e-Arc-Helix

Volume I: Geometry, Rational-Closure Sectors, and Upper Continuations

Arcman
INTERNATIONAL INSTITUTE OF ARC THEORY

April 28, 2026

Abstract

We study the *e-Arc-Helix*, a periodically modulated helix whose local differential geometry and global rational-closure arithmetic interact nontrivially. In aligned series-facing usage, *e-Arc-Helix* is the preferred English rendering; the earlier forms *Arc Helix* and *Arc-Helix* are retained only as historical identifiers where needed. Starting from a radially modulated helical embedding, we derive curvature, torsion, and the associated arc-helicity and curvature invariants, and show that rational commensurability of the modulation and azimuthal frequencies determines both a torus-knot type $T(p, q)$ and a corresponding closure sector $\mathcal{T}_{p,q}$. The minimal doubled sector $\mathcal{T}_{1,2}$ is thereby singled out as the first nontrivial doubled closure class, even though its standard torus representative $T(1, 2)$ is unknotted in the classical knot-theoretic sense.

On that geometric-topological foundation we introduce an explicitly *effective* sectorwise organization based on the curvature-torsion functional $E_A[X] = \lambda \int (\kappa^2 + \tau^2)\, ds$, formulate a reference-sector structure with the unmodulated helix as baseline configuration, and prove an ansatz-level recovery theorem for the modulated arc-helix family. Appendix A first fixes the downstream interface, the normalization rules, and the claim ladder, while Appendices B–D then develop the longer continuation chain: residual normalization, non-perturbative canonical evaluation, restricted variational relaxations, selector and barrier mechanisms, a uniquely characterized admissible reduced measure within the manuscript's intrinsic class, finite-family existence theorems, released-ladder compactness, finite microscopic lifts, projective infinite released-mode limits, a canonical intrinsic continuum support class with projective-to-continuum uniqueness on that class, and an odd-sector stochastic-field embedding with finite-Galerkin reversible generators, cylinder-level microscopic-to-field consistency, a unique stationary weak interaction-field law, canonical strong cylinder lifts, a canonical continuum Hunt semigroup that is ergodic on all L^2 observables inside the intrinsic support class, coordinate-wise Fukushima decompositions for the continuum interaction field, a conditional lift of spectral-gap and log-Sobolev control from finite observable cylinders to the full intrinsic continuum semigroup, and bounded or integrably singular renormalized tail reweightings on the same intrinsic support universe, together with a coherent cylindrical Brownian realization, a full-support pathwise strong continuum realization on the intrinsic compact support driven by that same countable Brownian family, a triangular reflected infinite-system representation of that strong continuum lift, a closed reflected cylinder martingale formulation characterizing that strong continuum lift on the full intrinsic cylinder algebra, an oscillation-based criterion reducing intrinsic quantitative coercivity to an explicit bounded-perturbation estimate on the cylinder interaction potentials, a summably local-block criterion making that coercivity automatic whenever the cylinder interactions admit a compatible quasilocal decomposition, a one-step local-increment criterion reducing existence of such a quasilocal decomposition to the actual released-mode interaction increments, a canonical tail-envelope criterion reducing automatic coercivity further to local-envelope oscillation plus released-mode tail memory, and, under summably controlled envelope/tail errors, a uniformly convergent renormalized intrinsic continuum interaction potential built from the genuine released-mode increments, together with a single weak reflected continuum field equation on the intrinsic finitely supported test space, a derivative-budget criterion reducing those renormalized interaction hypotheses to explicit summable partial-derivative bounds on the genuine released-mode increments themselves, a renormalized continuum interaction potential with continuous absolutely summable cylinder gradient, a weak reflected gradient-field formulation driven by that potential, a finite-cylinder derivative-Cauchy bridge reducing verification of that gradient package to the interaction potentials themselves, an explicit reflected gradient cylinder martingale formulation on the intrinsic cylinder core, a weighted cylinder-Lipschitz upgrade under summably controlled second-derivative budgets on the genuine released-mode increments, a finite-cylinder Hessian-Cauchy bridge reducing that stronger local regularity package to the interaction potentials themselves, a weighted cylinder-$C^{1,1}$ linearization and quadratic local-freezing estimate for the explicit continuum gradient drift under summably controlled third-derivative budgets, a finite-cylinder third-derivative-Cauchy bridge reducing that affine local-closure package to the interaction potentials themselves, a weighted cylinder-$C^{2,1}$ quadratic local model with cubic local-freezing error for the explicit continuum gradient drift under summably controlled fourth-derivative budgets, a finite-cylinder fourth-derivative-Cauchy bridge reducing that quadratic local-model package to the interaction potentials themselves, a weighted cylinder-$C^{3,1}$ cubic local model with quartic local-freezing error for the explicit continuum gradient drift under summably controlled fifth-derivative budgets, a finite-cylinder fifth-derivative-Cauchy bridge reducing that cubic local-model package to the interaction potentials themselves, an intrinsic cubically frozen local reflected-gradient martingale atlas with quartic chart defect, a finite-cylinder fifth-derivative-Cauchy criterion already sufficient for that quartic local atlas, a vanishing-radius exact local reflected-gradient generator germ extracted from that quartic atlas on the intrinsic cylinder core, a concrete tensor-majorant criterion reducing finite-cylinder fifth-derivative verification further to summable mode-decay envelopes, a graph-closed non-cylinder extension of that exact local generator germ on intrinsic balls, a weighted row-sum kernel criterion reducing fifth-order atlas activation further to per-row mode envelopes, an exact local-generator sheaf patching those ballwise graph-closed operators across intrinsic open sets, a banded-plus-tail criterion reducing weighted row-sum verification further to finite local interaction windows plus one ambient summable tail profile, a chartwise differential realization of that exact local-generator sheaf on a natural local Neumann test sheaf, a banded-plus-tail jet criterion reducing concrete recovery of the resulting local differential drift further to first/second finite-cylinder jet budgets, a single natural non-cylinder local martingale domain obtained as the graph closure of that chartwise differential sheaf, a finite-cylinder local-window C^2-Cauchy criterion reducing activation of that single-domain package further to direct first/second derivative stabilization of the interaction potentials themselves, a chart-free local Itô/carré-du-champ package on a canonical square-gradient subdomain of that same natural domain driven by the already-constructed Brownian family, a one-profile finite-cylinder local-window/tail C^2 criterion reducing activation of that Itô package further to one radius sequence, one tail profile, and one amplitude budget on the interaction potentials themselves, a chart-free dual local reflection functional on the intrinsic finitely supported test space compatible with the same common Brownian family and the coordinate reflected local times, a graph-closed chart-free reflection operator obtained as the local Itô-graph closure of that finitely supported dual core, a chart-free first-order reflection calculus on a canonical multiplicative natural-Itô core generated by bounded cylinder compositions and closed under product and scalar chain rules, a chart-free common-contact-clock factorization of that closed first-order reflection calculus on the same natural-Itô domain, a chart-free finitely supported cylinder-gradient representation of the resulting common-contact-clock reflection density package, a fixed-template telescoping C^2 criterion on the actual finite-cylinder interaction family compressing the same verification further to one common local window, one common tail profile, and one template seminorm sequence, a local-window/common-tail decomposition criterion reducing that same fixed-template verification further to one windowed shell core plus one common tail remainder, a chart-free coordinate-density realization of that same common-contact-clock gradient module, a chart-free algebraic contact-cone selector carried by that same common contact clock, a chart-free algebraic complementarity package for that same selector on the canonical coordinate contact wedge, a canonical frozen-tail criterion reducing the same fixed-window/common-tail verification further to one explicitly frozen local core plus its complementary tail remainder, a canonical shellwise frozen-tail seminorm criterion reducing that same frozen-tail verification further to one explicit seminorm of each telescoping finite-cylinder shell, a direct frozen-trace/difference shell seminorm criterion reducing that same frozen-tail verification further to one direct seminorm of each actual telescoping finite-cylinder shell together with its canonical frozen trace, an ambient coordinate-box normal-cone realization of that same chart-free algebraic complementarity selector carried by the common contact clock, a canonical frozen-segment oscillation criterion reducing the same direct frozen-trace verification further to one shellwise seminorm of the actual finite-cylinder shell along its canonical frozen-tail homotopies, a projected ambient-box Skorokhod decomposition for every finite observable cylinder driven by that same common contact clock, a chart-free intrinsic-support Skorokhod closure on the natural Itô domain's finitely supported cylinder core, a terminal canonical frozen-segment envelope criterion leaving no further abstract reduction on the interaction side beyond one shellwise segmentwise derivative envelope on each actual telescoping finite-cylinder shell, explicit tail-integrated actual e-Arc-Helix finite-cylinder interaction representatives extracted from the reduced infinite field law, a log-Laplace shell bridge reducing the terminal activation problem further to canonical frozen-segment C^2 estimates for the genuine infinite reduced free-energy shells, together with an exact posterior-cumulant reformulation of that same terminal shellwise verification under the actual tilted tail laws, and a posterior-variance-majorant simplification eliminating the remaining mixed-covariance bookkeeping, a posterior-quadratic-moment simplification absorbing the remaining posterior mean/variance split into weighted posterior L^2 shell budgets, a posterior-support essential-sup route replacing those remaining posterior moments by deterministic tilted-tail support envelopes, a frozen-tail support-interpolation majorant reducing those support envelopes further to canonical frozen-tail anchor values plus weighted tail-derivative transport budgets across the unreleased box, an anchor-plus-tail-transport compression reducing that stronger route further to four shellwise frozen-tail anchor and tail-transport budgets for the actual reduced free-energy shells, a tail-mass factorization compressing those same weighted transport budgets further to one explicit tail mass multiplying one m-uniform unreleased mixed-derivative envelope for each derivative order, a tail-profile factorization refining that same deterministic route through profile-weighted tail masses and profile-normalized unreleased mixed-derivative envelopes, an explicit verification skeleton organizing the remaining actual-shell burden into four shellwise target quantities, a menu of candidate profiles, and one terminal summability test, a one-step released-coordinate reduction compressing the first two rows of that dossier further to frozen-tail jets of the underlying infinite log-weight, and a generator-level quasilocal interface reducing the remaining locality question to genuinely external relativistic or gauge-theoretic dictionaries. Throughout, the main text is confined to differential-geometric, closure, and effective organizational claims, while the appendices keep candidate interpretations, microscopic lifts, and field-theoretic embeddings subordinate to those core results. No direct identification with QED couplings, the Standard Model, general relativity, or local quantum field theory is claimed.

我相信：于人而言，宇宙是几何的，且其言仅一。

I believe: for humankind, the universe is geometric, and it speaks but one language.

Series Alignment Note

This synchronized working build is part of the *Arc Theory first set*, whose normative anchor is *Arc Geometry Baseline Reference Text (Volume 0)*. The present e-Arc-Helix volumes remain the geometric and downstream continuation dossiers, but their series-level terminology and reference posture are now controlled externally by Volume 0: Appendix D is the preferred terminology authority, and Appendix E is the preferred reference-and-positioning authority. For terminology lock purposes, the preferred English rendering is *e-Arc-Helix*; the earlier forms *Arc Helix* and *Arc-Helix* are retained only as historical identifiers where needed. Likewise, *Quiescent State* is the preferred state label in aligned series-facing matter.

Accordingly, this build does not reopen the inherited mathematical core of the upper volume. It performs a series-level editorial synchronization only: the title-bearing *e-Arc-Helix* object is preserved, claim-strength boundaries remain unchanged, and later downstream continuations remain subordinate to that title-bearing geometric object. Any future wording cleanup across Volumes I/II should therefore be read as alignment to Volume 0 rather than as a change in theorem-level mathematical content.

The lower-volume continuation should also be read with the same hierarchy in mind: the external-realization completion layer belongs to the later completed witness package, while the present upper volume remains the archival geometric/effective front half of the first-set chain. Readers new to the series are advised to treat Volume 0 as the glossary-and-reference control layer, then read the present upper volume, and only then continue into the synchronized lower volume.

Contents

1 Introduction

This manuscript studies the *e-Arc-Helix*: a periodically modulated helical curve whose local curvature-torsion geometry and global rational-closure arithmetic are coupled through a two-frequency phase lift. The central aim is to isolate, in a single rigorous framework, the differential geometry of the curve, the resulting closure sectors, the effective sectorwise organization induced by a curvature-torsion energy, and a logically downstream family of microscopic and field-theoretic continuations.

Within the broader arc-geometric research sequence developed by the author, the present paper interfaces with the foundational non-spacetime formulation of Arc Theory [1], with the differential-topological study of arc helices [2], with the variational analysis of arc-helical energies [3], and with the spectral-stability and resonant-bifurcation viewpoint arising from the q-fold circle [4]. The longer-term ambition is to articulate, in a modern mathematical idiom and with deliberately conservative claim control, the geometric and physical significance of arc geometry itself. In the present manuscript, however, the emphasis is narrower: the title-bearing object is the e-Arc-Helix itself, and every later continuation is required to remain subordinate to that core geometric object.

1.1 Terminology and scope

A notational and terminological distinction will be used throughout. The symbol $T(p,q)$ denotes the standard torus-knot type, whereas $\mathcal{T}_{p,q}$ denotes the corresponding rational-closure sector in the present e-Arc-Helix framework. The sector $\mathcal{T}_{1,2}$ is mathematically distinguished as the minimal doubled closure sector, but it is not used here as the name of the overall theory. Earlier drafts employed the label "minimal doubled representative" for that doubled branch; the present revision deliberately retires that usage in order to keep the title-bearing object, the main geometric theory, and the downstream doubled-sector analysis cleanly separated.

Within the broader arc-theoretic background one may also study more composite internal-field constructions built from arc-helical constituents. The present paper does not develop those composite objects. Its title-bearing object remains the e-Arc-Helix itself. Accordingly, whenever later sections use the model-internal expressions *inertial plane*, *conjugate-time axis*, *light-end*, *magnetic-end*, *particle-type interaction*, or *wave-type interaction*, those expressions are to be read as interpretive descriptors attached to the e-Arc-Helix or to its centered doubled state, not as the introduction of a second primary theory object.

The manuscript is organized around three logical layers. The first layer, developed in Sections 2–3, is differential-geometric and topological: it defines the arc-helix family, derives curvature and torsion, and identifies the closure sectors produced by rational commensurability. The second layer, developed in Sections 4–6, is effective and ansatz-level: it introduces a curvature-torsion energy, a reference-sector structure, and a recovery theorem for the modulated arc-helix family. The third layer, carried by Appendices A–D, is explicitly downstream: Appendix A fixes the interface and first candidate, while Appendices B–D study normalization shifts, variational releases, microscopic lifts, infinite-mode limits, and stochastic-field continuations.

1.2 Claim ladder and external interfaces

The logical status of statements in the manuscript is tiered. The main text contains theorem-level geometric and closure statements, together with explicitly labeled effective organizational statements. Appendix A then fixes a four-step ladder — definition, effective relation, candidate interpretation, and conjectural identification — and Appendices B–D inherit that ladder so that downstream continuations cannot feed hidden assumptions back into Sections 2–6.

This distinction matters especially at the interface with existing mathematical and physical theories. The present manuscript interfaces directly with differential geometry of curves, torus-knot topology, elastic-curve variational language, reversible stochastic dynamics, and stochastic field embedding [7–10]. It does not claim a direct derivation of QED couplings, a general-relativistic embedding, or equivalence with a local quantum field theory. Likewise, whenever spectral, vacuum, microscopic, or field-theoretic language is used below, it is either explicitly effective or explicitly model-internal unless stated otherwise.

1.3 Logical layer matrix and reading order

Table 1 records the manuscript's logical layers, their highest permissible claim strengths, and the external mathematical or physical interfaces they directly touch. The practical reading rule is that the paper should be read from left to right across that matrix rather than from the later continuations backward into the main text.

Layer	Core objects	Highest claim status	Direct interface	Explicit non-claim boundary
Sections 2–3	Arc-helix embedding, Frenet invariants, primitive closure period, sectors $\mathcal{T}_{p,q}$	Definition / theorem-level geometry and closure	Differential geometry of curves; torus-knot and closure arithmetic	No spectral, microscopic, or field-theoretic claim enters here.
Sections 4–6	Curvature-torsion functional E_A, effective branch organization, reference sector, ansatz-level recovery	Effective relation / ansatz-level theorem	Elastic-curve variational language; constrained sector minimization; geometric holonomy	Not a Hamiltonian quantization, not a microscopic operator theory, and not a global inverse theorem for arbitrary space curves.
Appendix A	Downstream interface, normalization rules, claim ladder, first doubled-sector candidate	Candidate interpretation with explicit obstruction statements	Dimensionless normalization analysis and controlled candidate insertion	No completed constant identification is claimed, and no hidden assumption may feed back into Sections 2–6.
Appendices B–D	Variational releases, barrier mechanism, microscopic stochastic lift, projective limit, odd-sector field embedding	Model-internal microscopic / stochastic-field continuation	Reversible stochastic dynamics, coarse-grained measures, projective limits, reflected odd-sector field modeling	No direct equivalence with QED, the Standard Model, general relativity, condensed-matter roton physics, or local quantum field theory is claimed.

Table 1: Logical layers, claim status, and external interfaces.

The resulting master dependency chain is

$$\text{Sections 2–3} \Longrightarrow \text{Sections 4–6} \Longrightarrow \text{Appendix A} \Longrightarrow \text{Appendices B–D}, \tag{1}$$

and every later continuation in the manuscript is required to respect that one-way direction of dependence.

The paper proceeds as follows. Section 2 develops the differential geometry of the e-Arc-Helix and fixes the local invariants. Section 3 establishes rational closure, closure periods, and the sector labeling $\mathcal{T}_{p,q}$. Section 4 explains how a curvature-torsion energy organizes an effective sectorwise branch structure without invoking a full Hamiltonian quantization. Section 5 introduces the reference-sector language and singles out the minimal doubled branch above it. Section 6 formulates the corresponding ansatz-level recovery theorem. Appendix A then records the downstream interface and the first candidate, while Appendices B–D continue with normalization shifts, non-perturbative evaluations, restricted variational releases, selector and barrier mechanisms, finite microscopic lifts, infinite released-mode limits, and an odd-sector stochastic-field embedding.

2 Differential Geometry of Arc Helices

In this section we derive the differential-geometric structure of arc helices. The central goal is to obtain explicit expressions for curvature and torsion, and to define the basic invariants that will organize the later spectral and topological theory [2, 7].

2.1 Definition of the e-Arc-Helix

We consider the parametrized curve

$$X(t) = (R(t)\cos(\omega t), R(t)\sin(\omega t), vt), \tag{2}$$

where

$$R(t) = R_0 + \epsilon \sin(\Omega t) \tag{3}$$

represents a periodic radial modulation.

The parameters satisfy

$$R_0 > 0, \qquad 0 < \epsilon < R_0, \qquad \omega > 0, \qquad \Omega > 0, \qquad v > 0. \tag{4}$$

It is convenient to introduce the dimensionless weak-modulation parameter

$$\eta := \frac{\epsilon}{R_0}. \tag{5}$$

Definition 2.1 (Admissible e-Arc-Helix regime). The parameter tuple $(R_0, \epsilon, \omega, \Omega, v)$ is called *admissible* if $R_0 > 0$, $0 < \epsilon < R_0$, $\omega > 0$, $\Omega > 0$, $v > 0$, and the associated curve (2)–(3) satisfies $|X'(t) \times X''(t)| > 0$ on the interval under consideration. Whenever asymptotic formulas are invoked, one further assumes the weak-modulation regime $\eta \ll 1$.

Remark 2.2. The inequality $0 < \epsilon < R_0$ guarantees $R(t) \geq R_0 - \epsilon > 0$ for all t. Since $v > 0$, one has $|X'(t)| \geq v$, so the curve is regular and the arclength element $ds = |X'(t)|\,dt$ is always defined. The nonvanishing of $|X'(t) \times X''(t)|$ is automatic in the constant-radius limit and therefore persists for all sufficiently small ϵ by continuity.

2.2 Projection circle, inertial plane, and centered doubled state

For the later doubled-sector continuations it is convenient to fix a centered reference geometry attached to (2). Let

$$\Pi_{\text{in}} := \{(x, y, 0) \in \mathbb{R}^3\}, \qquad \pi_{\text{in}}(x, y, z) := (x, y, 0), \tag{6}$$

and let

$$\mathfrak{t} := \{(0, 0, z) : z \in \mathbb{R}\} \tag{7}$$

be the oriented axis orthogonal to Π_{in}. The mean-radius projection circle is

$$C_{\text{proj}} := \{(R_0 \cos\theta, R_0 \sin\theta, 0) : \theta \in [0, 2\pi)\} \subset \Pi_{\text{in}}. \tag{8}$$

The same reference circle will also be called the *traversal circle*. For any admissible representative Y with $\|\pi_{\text{in}} Y(u)\| > 0$, define its normalized inertial-plane traversal by

$$\widehat{\pi}_{\text{in}} Y(u) := \frac{R_0}{\|\pi_{\text{in}} Y(u)\|} \pi_{\text{in}} Y(u) \in C_{\text{proj}}. \tag{9}$$

This removes the instantaneous radial excursion and records only the oriented passage of the state around the inertial circle. In the arc-geometric terminology used here, this is the precise sense in which the projection circle is the *traversal*: C_{proj} is the reference carrier of the projected traversal.

Definition 2.3 (Centered doubled state; self-interaction and interaction branches). A 4π-parametrized representative $Y : [0, 4\pi] \to \mathbb{R}^3$ is called a *centered doubled state* if, after a phase-centering rigid motion, there exists an involutive reflection $\mathcal{J}_{\text{in}} : \Pi_{\text{in}} \to \Pi_{\text{in}}$ with fixed line through the origin such that

$$\widehat{\pi}_{\text{in}} Y(u + 2\pi) = \mathcal{J}_{\text{in}}(\widehat{\pi}_{\text{in}} Y(u)), \qquad u \in [0, 2\pi]. \tag{10}$$

Thus the two half-trajectories share the same inertial plane Π_{in}, viewed as the common spacetime reference background, while their projection-traversal circles in that plane form a mirror-symmetric pair. The doubled state is called

$$\textit{self-interaction type} \quad \text{if} \quad z_Y(u + 2\pi) = z_Y(u), \tag{11}$$

$$\textit{interaction type} \quad \text{if} \quad z_Y(u + 2\pi) = -z_Y(u), \tag{12}$$

for $u \in [0, 2\pi]$, where z_Y denotes the third coordinate of Y.

Proposition 2.4 (Projection-traversal symmetry in a centered doubled state). *Let Y be a centered doubled state and assume $\|\pi_{\text{in}} Y(u)\| > 0$ on $[0, 4\pi]$. Then its normalized projected traversal obeys*

$$\widehat{\pi}_{\text{in}} Y(u + 2\pi) = \mathcal{J}_{\text{in}}(\widehat{\pi}_{\text{in}} Y(u)), \qquad u \in [0, 2\pi]. \tag{13}$$

Hence the two half-trajectories of the doubled state share one inertial reference plane, while their projection-traversal circles form a mirror-symmetric pair in that plane. The projection circle is therefore not merely a static shadow of the doubled state but the normalized carrier of its inertial traversal, with each half-trajectory carrying its own mirror partner on the common inertial background.

Proof. This is exactly the defining relation (10). It states that the second half-traversal is the in-plane mirror image of the first after normalization to the reference circle C_{proj}. Thus the two half-trajectories occupy the same inertial background Π_{in} while their projected traversals are mirror-symmetric there. □

Proposition 2.5 (Antiperiodic normal deviation for the interaction doubled branch). *Let Y be a centered doubled state of interaction type and define its signed normal deviation by $\delta_Y(u) := z_Y(u)$. Then δ_Y is 4π-periodic and 2π-antiperiodic:*

$$\delta_Y(u + 4\pi) = \delta_Y(u), \qquad \delta_Y(u + 2\pi) = -\delta_Y(u). \tag{14}$$

Consequently δ_Y has an $L^2([0, 4\pi])$ Fourier expansion involving only odd half-integer harmonics,

$$\delta_Y(u) = \sum_{n \geq 0} \left(\alpha_n \cos\left(\frac{(2n+1)u}{2}\right) + \beta_n \sin\left(\frac{(2n+1)u}{2}\right) \right). \tag{15}$$

If the doubled representative is phase-centered so that the cosine sector vanishes, then

$$\delta_Y(u) = \sum_{n\geq 0} \beta_n \sin\left(\frac{(2n+1)u}{2}\right), \tag{16}$$

which is exactly the odd half-mode form later used in the released-mode continuations.

Proof. The interaction-branch condition (12) gives $\delta_Y(u+2\pi) = -\delta_Y(u)$ directly, and applying it twice gives $\delta_Y(u+4\pi) = \delta_Y(u)$. For a 4π-periodic function, Fourier frequencies are integer multiples of $1/2$. The antiperiodicity under a shift by 2π eliminates the even half-integer harmonics and leaves only frequencies $(2n+1)/2$, which yields (15). The final statement is the usual phase-centering normalization that removes the cosine sector. □

Remark 2.6 (Model-internal interpretive conventions)**.** In the model-internal physical language used later, Π_{in} is also called the *inertial plane* or *inertial field*, C_{proj} is the *projection circle*, equivalently the *traversal circle*, and $\mathfrak{t}$ is the *conjugate-time axis*. Relative to the positive orientation of $\mathfrak{t}$, the projected traversal from the light branch to the magnetic branch is taken counterclockwise on C_{proj}, while the reverse traversal is clockwise. When both oriented branches are active simultaneously, the dominant local reading is assigned by their relative arc-energy strengths. These conventions are interpretive labels attached to the centered doubled state; they are not new theorem-level primitives.

Remark 2.7 (Field partition and state taxonomy)**.** For a centered self-mirror doubled state Y, the regions between Π_{in} and the two branches define the model-internal *light-end domain* and *magnetic-end domain*; the thickness of the reference plane itself is read as the *inertial field.* The undeformed single arc helix is the basic eigenstate of the present paper. Its self-mirror doubled state supplies two opposite oriented generators, conventionally denoted $L \to M$ and $M \to L$. Equation (14) is the precise content of the informal cancellation statement: the signed normal deviations of those two opposite generators add to zero after a half-period shift on the doubled circle. The further expressions *simultaneous coupling*, *coherent coupling*, *particle-type interaction*, and *wave-type interaction* are retained only as model-internal interpretive shorthand for different couplings of these oriented generators; no literal identification with external antiparticle, dark-matter, electromagnetic, or wave-mechanical theories is asserted in this manuscript.

2.3 First Derivative

The velocity vector is

$$X'(t) = \frac{dX}{dt}. \tag{17}$$

Using the product rule, we obtain

$$X'(t) = \big(R'(t)\cos(\omega t) - R(t)\omega\sin(\omega t), R'(t)\sin(\omega t) + R(t)\omega\cos(\omega t), v\big), \tag{18}$$

where

$$R'(t) = \epsilon\Omega\cos(\Omega t). \tag{19}$$

2.4 Second Derivative

Differentiating again yields

$$X''(t) = \big(R''\cos\omega t - 2R'\omega\sin\omega t - R\omega^2\cos\omega t, R''\sin\omega t + 2R'\omega\cos\omega t - R\omega^2\sin\omega t, 0\big), \tag{20}$$

where

$$R''(t) = -\epsilon\Omega^2\sin(\Omega t). \tag{21}$$

2.5 Third Derivative

The third derivative takes the form

$$X_x''' = R''' \cos \omega t - 3R''\omega \sin \omega t - 3R'\omega^2 \cos \omega t + R\omega^3 \sin \omega t, \tag{22}$$

$$X_y''' = R''' \sin \omega t + 3R''\omega \cos \omega t - 3R'\omega^2 \sin \omega t - R\omega^3 \cos \omega t, \tag{23}$$

$$X_z''' = 0, \tag{24}$$

where

$$R'''(t) = -\epsilon\Omega^3 \cos(\Omega t). \tag{25}$$

Proposition 2.8 (Regularity and admissible arclength geometry). *In the admissible e-Arc-Helix regime, the curve X is C^∞, the speed satisfies*

$$|X'(t)|^2 = R'(t)^2 + (R(t)\omega)^2 + v^2 \geq v^2 > 0, \tag{26}$$

and the arclength parameter is globally well defined. Moreover, $\kappa(t)$ and $\tau(t)$ are smooth on every interval on which $|X'(t) \times X''(t)| > 0$.

Proof. The modulation $R(t) = R_0 + \epsilon \sin(\Omega t)$ is smooth, hence so is $X(t)$. The explicit speed identity follows directly from (18); because $v > 0$, the speed never vanishes. Smoothness of κ and τ then follows from the defining formulas once the denominator $|X'(t) \times X''(t)|$ is assumed nonzero. □

2.6 Curvature

The curvature is defined by

$$\kappa(t) = \frac{|X'(t) \times X''(t)|}{|X'(t)|^3}. \tag{27}$$

The velocity magnitude is

$$|X'(t)|^2 = R'(t)^2 + (R(t)\omega)^2 + v^2. \tag{28}$$

A direct computation gives

$$|X'(t) \times X''(t)|^2 = v^2\left((R\omega^2 - R'')^2 + 4\omega^2(R')^2\right) + \omega^2\left(R^2\omega^2 - RR'' + 2(R')^2\right)^2. \tag{29}$$

Hence the curvature admits the exact representation

$$\kappa(t) = \frac{\sqrt{v^2((R\omega^2 - R'')^2 + 4\omega^2(R')^2) + \omega^2(R^2\omega^2 - RR'' + 2(R')^2)^2}}{((R')^2 + (R\omega)^2 + v^2)^{3/2}}. \tag{30}$$

All later weak-modulation expansions of κ are obtained by expanding (30) in powers of $\eta = \epsilon/R_0$. No separate approximate definition of curvature is introduced.

2.7 Torsion

The torsion is defined by

$$\tau(t) = \frac{(X'(t) \times X''(t)) \cdot X'''(t)}{|X'(t) \times X''(t)|^2}. \tag{31}$$

Using (29), one obtains the exact formula

$$\tau(t) = \frac{\omega v \left(R^2\omega^4 - 4RR''\omega^2 + 6(R')^2\omega^2 - 2R'R''' + 3(R'')^2\right)}{v^2((R\omega^2 - R'')^2 + 4\omega^2(R')^2) + \omega^2(R^2\omega^2 - RR'' + 2(R')^2)^2}. \tag{32}$$

As with curvature, every later asymptotic torsion formula is understood as an expansion of (32) in the weak-modulation regime.

2.8 Constant-Radius Limit

When the modulation vanishes, $\epsilon = 0$, the arc helix reduces to the standard helix

$$X_0(t) = (R_0 \cos \omega t, R_0 \sin \omega t, vt). \tag{33}$$

In this case the curvature and torsion are constant:

$$\kappa_0 = \frac{R_0 \omega^2}{R_0^2 \omega^2 + v^2}, \qquad \tau_0 = \frac{v\omega}{R_0^2 \omega^2 + v^2}. \tag{34}$$

2.9 Arc Helicity and Curvature Invariant

We define the *arc helicity*

$$H_A = \frac{\tau}{\kappa}. \tag{35}$$

This ratio is understood on the nondegenerate part of the admissible regime where $\kappa > 0$; in particular, it is well defined on the constant-radius baseline and on all sufficiently small modulations around it. In the constant-radius limit,

$$H_A = \frac{v}{R_0 \omega}. \tag{36}$$

We also define the *arc curvature invariant*

$$I_A = \kappa^2 + \tau^2. \tag{37}$$

In the constant-radius limit,

$$I_A = \frac{\omega^2}{R_0^2 \omega^2 + v^2}. \tag{38}$$

These quantities will serve as the basic geometric observables in the later spectral construction.

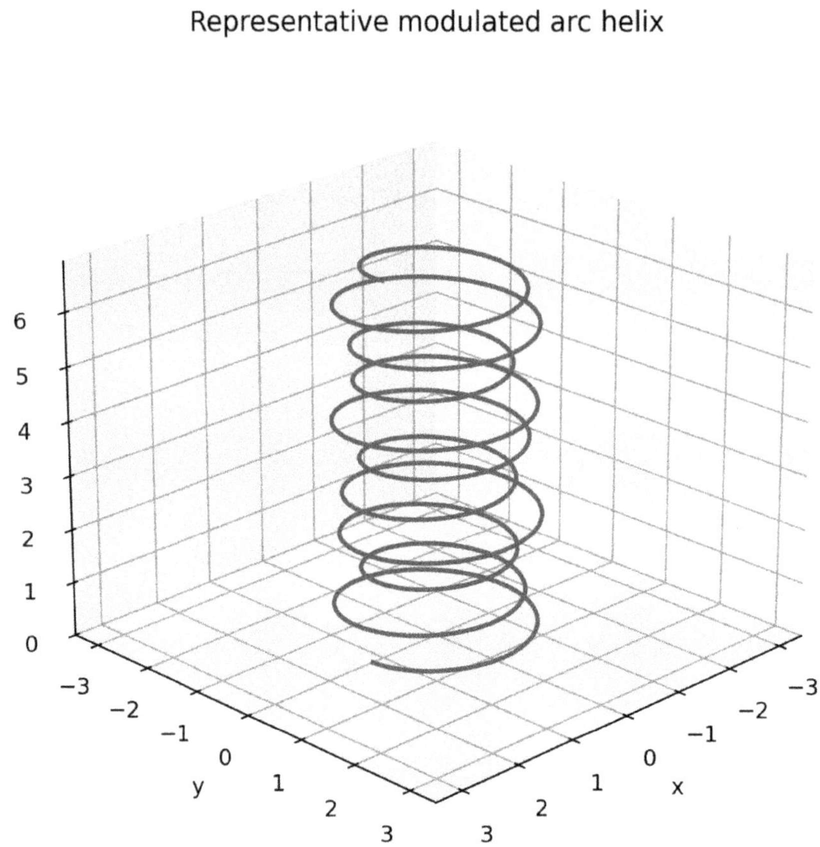

Figure 1: Representative modulated arc helix generated by the ansatz (2)–(3). The periodic radial modulation deforms the standard helix while preserving the two-frequency structure that later governs closure and sector formation.

3 Rational Closure and Sector Structure

In this section we establish the topological classification of periodically identified arc helices, using the standard torus-knot viewpoint as background [8].

3.1 Periodic Structure

The arc helix carries two independent phases:

$$\theta(t) = \omega t, \qquad \phi(t) = \Omega t. \tag{39}$$

Since both variables are angular modulo 2π, they define a lifted phase curve

$$\Gamma(t) = \big(\theta(t), \phi(t)\big) = \big(\omega t, \Omega t\big) \mod 2\pi \tag{40}$$

on the phase torus

$$T^2 = S^1 \times S^1. \tag{41}$$

The global closure properties are therefore controlled by the ratio

$$\frac{\Omega}{\omega}. \tag{42}$$

The relevant notion of closure is closure of the lifted phase curve on T^2, or equivalently closure in a periodically identified geometry, rather than literal Euclidean closure of $X(t)$ in $\mathbb{R}^3$.

3.2 Rational Closure Condition

Definition 3.1 (Rational closure). The arc helix satisfies the rational closure condition if

$$\frac{\Omega}{\omega} = \frac{p}{q}, \qquad p, q \in \mathbb{Z}, \qquad \gcd(p, q) = 1. \tag{43}$$

Definition 3.2 (Primitive closure sector). For a coprime pair (p, q) with $q > 0$, the symbol

$$\mathcal{T}_{p,q} \tag{44}$$

denotes the class of admissible Arc Helices satisfying (43), modulo rigid Euclidean motions and periodic torus identification of the lifted phase curve. The symbol records closure arithmetic and admissible toroidal isotopy data; it is not, by itself, a complete replacement for the full package of standard knot invariants.

3.3 Closure Period

Lemma 3.3. *If the rational closure condition* (43) *holds, then the lifted phase curve associated with the arc helix closes after the common period*

$$T = \frac{2\pi q}{\omega}. \tag{45}$$

Proof. After time T, we have

$$\omega T = 2\pi q, \qquad \Omega T = \frac{p}{q}\omega T = 2\pi p. \tag{46}$$

Hence the azimuthal angle and the radial-modulation phase both return modulo 2π, so that

$$\Gamma(T) = \Gamma(0) \tag{47}$$

on T^2. Therefore the lifted phase curve is globally periodic in the identified geometry. □

Proposition 3.4 (Primitive closure period). *Under the coprime rational-closure condition* (43), *the period*

$$T_* = \frac{2\pi q}{\omega} \tag{48}$$

is the least positive period of the lifted phase curve.

Proof. Suppose $T' > 0$ also satisfies $\Gamma(T') = \Gamma(0)$. Then $\omega T' = 2\pi m$ for some integer $m \geq 1$. Compatibility with the second phase gives

$$\Omega T' = \frac{p}{q}\omega T' = 2\pi \frac{pm}{q} \in 2\pi\mathbb{Z}. \tag{49}$$

Since $\gcd(p, q) = 1$, this implies $q \mid m$. Hence $m \geq q$, so $T' \geq 2\pi q/\omega = T_*$. Therefore T_* is primitive. □

3.4 Toroidal Embedding

The oscillation of $R(t)$ around R_0 generates an effective toroidal envelope. To interpret the closed phase orbit geometrically, we introduce the standard torus embedding

$$\mathcal{E}(\theta, \phi) = \big((R_0 + a\cos\phi)\cos\theta, (R_0 + a\cos\phi)\sin\theta, a\sin\phi\big), \qquad a > 0. \tag{50}$$

The lifted phase curve then induces the torus curve

$$\widetilde{X}(t) = \mathcal{E}\big(\omega t, \Omega t\big). \tag{51}$$

Over one primitive period, $\widetilde{X}(t)$ winds

- q times in the longitudinal direction,
- p times in the meridional direction.

Thus the periodically identified arc helix may be regarded as a curve on an effective torus with winding data (p, q).

3.5 Rational-closure torus-sector theorem

Theorem 3.5 (e-Arc-Helix Torus Knot Theorem). *Let*

$$X(t) = (R(t)\cos(\omega t), R(t)\sin(\omega t), vt), \qquad R(t) = R_0 + \epsilon \sin(\Omega t), \tag{52}$$

with $0 < \epsilon < R_0$. *If*

$$\frac{\Omega}{\omega} = \frac{p}{q}, \qquad \gcd(p, q) = 1, \tag{53}$$

then the periodically identified arc helix determines a primitive torus curve with winding numbers (p, q), *and is therefore topologically equivalent to the torus knot*

$$T(p, q). \tag{54}$$

Proof. By the rationality condition, the lifted phase curve $\Gamma(t)$ closes on T^2 after the primitive period $T = 2\pi q/\omega$. During this period, the first phase winds q times and the second phase winds p times, so the corresponding torus curve $\widetilde{X}(t) = \mathcal{E}(\omega t, \Omega t)$ has longitudinal winding number q and meridional winding number p. Since p and q are coprime, the curve is primitive and connected. By the standard classification of simple closed curves on the torus, $\widetilde{X}(t)$ is isotopic to the torus knot $T(p, q)$. Because the periodically identified arc helix carries the same phase winding data, it determines the same toroidal isotopy class. □

Proposition 3.6 (Primitive sectors and coprimality). *If integers* p, q *satisfy* $d = \gcd(p, q) > 1$, *then the associated torus representative is a d-fold repetition of a primitive curve with winding data* $(p/d, q/d)$. *Consequently, primitive closure sectors are naturally indexed by coprime pairs.*

Proof. The torus representative is determined by the phase map $t \mapsto (\omega t, \Omega t)$ with $\Omega/\omega = p/q$. If $p = dp_0$ and $q = dq_0$ with $\gcd(p_0, q_0) = 1$, then one full period for the nonreduced pair decomposes into d repeats of the primitive phase orbit associated with (p_0, q_0). Thus the primitive topological content is already captured by the coprime pair. □

Remark 3.7 (Sector labels versus torus-knot types). The passage

$$\mathcal{T}_{p,q} \longmapsto T(p, q) \tag{55}$$

forgets some of the bookkeeping that matters later in the e-Arc-Helix framework. In particular, $\mathcal{T}_{p,q}$ retains the closure arithmetic, primitive period, and admissible toroidal realization used in the effective branch organization, whereas the ordinary torus-knot type records only the resulting knot class. This distinction is essential when $T(p, q)$ is classically unknotted but the closure arithmetic remains nontrivial.

3.6 Minimal doubled sector and its distinguished representative

We distinguish carefully between the torus-knot type $T(p, q)$ and the corresponding closure sector $\mathcal{T}_{p,q}$ in the e-Arc-Helix framework.

The simplest nontrivial closed sector is

$$(p, q) = (1, 2). \tag{56}$$

This is the minimal doubled toroidal configuration.

Definition 3.8 (Minimal doubled closure sector)**.** The closure sector

$$\mathcal{T}_{1,2} \tag{57}$$

is called the *minimal doubled closure sector.*

Remark 3.9. In standard knot-theoretic terminology, the torus curve $T(1,2)$ is isotopic to the unknot [8]. In the present manuscript, however, the sector $\mathcal{T}_{1,2}$ is distinguished not by knot complexity alone, but by being the first sector whose global closure period is doubled relative to the local helical period. The manuscript therefore treats $\mathcal{T}_{1,2}$ as a distinguished closure sector rather than as a title-bearing object in its own right.

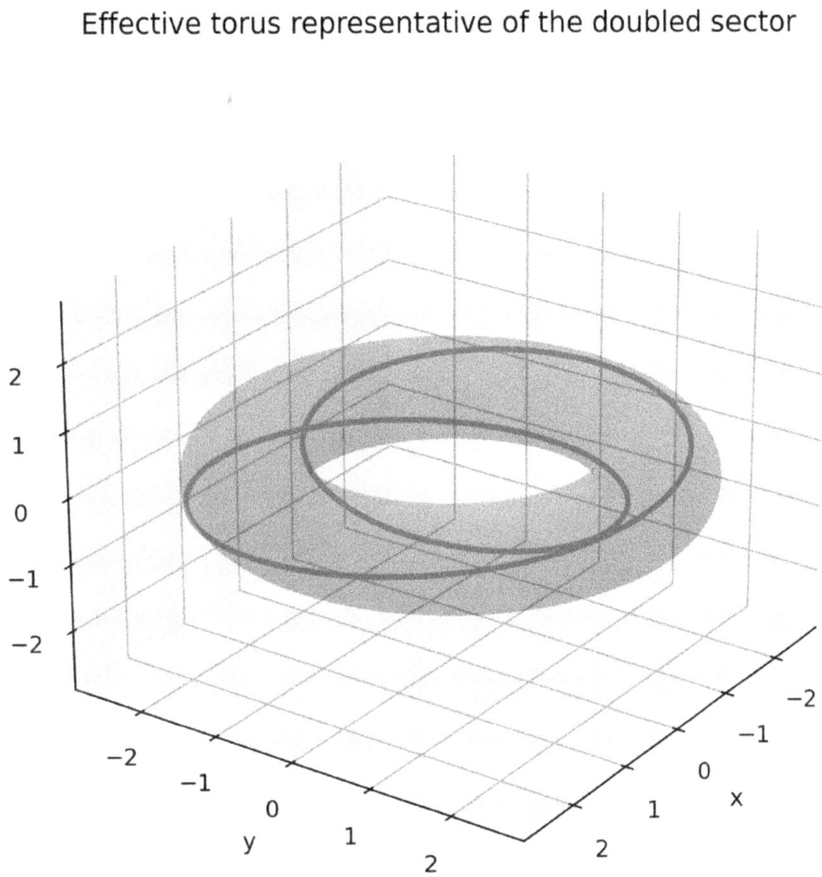

Figure 2: Effective torus representative of the minimal doubled sector $\mathcal{T}_{1,2}$. The sector is singled out by its doubled closure arithmetic and phase winding data, even though the corresponding torus curve $T(1,2)$ is unknotted in the standard knot-theoretic sense.

3.7 Spontaneous Doubling

The emergence of the $(1,2)$ sector motivates the concept of *spontaneous doubling.* A locally helical motion reorganizes into a globally doubled closed orbit once the rational closure condition is imposed. In the present manuscript this is read as the first nontrivial *closure reorganization* above the reference sector, not as an independently quantized particle-like excitation. That doubled reorganization connects directly with the broader doubling-based interpretation developed in [4, 5].

4 Effective Spectral Organization of e-Arc-Helix Sectors

We now introduce an effective sectorwise organization of the e-Arc-Helix framework. The goal of the present section is not to derive a full microscopic quantization rule, but to show how the differential-geometric invariants of the arc helix organize a natural hierarchy of closure-constrained branches. This viewpoint is guided both by curvature-based functionals familiar from the geometry of elastic curves [9, 10] and by the author's broader variational and spectral arc-geometry program [3, 4].

4.1 From geometric energy to effective sectorwise branching

The passage from the continuous functional $E_A[X]$ to a discrete family of sectorwise branches is not taken here as a literal Hamiltonian quantization of the classical curve. What is used instead is the joint effect of rational closure and constrained minimization. Once $\Omega/\omega = p/q$ with $\gcd(p, q) = 1$, admissible configurations split into closure classes $\mathcal{C}_{p,q}$; minimizing a common geometric functional separately on those classes yields a discrete family of sectorwise reference levels $E^{\min}_{p,q}$. The resulting discreteness is therefore a discreteness of topological constraint classes and constrained extrema, not yet of operator spectra.

A complementary geometric picture comes from the Frenet–Serret frame. Along a primitive closed cycle, the frame defines an $SO(3)$-bundle over the base circle, and the torsion integral

$$\Phi_{\text{geom}} = \oint_{\mathcal{C}_{p,q}} \tau(s)\, ds \tag{58}$$

acts as a sector-sensitive holonomy quantity. Under rational closure, admissible relaxations must remain compatible with the global winding data encoded by $\mathcal{T}_{p,q}$. In that limited but rigorous sense, closure arithmetic and frame holonomy jointly organize a discrete family of admissible sectorwise minima.

Proposition 4.1 (Holonomy-compatible constrained sector selection). *Let $X \in \mathcal{C}_{p,q}$ be a primitive closed e-Arc-Helix in the admissible regime of Section 2. Then over the primitive closure cycle the Frenet–Serret frame determines a holonomy element $H_{p,q}(X) \in SO(3)$, well defined up to conjugacy, whose normal-plane rotation angle contains the integrated torsion Φ_{geom} modulo the sector closure data. Consequently, sectorwise reference levels arise from two simultaneous constraints: the discrete closure arithmetic (p, q) and the requirement that admissible variations remain compatible with the corresponding frame holonomy class along the primitive cycle.*

Proof. Because X is C^3 and nondegenerate in the admissible regime, the Frenet–Serret frame (T, N, B) exists along the whole primitive cycle. Writing the frame as an $SO(3)$-valued path $F(s)$, one has $F(L_{p,q}) = F(0)H_{p,q}(X)$ after one primitive closure period, which defines the holonomy element up to conjugacy. The torsion enters this rotation through the normal-plane part of the Frenet–Serret connection, so the integrated torsion (58) records the associated geometric phase modulo the global closure data.

Now fix a primitive sector $\mathcal{C}_{p,q}$. Admissible variations inside that class preserve the primitive closure arithmetic and therefore remain inside the same disconnected closure family while changing E_A continuously. Passing to a different primitive sector requires changing the coprime closure data and hence leaving the original closure class. Minimizing the same functional on each disconnected family separately therefore produces sectorwise reference levels indexed jointly by closure arithmetic and holonomy-compatible admissibility. This explains the effective branch bookkeeping without invoking a prior operator quantization rule. □

Accordingly, the words "spectral", "branch", and "reference energy" in this section should be read as effective organizational language. The section does not claim a canonical quantization rule, an operator algebra, or a microscopic gap theorem.

4.2 Geometric Spectral Principle

The invariant

$$I_A(t) = \kappa(t)^2 + \tau(t)^2 \tag{59}$$

provides a natural nonnegative geometric density from which an effective energy may be constructed. In the author's program, this invariant plays the role of the leading local observable linking the differential description of arc helices to their variational and spectral organization [3,4].

4.3 Curvature Energy Functional

We define the effective geometric energy

$$E_A[X] = \lambda \int_0^T \left(\kappa(t)^2 + \tau(t)^2\right) ds, \qquad ds = \left|X'(t)\right| dt, \tag{60}$$

where $\lambda > 0$ is a coupling constant. For a nontrivial closed sector, T denotes the primitive closure period determined by the rationality condition (43); for the vacuum sector it is understood as the corresponding reference period of the unmodulated helix. This functional should be understood as a minimal effective geometric energy capturing the leading bending-twisting cost of the arc helix, rather than as a uniquely determined microscopic Hamiltonian [9, 10].

In the weak-modulation regime,

$$E_A \approx \lambda \int_0^T I_A(t)\left|X'(t)\right| dt. \tag{61}$$

4.4 Topological Spectral Branches

The closed sectors define an effective branch family

$$E_{p,q}, \qquad \mathcal{S} = \bigcup_{(p,q)=1} \mathcal{S}_{p,q}. \tag{62}$$

More precisely, if $\mathcal{C}_{p,q}$ denotes the admissible periodically identified e-Arc-Helix configurations belonging to sector $\mathcal{T}_{p,q}$, one may regard

$$E_{p,q}^{\min} := \inf_{X \in \mathcal{C}_{p,q}} E_A[X] \tag{63}$$

as the sectorwise reference energy.

Proposition 4.2 (Constraint-class origin of branch discreteness). *Assume each primitive sector class $\mathcal{C}_{p,q}$ is nonempty. Then the family $\{E_{p,q}^{\min}\}_{(p,q)=1}$ is indexed by the discrete set of primitive closure sectors. Any discreteness appearing at this stage arises from rational closure and separate constrained minimization over disconnected sector classes, not from a prior operator quantization rule.*

Proof. Primitive sectors are labeled by coprime integer pairs (p, q), hence by a discrete index set. For each such sector, the value $E_{p,q}^{\min}$ is defined by a separate infimum over the corresponding admissible class $\mathcal{C}_{p,q}$. The resulting collection of numbers is therefore discrete by indexing and constraint separation alone, irrespective of whether the infimum is later attained by a unique minimizer. □

The branch $\mathcal{S}_{p,q}$ is therefore only a bookkeeping label for configurations organized around the constrained minimum $E_{p,q}^{\min}$ and its admissible variational descendants within the same sector. No operator-valued excitation interpretation is required at the present stage.

4.5 The doubled-sector branch

The minimal nontrivial spectral branch is

$$\mathcal{S}_{1,2}. \tag{64}$$

We denote its sectorwise reference energy by

$$E_R := E_{1,2}^{\min}, \tag{65}$$

and refer to the corresponding branch as the doubled-sector branch.

4.6 Small-Amplitude Expansion

For $\epsilon \ll R_0$, the invariant admits an expansion

$$I_A(t) = I_0 + \epsilon I_1(t) + \epsilon^2 I_2(t) + \cdots, \qquad I_0 = \frac{\omega^2}{R_0^2\omega^2 + v^2}. \tag{66}$$

Correspondingly,

$$E_A = E_0 + \epsilon E_1 + \epsilon^2 E_2 + \cdots. \tag{67}$$

For centered sinusoidal modulation and primitive closed sectors, the first-order averaged correction vanishes over one full closure period in the generic symmetric case. Accordingly, the leading nontrivial spectral shift is typically quadratic:

$$E_A \approx E_0 + \epsilon^2 \Delta E_{p,q}. \tag{68}$$

4.7 Effective Spectral Law

At the structural level, the sectorwise spectrum may be written schematically as

$$E_{p,q}^{\text{eff}} = E_{\text{geom}}^{\min}(p,q) + E_{\text{top}}(p,q) + \delta E_{p,q}, \tag{69}$$

where $E_{\text{geom}}^{\min}$ represents the sectorwise minimum of the integrated curvature-torsion contribution, E_{top} encodes the energetic dependence on the winding sector (p,q), and $\delta E_{p,q}$ denotes higher-order corrections or branch-level corrections. Equivalently, one may write schematically,

$$E_{p,q}^{\text{eff}} \sim \lambda \int_0^T (\kappa^2 + \tau^2)\, ds + \mu\, \Phi(p,q) + \delta E_{p,q}, \tag{70}$$

where $\Phi(p,q)$ is a sector function encoding topological cost. At the present level of the theory, the topological contribution should be read as an effective bookkeeping term rather than as an independently derived action term.

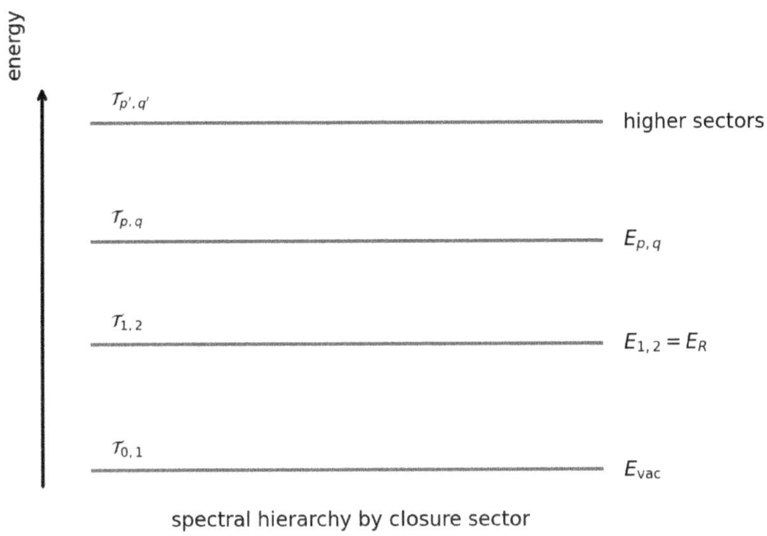

Figure 3: Schematic effective branch hierarchy of the e-Arc-Helix framework. The reference sector $\mathcal{T}_{0,1}$ serves as the baseline state, the doubled sector $\mathcal{T}_{1,2}$ defines the doubled-sector branch, and higher winding sectors populate the upper part of the hierarchy.

5 Reference Sector and Emergent Doubled Branch

We now introduce the reference-sector structure of the framework, in a form compatible with the variational viewpoint of [3]. Throughout this section, the notation E_{vac} is retained for continuity with earlier drafts, but it is interpreted strictly as the energy of the internal *reference sector* $\mathcal{T}_{0,1}$, not as a claim about an absolute physical vacuum in a more complete underlying theory.

Definition 5.1 (Reference-sector notation)**.** We write

$$\mathcal{T}_{\text{ref}} := \mathcal{T}_{0,1}, \qquad E_{\text{ref}} := E_{\text{vac}} := E_0, \tag{71}$$

where the subscript "ref" records the logical role of the unmodulated helix as the internal baseline sector. The notation "vac" is retained only for continuity with earlier drafts.

5.1 Geometric reference configuration

The natural reference configuration is the unmodulated helix,

$$R(t) = R_0. \tag{72}$$

We denote the corresponding state by

$$|0_A\rangle\,. \tag{73}$$

Its energy is

$$E_{\text{vac}} = E_0. \tag{74}$$

Within the present effective theory, this state serves as the baseline configuration relative to which the doubled and higher winding sectors are compared.

5.2 Reference sector

The reference configuration corresponds to the trivial sector

$$\mathcal{T}_{\text{ref}} = \mathcal{T}_{0,1}. \tag{75}$$

This sector has no nontrivial meridional winding and no doubled closure. It is therefore the natural baseline sector for the hierarchy introduced in Section 4.

5.3 Emergence of the doubled branch above the reference sector

When the modulation becomes commensurate with

$$\frac{\Omega}{\omega} = \frac{1}{2}, \tag{76}$$

the system enters the first doubled topological branch

$$\mathcal{T}_{1,2}. \tag{77}$$

This is the minimal doubled branch above the reference sector. In the language of the effective sectorwise description, it is the first sector whose global closure arithmetic differs essentially from that of the unmodulated reference helix.

Remark 5.2 (Self-mirror normal form of the minimal doubled branch)**.** In the centered gauge of Section 2, the sector $\mathcal{T}_{1,2}$ is the first closure class whose primitive period naturally occupies a doubled angular interval $u \in [0, 4\pi]$. The two half-traversals then define a centered doubled normal form on a common inertial plane, with mirror-symmetric projection-traversal circles and, in the interaction branch, antiperiodic signed normal deviation, exactly as in Propositions 2.4 and 2.5. This is the geometric reason the later released-mode and field-theoretic continuations are organized on the doubled circle and in odd half-modes.

5.4 Doubling order parameter

A simple order parameter for spontaneous doubling is

$$D = q - 1. \tag{78}$$

Then

$$D = 0 \quad \text{for the reference sector,} \qquad D = 1 \quad \text{for the minimal doubled sector.} \tag{79}$$

Equivalently, one may define the global-period mismatch

$$\Delta_T = \frac{T_{\text{global}}}{T_{\text{local}}} - 1 = q - 1. \tag{80}$$

Thus the onset of the doubled-sector branch may be read directly as the first nonzero value of the doubling order parameter.

5.5 Effective reference functional

The sectorized reference structure is organized by the functional

$$\mathcal{F}[X] = \lambda \int (\kappa^2 + \tau^2)\, ds + \mu\, \Phi_{\text{top}}[X]. \tag{81}$$

As in Section 4, this should be understood as an effective organizing functional. The relevant reference data are obtained by minimizing $\mathcal{F}$ subject to the sector constraint. Accordingly, one may regard

$$E_{\text{vac}}^{\text{eff}} := \inf_{X \in \mathcal{C}_{0,1}} \mathcal{F}[X], \qquad E_R^{\text{eff}} := \inf_{X \in \mathcal{C}_{1,2}} \mathcal{F}[X], \tag{82}$$

where $\mathcal{C}_{0,1}$ and $\mathcal{C}_{1,2}$ denote admissible representatives of the corresponding sectors. The reference sector minimizes $\mathcal{F}$ in the trivial class, while the minimal doubled branch is the first local minimum in the doubled class.

6 Recovery Structure within the e-Arc-Helix Ansatz

We now formulate the recovery principle of the model, completing the differential-topological reconstruction viewpoint initiated for arc helices in [2]. In the present effective setting, recovery means reconstruction of the *configuration class* determined by the modulated-helical ansatz, up to rigid motion and sector label, from a specified set of local invariants and closure data. It does not yet amount to an inversion theorem for arbitrary space curves, nor to recovery from a full microscopic field theory. The claim is therefore intentionally confined to the present ansatz-level effective description.

6.1 Recovery Data

The minimal doubled branch is specified, at the ansatz level, by three layers of data:

1. local differential data: $\kappa(t), \tau(t)$,
2. reduced closure data: the commensurability ratio $\Omega/\omega = p/q$ together with equivalent phase information,
3. topological sector label: $\mathcal{T}_{p,q}$.

Remark 6.1. The recovery problem considered here is restricted to the modulated arc-helix family introduced in Section 2. Within this family, the pair (κ, τ) together with reduced closure data is intended to determine an equivalence class of admissible configurations. No claim is made here about global identifiability for arbitrary space curves lying outside the present ansatz.

6.2 Frenet–Serret Recovery

By the fundamental theorem of space curves, a regular space curve is determined up to rigid motion by its curvature and torsion as functions of arc length [7]. Thus (κ, τ) determine the local shape through the Frenet–Serret system

$$\frac{dT}{ds} = \kappa N, \tag{83}$$

$$\frac{dN}{ds} = -\kappa T + \tau B, \tag{84}$$

$$\frac{dB}{ds} = -\tau N. \tag{85}$$

6.3 Recovery of Modulation Parameters

In the e-Arc-Helix ansatz, the invariants arise from the specific modulated structure

$$R(t) = R_0 + \epsilon \sin(\Omega t). \tag{86}$$

Hence the periodic content of $\kappa(t)$ and $\tau(t)$ contains, in principle, recoverable information about the parameters

$$R_0,\ \epsilon,\ \omega,\ \Omega,\ v. \tag{87}$$

This statement should be read as an ansatz-level identifiability claim: once one restricts attention to the family (2)–(3), the dominant harmonics of κ and τ encode the underlying frequency and amplitude data. A complete inversion formula is beyond the scope of the present paper.

6.4 Recovery of the Topological Sector

Once the ratio Ω/ω is reduced to lowest terms,

$$\frac{\Omega}{\omega} = \frac{p}{q}, \tag{88}$$

the global closure class is recovered as

$$\mathcal{T}_{p,q}. \tag{89}$$

6.5 Recovery Theorem

Theorem 6.2 (Ansatz-level recovery theorem for e-Arc-Helix sector data). *Let $X(t)$ belong to the modulated arc-helix family (2)–(3), and assume that it satisfies the rational closure condition*

$$\frac{\Omega}{\omega} = \frac{p}{q}, \qquad \gcd(p, q) = 1. \tag{90}$$

Then its arc-helix configuration class is recoverable, within this ansatz and up to rigid motion and toroidal isotopy, from the data

$$\big(\kappa(t), \tau(t), p, q\big). \tag{91}$$

Proof. By the fundamental theorem of space curves, the pair (κ, τ) determines the local space-curve geometry up to Euclidean motion [7]. Restricting to the modulated-helical ansatz (2)–(3) identifies the admissible family in which the recovered configuration is to lie. The reduced ratio p/q determines the closure arithmetic, the primitive period, and the associated topological sector $\mathcal{T}_{p,q}$, equivalently the torus-knot type $T(p, q)$. Together, these data determine the configuration class within the ansatz up to rigid motion and toroidal isotopy. □

Corollary 6.3 (Recovery of the minimal doubled sector). *Within the modulated arc-helix ansatz, the data $\big(\kappa(t), \tau(t), 1, 2\big)$ determine the minimal doubled sector $\mathcal{T}_{1,2}$ up to rigid motion and toroidal isotopy.*

Remark 6.4 (Scope of the recovery theorem). The recovery theorem is intentionally conditional. It does not assert that arbitrary space curves with the same invariants must lie in the e-Arc-Helix family, nor that the modulated ansatz is uniquely selected among all sectorwise competitors. What it proves is an ansatz-level reconstruction statement: once one works inside (2)–(3) and records the primitive closure data, the corresponding e-Arc-Helix configuration class is fixed up to the stated symmetries.

7 Structural Summary, Boundaries, and Outlook

Proposition 7.1 (Master structural synthesis of the e-Arc-Helix framework). *The manuscript establishes a one-way logical chain*

$$\text{differential geometry and primitive closure} \Longrightarrow \text{effective sectorwise organization and ansatz-level recovery} \Longrightarrow \text{downstream residual, variational, microscopic, and field continuations.} \tag{92}$$

More precisely:

- *(i) Sections 2–3 define the admissible e-Arc-Helix family, derive its Frenet invariants, and identify the primitive rational-closure sectors $\mathcal{T}_{p,q}$, with $\mathcal{T}_{1,2}$ singled out as the minimal doubled closure sector.*
- *(ii) Sections 4–6 construct an explicitly effective sectorwise organization from the curvature-torsion functional $E_A[X]$, identify the reference sector, and prove an ansatz-level recovery theorem inside the modulated e-Arc-Helix family.*

(iii) *Appendices A–D develop only logically downstream continuations of that structure: normalization maps, candidate families, variational releases, barrier-regularized selections, finite microscopic lifts, projective infinite released-mode closure, and odd-sector field embeddings.*

(iv) *No statement from Appendices A–D is used to strengthen, reinterpret, or silently modify the theorem-level content of Sections 2–6.*

Proof. Items (i)–(ii) simply collect the definitions, propositions, and theorems proved in Sections 2–6: the admissible-regime results of Section 2, the primitive closure and sector structure of Section 3, the effective branch organization of Sections 4–5, and the ansatz-level recovery theorem of Section 6. Item (iii) is exactly the reading protocol fixed in the Introduction and Appendix A, and item (iv) restates the one-way dependency rule (1). No additional argument is required beyond the organizational hypotheses already imposed on the manuscript. □

Proposition 7.1 is the correct reading key for the full paper. The main text fixes the geometric, closure, effective, and ansatz-level recovery architecture; the appendices continue that architecture downstream without being allowed to feed hidden assumptions back into it.

The sharpest purely structural consequence inside the main text is the distinguished role of the sector $\mathcal{T}_{1,2}$. In the periodically identified toroidal picture its standard torus representative is $T(1,2)$, which is unknotted in the classical knot-theoretic sense. What is nontrivial here is therefore not classical knot complexity by itself, but the appearance of the first doubled global closure period within the rational-closure classification.

Interpretive boundaries. Four boundaries govern how the downstream chain should be read. First, the passage from the continuous functional $E_A[X]$ to discrete branch language is an effective closure-and-holonomy construction, not a full operator quantization. Second, the weak-modulation asymptotics used at the start of Appendix A are diagnostic only: once the candidate-scale crossings move into the moderate regime in η, the evidence comes from exact quadrature or non-perturbative scans rather than from truncated perturbation theory. Third, the selector and barrier mechanisms developed in Appendices B–D now possess reduced-measure, simplex-pushforward, and Gamma-microcell origins inside the model, and the admissible reduced measure is uniquely characterized within the manuscript's intrinsic reduced class by its endpoint geometry and reversible balance law. Fourth, the successive released-mode tests exhibit damped rather than runaway drift. The appendices now close the projective-to-continuum passage on the manuscript's intrinsic tail-controlled support class, but genuine ansatz-independence still requires control beyond that intrinsic continuum universe rather than a short numerical tail fit alone.

External interfaces. These same boundaries clarify the manuscript's external interfaces. The work speaks directly to differential geometry, geometric topology, elastic-curve variational theory, reversible stochastic dynamics, and stochastic-field modeling. The model-internal descriptors *inertial plane, conjugate-time axis, light-end, magnetic-end, self-interaction, interaction, particle-type interaction*, and *wave-type interaction* belong to this internal dictionary and are not themselves external identifications. By contrast, any relation to QED or the Standard Model, to general relativity, to condensed-matter roton physics, or to local quantum field theory should presently be read as an open dictionary problem rather than as a completed identification.

Next mathematical steps. Natural next steps divide into two families. In the geometric direction, one would like sharper regularity and sector-minimization results for $E_{p,q}^{\min}$, together with stronger control of self-intersection, admissible closure classes, and ansatz-level inversion. In the downstream direction, the present revision now supplies finite-family existence, released-ladder compactness, intrinsic attainment, projective closure, a canonical intrinsic continuum support law, finite-Galerkin reversible generators, canonical strong realizations on every finite observable cylinder, exponential cylinder mixing, a unique stationary weak interaction-field law inside the intrinsic cylinder-Markov class, a canonical continuum Hunt semigroup that is ergodic on the full intrinsic support, coordinate-wise Fukushima decompositions for the continuum interaction field, a conditional full-continuum spectral-gap/log-Sobolev package under uniform cylinder coercivity, and bounded or integrably singular renormalized tail closure on the same intrinsic support universe, a coherent cylindrical Brownian realization driven by a common countable noise family, a full-support pathwise strong continuum realization on the intrinsic compact support, a triangular reflected infinite system driven by that same Brownian family, a closed reflected cylinder martingale problem characterizing that strong continuum lift on the intrinsic cylinder algebra, an oscillation-based criterion reducing intrinsic continuum coercivity to a uniform bounded-perturbation estimate on the cylinder interaction potentials, a summably local-block criterion under which that coercivity becomes automatic, a one-step local-increment criterion reducing exact quasilocal existence to the actual released-mode interaction increments, a canonical tail-envelope criterion reducing automatic coercivity further to local-envelope oscillation plus summably small released-mode tail memory, a renormalized intrinsic continuum interaction potential obtained from the genuine released-mode increments under summable envelope/tail control, a single weak reflected continuum field equation on the intrinsic finitely supported test space, a derivative-budget criterion reducing those renormalized interaction hypotheses to explicit summable partial-derivative bounds on the genuine released-mode increments themselves, a renormalized continuum interaction potential with continuous absolutely summable cylinder gradient, a weak reflected gradient-field formulation driven by that potential, a finite-cylinder derivative-Cauchy bridge reducing verification of that gradient package to the interaction potentials themselves, an explicit reflected gradient cylinder martingale formulation on the intrinsic cylinder core, a weighted cylinder-Lipschitz upgrade under summably controlled second-derivative budgets on the genuine released-mode increments, a finite-cylinder Hessian-Cauchy bridge reducing that stronger local regularity package to the interaction potentials themselves, a weighted cylinder-$C^{1,1}$ linearization and quadratic local-freezing estimate under summably controlled third-derivative budgets, a finite-cylinder third-derivative-Cauchy bridge reducing that affine local-closure package to the interaction potentials themselves, a weighted cylinder-$C^{2,1}$ quadratic local model with cubic local-freezing error under summably controlled fourth-derivative budgets, a finite-cylinder fourth-derivative-Cauchy bridge reducing that quadratic local-model package to the interaction potentials themselves, a weighted cylinder-$C^{3,1}$ cubic local model with quartic local-freezing error under summably controlled fifth-derivative budgets, a finite-cylinder fifth-derivative-Cauchy bridge reducing that cubic local-model package to the interaction potentials themselves, an intrinsic cubically frozen local reflected-gradient martingale atlas with quartic chart defect, a finite-cylinder fifth-derivative-Cauchy criterion already sufficient for that quartic local atlas, a vanishing-radius exact local reflected-gradient generator germ on the intrinsic cylinder core extracted from that quartic atlas, a concrete tensor-majorant criterion reducing fifth-order atlas activation further to summable mode-decay envelopes, a graph-closed non-cylinder extension of that exact local generator germ on intrinsic balls, a weighted row-sum kernel criterion reducing fifth-order atlas activation further to per-row mode envelopes, an exact local-generator sheaf patching those ballwise graph-closed operators across intrinsic open sets, a banded-plus-tail criterion reducing weighted row-sum verification further to finite local interaction windows plus one ambient summable tail profile, a chartwise differential realization of the resulting exact local-generator sheaf on a natural local Neumann test sheaf, a banded-plus-tail jet criterion reducing concrete identification of the resulting local differential drift further to first/second finite-cylinder jet budgets, a single natural non-cylinder local martingale domain obtained as the graph closure of that chartwise differential sheaf, a finite-cylinder local-window C^2-Cauchy criterion reducing activation of that single-domain package further to direct first/second derivative stabilization of the interaction potentials themselves, a chart-free local Itô/carré-du-champ package on a canonical square-gradient subdomain of that same natural domain driven by the already-constructed Brownian family, a one-profile finite-cylinder local-window/tail C^2 criterion reducing activation of that Itô package further to one radius sequence, one tail profile, and one amplitude budget on the interaction potentials themselves, a chart-free dual local reflection functional on the intrinsic finitely supported test space compatible with the same common Brownian family and the coordinate reflected local times, a graph-closed chart-free reflection operator obtained as the local Itô-graph closure of that finitely supported dual core, a chart-free first-order reflection calculus on a canonical multiplicative natural-Itô core generated by bounded cylinder compositions and closed under product and scalar chain rules, a chart-free common-contact-clock factorization of that closed first-order reflection calculus on the same natural-Itô domain, a chart-free finitely supported cylinder-gradient representation of the resulting common-contact-clock density package, a fixed-template telescoping C^2 criterion on the actual finite-cylinder interaction family compressing the same verification further to one common local window, one common tail profile, and one template seminorm sequence, a local-window/common-tail decomposition criterion reducing that same fixed-template verification further to one windowed shell core plus one common tail remainder, a chart-free coordinate-density realization of the same common-contact-clock gradient module, a chart-free algebraic contact-cone selector carried by that same common contact clock, a chart-free algebraic complementarity package for that same selector on the canonical coordinate contact wedge, a canonical frozen-tail criterion reducing that same fixed-window/common-tail verification further to one explicitly frozen local core plus its complementary tail remainder, a canonical shellwise frozen-tail seminorm criterion reducing that same frozen-tail verification further to one explicit seminorm of each telescoping finite-cylinder shell, a direct frozen-trace/difference shell seminorm criterion reducing that same frozen-tail verification further to one direct seminorm of each actual telescoping finite-cylinder shell together with its canonical frozen trace, an ambient coordinate-box normal-cone realization of that same chart-free algebraic complementarity selector carried by the common contact clock, a canonical frozen-segment oscillation criterion reducing the same direct frozen-trace verification further to one shellwise seminorm of the actual finite-cylinder shell along its canonical frozen-tail homotopies, a projected ambient-box Skorokhod decomposition for every finite observable cylinder driven by that same common contact clock, a chart-free intrinsic-support Skorokhod closure on the natural Itô domain's finitely supported cylinder core, a terminal canonical frozen-segment envelope criterion leaving no further abstract reduction on the interaction side, explicit tail-integrated actual e-Arc-Helix finite-cylinder interaction representatives extracted from the reduced infinite field law, a log-Laplace shell bridge reducing the remaining terminal verification to concrete frozen-segment C^2 estimates for the genuine infinite reduced free-energy shells, together with an exact posterior-cumulant reformulation of that same terminal shellwise verification under the actual tilted tail laws, and a posterior-variance-majorant simplification eliminating the remaining mixed-covariance bookkeeping, a posterior-quadratic-moment simplification absorbing the remaining posterior mean/variance split into weighted posterior L^2 shell budgets, a posterior-support essential-sup route replacing those remaining posterior moments by deterministic tilted-tail support envelopes, a frozen-tail support-interpolation majorant reducing those support envelopes further to canonical frozen-tail anchor values plus weighted tail-derivative transport budgets across the unreleased box, an anchor-plus-tail-transport compression reducing that stronger route further to four shellwise frozen-tail anchor and tail-transport budgets for the actual reduced free-energy shells, a tail-mass factorization compressing those same weighted transport budgets further to one explicit tail mass multiplying one m-uniform unreleased mixed-derivative envelope for each derivative order, a tail-profile factorization refining that same deterministic route further through profile-weighted tail masses and profile-normalized unreleased mixed-derivative envelopes, an explicit verification skeleton organizing the remaining actual-shell burden into four shellwise target quantities, a menu of candidate profiles, and one terminal summability test, a one-step released-coordinate anchor reduction compressing the first two rows of that dossier further to frozen-tail jets of the underlying infinite log-weight, a frozen-tail anchor-trace/self-transport reduction splitting those same one-step anchor jets further into frozen-trace values plus one released-coordinate self-transport envelope for each derivative order, and an n-truncated frozen-core/pre-shell-tail-transport split of those same frozen-trace values into frozen anchor cores plus one pre-shell transport envelope for each derivative order, and an axis-frozen/finite-core-transport split of those same frozen anchor cores into axis traces plus one low-core transport envelope for each derivative order, and an origin-axis/axial-height split of those same axis-frozen traces into origin-axis traces plus one axial-height transport envelope weighted by the chartwise axis height for each derivative order, and a generator-level quasilocal interface for the intrinsic odd-sector field; what remains is construction — or a sharp obstruction — of monotone intrinsic reductions for genuinely broader external classes, direct filling of that verification skeleton for the actual tail-integrated e-Arc-Helix shell family — choose one candidate profile family, bound the sector-zero vacuum traces together with the family-index transport envelopes, the sector-parameter transport envelopes, the axial-height transport envelopes, the low-core transport envelopes, the pre-shell tail-transport envelopes, the released-coordinate self-transport envelopes, and the two profile-normalized unreleased mixed-derivative envelopes, and close the resulting shell summability test of the lower-volume reference-vacuum terminal proposition — or, failing that, retreat first to the raw v91 vacuum-origin route, then through the v90 origin-axis route and the coarser tail-profile, tail-mass, frozen-tail anchor-plus-tail-transport, frozen-tail support-interpolation, posterior-support essential-sup, or posterior-quadratic-moment routes under their own tilted tail laws (hence also the posterior mean/variance-majorant criterion, the posterior-cumulant criterion, the direct log-Laplace criterion, the terminal canonical frozen-segment-envelope criterion, and all prior frozen-trace, frozen-tail, fixed-template, fixed-profile, one-seminorm, one-step, one-profile, local-window, jet, row-sum, tensor-majorant, and fifth-derivative criteria feeding the same natural-domain package), genuinely external relativistic/gauge-theoretic dictionary building beyond the intrinsic quasilocal interface, and extension to non-box or noncompact renormalized tail universes for the reflected odd-sector field dynamics.

The present manuscript should therefore be read as a master rigorous edition of the e-Arc-Helix framework. In this form, the e-Arc-Helix itself — not any single downstream label — remains the organizing object of the theory.

A Interface, Reading Protocol, and First Candidate

These appendices are designed to separate the downstream continuation program into explicit structural layers. Appendix A carries the interface, the reading protocol, and the first candidate. Appendix B records normalization shifts and variational selection rules. Appendix C collects stability tests and broader released-mode continuations. Appendix D contains the microscopic lift, the infinite released-mode closure, and the odd-sector field embedding. No derivation in Appendices A–D is used in any theorem or effective construction of the main text, and every downstream statement must remain subordinate to the kinematic, topological, spectral, reference-sector, and recovery structures already established in Sections 2–6 [4, 5].

A.1 Reading protocol, risk register, and block guide

Conceptually, the downstream continuation program is divided into four appendices. The logic always runs forward through them and never backward into the main text. Four cautions govern the reading. First, whenever branch language is built from E_A, it inherits the effective status explained in Section 4 rather than a full quantization claim. Second, asymptotic formulas derived under $\eta \ll 1$ are used only as diagnostics; candidate-scale comparisons at moderate η rely on exact quadrature or other non-perturbative evaluations. Third, the selector and barrier constructions developed later now have endogenous reduced-measure and microcell realizations, and those realizations are eventually shown to determine a unique admissible reduced measure within the manuscript's intrinsic class. Fourth, successive mode releases show a damped but nonzero drift of the preferred crossing branch; the projective infinite-mode and field-theoretic layers are therefore part of the mathematical burden of the appendices, not decorative afterthoughts.

1. **Appendix A — Interface and first candidate:** setup, normalization rules, claim ladder, and the first doubled-sector candidate with its controlled weak-modulation obstruction.
2. **Appendix B — Normalization shift and variational selection:** residual normalization, exact canonical evaluation, restricted minimization, selector construction, and barrier regularization.
3. **Appendix C — Stability and broader released-mode tests:** higher-mode robustness, broader doubled-family continuation, and damped mode-release diagnostics.
4. **Appendix D — Microscopic lift, infinite released-mode limit, and field embedding:** reduced-measure origins, microscopic stochastic lift, projective infinite closure, and odd-sector field dynamics.

A.2 Downstream theory-interface matrix

Table 2 records the scope of each appendix, its direct input data, and the highest logical status permitted inside it.

Appendix	Primary role	Depends directly on	Highest claim status	Boundary condition
A	Interface, normalization map, claim ladder, first candidate	Sections 2–6 only	Candidate interpretation / obstruction statement	No successful constant identification is asserted.
B	Residual normalization, variational releases, selector and barrier construction	Appendix A plus explicit downstream normalizations	Effective relation / model-internal variational continuation	No reverse support to theorems in the main text is allowed.
C	Higher-mode robustness, broader families, damped drift diagnostics	Appendices A–B	Local stability and broader-family continuation	No ansatz-independence is claimed without stronger infinite-mode control.
D	Microscopic lift, projective limit, odd-sector field embedding	Appendices A–C	Model-internal microscopic and stochastic-field embedding	No equivalence with QED, SM, GR, or local QFT is claimed.

Table 2: Downstream appendix blocks, direct dependencies, and claim ceilings.

A.3 Purpose and scope

The purpose of Appendix A is not to present a completed derivation of physical constants from the present framework. Rather, it fixes the interface through which later constant candidates, if viable, may be inserted without forcing retrospective changes to the main paper. The governing direction of dependence is therefore

$$\text{Sections 2–6} \Longrightarrow \text{Appendix A interface} \Longrightarrow \text{Appendices B–D continuations,} \qquad (93)$$

which is the appendix-level portion of the master dependency chain (1). In particular, no statement in Appendix A is allowed to supply hidden assumptions for the torus-sector theorem, the effective spectral interpretation, the reference-sector construction, or the ansatz-level recovery theorem.

Accordingly, Appendix A is restricted to interface tasks: scope control, admissible inputs, normalization rules, claim-strength labels, and a minimal insertion template. A single worked candidate is included in Section A.9 to demonstrate this interface discipline, and that same subsection now records its first downstream asymptotic, sector-comparison, and numerical diagnostic layer. Appendices B–D then test a growing sequence of admissible continuations of that candidate — alternative normalizations, exact canonical evaluation, restricted sectorwise minimization, reduced-envelope analysis, interior helicity selection, selector-coupled variational relaxation, barrier-regularized local stability, higher-mode robustness checks, reduced-measure reinterpretations of the helicity barrier, microscopic lifts, projective infinite released-mode closure, and field embedding. All such additions remain explicitly non-retroactive: they are evidence about downstream candidates, not premises for Sections 2–7.

A.4 Dependency map to main-text quantities

Any future candidate inserted into Appendix A may depend only on quantities already defined in the frozen main text, together with explicit downstream normalizations built from them. The admissible dependency layers are as follows.

(i) Differential-geometric inputs from Section 2. The primary local quantities are

$$\kappa(t), \qquad \tau(t), \qquad H_A = \frac{\tau}{\kappa}, \qquad I_A = \kappa^2 + \tau^2, \tag{94}$$

as well as the constant-radius reference quantities κ_0, τ_0, and the corresponding reference invariant values introduced in the main text.

(ii) Closure and sector data from Section 3. The admissible global-topological inputs are

$$\frac{\Omega}{\omega} = \frac{p}{q}, \qquad T = \frac{2\pi q}{\omega}, \qquad T(p,q), \qquad \mathcal{T}_{p,q}, \qquad D = q - 1. \tag{95}$$

Here the notational distinction between torus-knot type $T(p,q)$ and closure sector $\mathcal{T}_{p,q}$ must be preserved exactly.

(iii) Effective spectral and vacuum quantities from Sections 4–5. Where a candidate requires sectorwise energetic language, it may use only the already declared effective objects

$$E_A[X], \qquad E_{p,q}^{\min}, \qquad E_R, \qquad E_{\text{vac}}, \qquad \mathcal{F}[X], \tag{96}$$

with the understanding that these remain effective organizing quantities rather than microscopic observables.

(iv) Recovery-layer restrictions from Section 6. No candidate in Appendix A may strengthen the recovery theorem beyond the modulated arc-helix ansatz. Recovery data may be cited only as a structural constraint on admissible sector labels and not as a new source of arbitrary-curve identifiability.

The practical rule is therefore simple: Appendix A may read from Sections 2–6, but Sections 2–6 may not read back from Appendix A.

A.5 Normalization map

Any later constant candidate should, whenever possible, be formulated first in terms of dimensionless variables. The preferred normalized interface begins with

$$h := H_A = \frac{\tau}{\kappa}, \qquad \rho := \frac{\Omega}{\omega} = \frac{p}{q}, \qquad d := D = q - 1. \tag{97}$$

When the curvature-torsion density itself is used, the preferred reference normalization is the leading reference value I_0 from the weak-modulation expansion of Section 4, namely

$$\widehat{I}_A(t) := \frac{I_A(t)}{I_0}, \tag{98}$$

provided the reference regime is stated explicitly.

If dimensional quantities enter at a later stage, they must first be converted into an explicitly named normalized quantity before any candidate identification is proposed. Thus a future dimensional input Q_{dim} should only appear through a ratio of the form

$$\widehat{Q} := \frac{Q_{\mathrm{dim}}}{Q_{\mathrm{ref}}}, \tag{99}$$

where the reference scale Q_{ref} is declared in the same subsection. Bare dimensional equalities are discouraged at the interface stage.

Finally, when a later candidate uses sectorwise minima or vacuum comparisons, the normalization choice must be named explicitly; for example, one must specify whether the comparison is made relative to E_0, E_{vac}, E_R, or another stated reference quantity. This requirement is meant to prevent hidden renormalizations by notation alone.

A.6 Claim-strength labels

To preserve consistency with the frozen main text, every future statement in Appendix A should be marked by one of the following logical-status labels.

- **definition**: an exact introduced quantity, normalization, or sector-restricted notation.
- **effective relation**: a model-level organizing formula compatible with the effective language of Sections 4–5.
- **asymptotic estimate**: a perturbative, weak-modulation, or limiting expression whose regime of validity is stated.
- **candidate interpretation**: a structurally motivated proposal that maps a normalized geometric expression to a selected constant candidate.
- **conjectural identification**: a proposed identification for which a full derivation has not yet been completed.

The label should appear before a candidate formula is used interpretively, so that a symbolic structural relation is not mistaken for a theorem-level consequence of the present paper.

A.7 Admissibility checklist for candidate constants

Before any future constant candidate is inserted, it should pass the following checklist.

1. It uses only quantities already available from Sections 2–6, together with explicit downstream normalizations introduced in this appendix.

2. Its target expression is dimensionless, or else is converted into an explicitly normalized dimensionless ratio before interpretation.
3. Its derivation path states the regime being used: exact, effective, asymptotic, sector-minimized, doubled-branch restricted, or otherwise.
4. It does not alter the scope of the torus-knot theorem, the effective spectral interpretation, the reference-sector construction, or the ansatz-level recovery theorem.
5. It preserves the notation $T(p,q)$ for torus-knot type and $\mathcal{T}_{p,q}$ for doubled sector.
6. It distinguishes clearly between equality, approximation, fit, interpretation, and conjecture.
7. If a specific sector such as $\mathcal{T}_{1,2}$ is singled out, the reason for that choice is stated structurally and not only by retrospective numerical matching.
8. Any numerical comparison, if later included, is downstream evidence only and is not allowed to function as a hidden premise for Sections 2–7.

A candidate that fails any of the above conditions should be regarded as outside the admissible scope of the present appendix architecture.

A.8 Minimal template for future candidate insertion

Each future worked candidate should be inserted using a minimal template of the following form.

1. **Target quantity.** Name the constant candidate being discussed.
2. **Status label.** Assign one of the labels from Section A.6.
3. **Input data.** List the exact main-text quantities being used, with section references if needed.
4. **Normalization step.** Convert all inputs to the normalized interface variables of Section A.5.
5. **Derivation path.** State whether the formula is exact, effective, asymptotic, or sector-restricted.
6. **Candidate expression.** Present the proposed formula with its equality or approximation symbol justified.
7. **Interpretive status.** Explain whether the result is a structural candidate, an asymptotic estimate, or only a conjectural identification.
8. **Non-retroactivity note.** State explicitly that the candidate is downstream of Sections 2–7 and does not modify their logical content.

At the purely schematic level, a later insertion should therefore look like

$$\mathcal{C}_{\text{cand}} = F\Big(h, \widehat{I}_A, \rho, d; \text{sector restriction}\Big), \tag{100}$$

with every symbol either already defined in the main text or declared as an explicit normalization in the present appendix.

A.9 First worked candidate: doubled-sector inverse-gap coupling

As a first fully labeled example, we record a downstream candidate built only from the vacuum reference state and the minimal doubled branch. The purpose of this subsection is not to claim a completed derivation of a physical constant, but to show how a concrete candidate can be inserted into Appendix A without feeding back into Sections 2–7.

Target quantity. A first inverse-coupling-style candidate attached to the doubled sector $\mathcal{T}_{1,2}$. In possible later applications one may ask whether this quantity should be compared with a fine-structure-type constant, but no such identification is assumed at the start of the present construction.

Status label. The normalized gap introduced below has status **definition**. The inverse-gap expression built from it has status **candidate interpretation**. Any later comparison with a named physical constant has status **conjectural identification**.

Input data. From Section 4 we use the sectorwise minimum $E_R = E_{1,2}^{\min}$. From Section 5 we use the vacuum reference energy $E_{\text{vac}} = E_0$. From Section 3 we use the closure data of the minimal doubled sector, namely $(p, q) = (1, 2)$, so that

$$\rho_R = \frac{1}{2}, \qquad d_R = 1, \qquad \Theta_R := 2\pi q = 4\pi. \tag{101}$$

Here Θ_R is only the primitive total phase advance of the doubled closure cycle and is introduced as a downstream normalization, not as a new main-text dynamical variable.

Normalization step. Define the dimensionless doubled-sector gap by

$$\widehat{\Delta}_R := \frac{E_R - E_{\text{vac}}}{E_{\text{vac}}}. \tag{102}$$

This is the preferred vacuum-normalized measure of the first doubled-sector energy-gap cost. More generally, one could define $\widehat{\Delta}_{p,q} := (E_{p,q}^{\min} - E_{\text{vac}})/E_{\text{vac}}$, but for the first worked candidate we stay with the minimal doubled branch.

Derivation path. The construction is *effective* and *sector-restricted*. It uses only the spectral and vacuum language of Sections 4–5 together with the closure arithmetic of Section 3. It is not an exact theorem about microscopic dynamics, nor does it add any new recovery input beyond the ansatz-level data already fixed in Section 6.

Candidate expression. Using the doubled-sector closure factor $\Theta_R = 4\pi$, define the closure-normalized inverse-gap quantity

$$\mathcal{A}_R := \frac{\Theta_R}{\widehat{\Delta}_R} = \frac{4\pi E_{\text{vac}}}{E_R - E_{\text{vac}}}. \tag{103}$$

Equivalently, the direct coupling-style form is

$$g_R := \mathcal{A}_R^{-1} = \frac{1}{4\pi}\frac{E_R - E_{\text{vac}}}{E_{\text{vac}}}. \tag{104}$$

At the level of a downstream named-constant proposal, one may then write

$$\alpha_{\text{cand}}^{-1} \stackrel{?}{=} \mathcal{A}_R, \qquad \alpha_{\text{cand}} \stackrel{?}{=} g_R, \tag{105}$$

where the question mark is essential: it signals a conjectural identification rather than a theorem-level consequence of the present paper.

Interpretive status. The structural reason for singling out $\mathcal{T}_{1,2}$ is not retrospective numerics, but the fact that it is the first nontrivial closed sector above vacuum and the first sector with doubled global closure. The quantity $\mathcal{A}_R$ should therefore be read as a doubled-sector inverse-gap candidate generated by the existing effective geometry. At the definitional stage no numerical fit or calibration is built in; any numerical check that appears below is a downstream diagnostic only, not a premise of the construction.

Non-retroactivity note. This worked candidate is downstream of Sections 2–7. It does not strengthen the torus-knot theorem, the effective spectral law, the reference-sector construction, or the ansatz-level recovery theorem. It merely packages already-declared quantities into a dimensionless candidate associated with the minimal doubled sector.

Asymptotic evaluation. Introduce the dimensionless modulation amplitude

$$\eta := \frac{\epsilon}{R_0}, \qquad h := \frac{v}{R_0 \omega}, \qquad \rho := \frac{\Omega}{\omega} = \frac{p}{q}. \tag{106}$$

Then the vacuum energy over one primitive local cycle is

$$E_{\text{vac}} = \frac{2\pi\lambda}{R_0\sqrt{1+h^2}}. \tag{107}$$

Evaluating the effective functional (60) on the canonical modulated branch with primitive closure period $T = 2\pi q/\omega$ gives the **asymptotic estimate**

$$E_{p,q} \approx \frac{2\pi q \lambda}{R_0\sqrt{1+h^2}} \left[1 + \eta^2 \mathcal{K}(h,\rho) + O(\eta^3)\right], \tag{108}$$

where

$$\mathcal{K}(h,\rho) := \frac{8h^4\rho^4 + 10h^2\rho^4 - 11h^2\rho^2 - h^2 + 2\rho^4 - 5\rho^2 + 2}{4(1+h^2)^2}. \tag{109}$$

Because the modulation is centered and sinusoidal, the integrated $O(\eta)$ contribution vanishes over the primitive closure cycle, so the first nontrivial correction is quadratic. Accordingly,

$$\widehat{\Delta}_{p,q} \approx (q-1) + q\eta^2 \mathcal{K}(h,\rho) + O(\eta^3). \tag{110}$$

For the doubled sector $\rho_R = 1/2$, this becomes

$$\widehat{\Delta}_R \approx 1 + \eta^2 \frac{4h^4 - 25h^2 + 7}{16(1+h^2)^2} + O(\eta^3), \tag{111}$$

and hence

$$\mathcal{A}_R \approx 4\pi \left[1 - \eta^2 \frac{4h^4 - 25h^2 + 7}{16(1+h^2)^2} + O(\eta^3)\right]. \tag{112}$$

Thus the current candidate is asymptotically of order 4π, not of parametrically large inverse-coupling size.

Sector comparison. For a general primitive nonvacuum sector, define the closure-normalized inverse-gap quantity

$$\mathcal{A}_{p,q} := \frac{2\pi q}{\widehat{\Delta}_{p,q}}. \tag{113}$$

This quantity has status **candidate interpretation**. Combining this with (110) yields the **effective relation**

$$\mathcal{A}_{p,q} \approx \frac{2\pi q}{q-1} \left[1 - \frac{q}{q-1}\eta^2 \mathcal{K}(h,\rho) + O(\eta^3)\right]. \tag{114}$$

At zeroth order one therefore finds

$$\mathcal{A}^{(0)}_{1,2} = 4\pi, \qquad \mathcal{A}^{(0)}_{p,3} = 3\pi, \qquad \mathcal{A}^{(0)}_{p,4} = \frac{8\pi}{3}, \qquad \mathcal{A}^{(0)}_{p,5} = \frac{5\pi}{2}, \ \ldots \tag{115}$$

Since $q \mapsto 2\pi q/(q-1)$ is strictly decreasing for $q \geq 2$, the doubled sector is already the largest inverse-gap candidate at leading order among primitive nonvacuum sectors. This gives a sector-comparison reason—independent of retrospective numerics—for testing the doubled sector first.

If one wishes to remove the trivial closure multiplicity before comparing sectors, the natural residual quantity is

$$\delta^{\mathrm{exc}}_{p,q} := \widehat{\Delta}_{p,q} - (q-1) \approx q\eta^2 \mathcal{K}(h,\rho) + O(\eta^3). \tag{116}$$

For the doubled-sector branch,

$$\delta^{\mathrm{exc}}_{R} \approx \eta^2 \frac{4h^4 - 25h^2 + 7}{16(1+h^2)^2} + O(\eta^3). \tag{117}$$

This residual is the first place where sector dependence survives after the trivial $(q-1)$ baseline is stripped off.

Numerical layer. For reference, NIST/CODATA 2022 gives $\alpha^{-1}_{\mathrm{CODATA}} = 137.035999177(21)$. [6] A literal identification $\alpha^{-1}_{\mathrm{cand}} \stackrel{?}{=} \mathcal{A}_R$ would require the gap size

$$\widehat{\Delta}^{\mathrm{req}}_{R} := \frac{4\pi}{\alpha^{-1}_{\mathrm{CODATA}}} \approx 9.17012 \times 10^{-2}. \tag{118}$$

By contrast, the asymptotic estimate (111) gives

$$\widehat{\Delta}_R = 1 + O(\eta^2), \qquad \mathcal{A}_R = 4\pi + O(\eta^2) \approx 12.5664 + O(\eta^2). \tag{119}$$

Thus the naive vacuum-normalized candidate undershoots the inverse fine-structure constant by the factor

$$\frac{\alpha^{-1}_{\mathrm{CODATA}}}{4\pi} \approx 10.905. \tag{120}$$

Within the present weak-modulation asymptotic regime, the first worked candidate therefore does *not* numerically reproduce the inverse fine-structure constant.

A weaker diagnostic question is whether the closure-subtracted residual scale (117) could at least be of the required order of magnitude. The coefficient of η^2 in (117) obeys

$$-\frac{57}{256} \le \frac{4h^4 - 25h^2 + 7}{16(1+h^2)^2} \le \frac{7}{16}. \tag{121}$$

Therefore even on the most favorable positive branch one would need

$$\eta \gtrsim \sqrt{\frac{4\pi/\alpha^{-1}_{\mathrm{CODATA}}}{7/16}} \approx 0.458 \tag{122}$$

to reach the scale in (118). That requirement is already outside the strict weak-modulation regime $\eta \ll 1$. The numerical layer therefore points to a clear conclusion: the present Appendix-A candidate is structurally well defined, but any successful fine-structure identification would need either a different normalization, an additional dynamical enhancement, or a non-perturbative treatment beyond the current asymptotic layer.

Controlled obstruction summary. The three downstream layers above can be compressed into one structural obstruction statement and one leading-order comparison statement.

Proposition A.1 (Weak-modulation obstruction for the vacuum-normalized doubled-sector candidate). *Assume the weak-modulation regime $\eta \ll 1$ and retain the vacuum-normalized doubled-sector definition*

$$\widehat{\Delta}_R = \frac{E_R - E_{\mathrm{vac}}}{E_{\mathrm{vac}}}, \qquad \mathcal{A}_R = \frac{4\pi}{\widehat{\Delta}_R}. \tag{123}$$

Then

$$\mathcal{A}_R = 4\pi + O(\eta^2). \tag{124}$$

Hence, within this regime, the candidate cannot reach a parametrically large inverse-coupling scale without an additional normalization or enhancement mechanism.

Proof. Equation (111) gives $\widehat{\Delta}_R = 1 + O(\eta^2)$. Substituting this into (123) and expanding $(1+x)^{-1} = 1+O(x)$ yields (124). The final sentence follows because the weak-modulation regime keeps $\mathcal{A}_R$ in an $O(1)$ neighborhood of 4π, not at an inverse-coupling scale of order 10^2. □

Corollary A.2 (Leading-order maximality of the doubled branch). *For the zeroth-order inverse-gap family*

$$\mathcal{A}^{(0)}_{p,q} = \frac{2\pi q}{q-1}, \qquad q \geq 2, \tag{125}$$

the maximum over primitive nonvacuum sectors occurs at $q = 2$. Thus the doubled sector $\mathcal{T}_{1,2}$ is the first sector to test within the present inverse-gap candidate family.

Proof. The function $q \mapsto q/(q-1)$ is strictly decreasing for $q > 1$. Applying this monotonicity to the zeroth-order form extracted from (114) proves the claim. □

Remark A.3. The first worked candidate now has three explicit downstream layers: a structural definition/candidate layer, an asymptotic sector-comparison layer, and a numerical diagnostic layer. What this extended analysis presently delivers is not a completed constant derivation but a controlled obstruction: under the current vacuum-normalized definition and weak-modulation asymptotics, the candidate remains of order 4π, so matching a fine-structure-type inverse coupling would require a further mechanism beyond the present layer.

B Normalization Shift and Variational Selection

This appendix replaces the obstructed vacuum-normalized route by a residual normalization, then studies how far that route can be stabilized by exact quadrature, restricted minimization, selector weights, and barrier regularization. The logical status rises from definition to effective relation and numerical diagnostic, but not to theorem-level main-text content.

B.1 Standing conventions for Appendices B–D

To standardize the downstream notation, Appendices B–D use a common released-mode bookkeeping.

Released-mode index. The integer $m \geq 0$ counts the number of noncanonical odd half-modes released beyond the canonical doubled profile. We write

$$\mathbf{a}^{[0]} := \varnothing, \qquad \mathbf{a}^{[1]} := (\sigma), \qquad \mathbf{a}^{[2]} := (\sigma, \chi), \qquad \mathbf{a}^{[3]} := (\sigma, \chi, \psi), \tag{126}$$

and more generally $\mathbf{a}^{[m]} = (a_1, \ldots, a_m)$, where a_j multiplies the odd half-mode $\sin\big((2j+1)\omega t/2\big)$.

Residual-gap notation. For a fixed released family, $\widetilde{\Delta}^{[m],\mathrm{raw}}_R(\eta, h, \mathbf{a}^{[m]})$ denotes the closure-subtracted residual gap before selector or measure corrections. Whenever the gap is positive, the associated inverse-residual proxy is

$$\mathcal{B}^{[m]}_R := \frac{4\pi}{\widetilde{\Delta}^{[m]}_R}. \tag{127}$$

When no confusion can arise, the superscript raw will be suppressed.

Regularized free energies. The barrier action and the measure-corrected free energy are written schematically as

$$\mathfrak{F}_R^{[m]} := \log \widetilde{\Delta}_R^{[m],\mathrm{raw}} - \log\left(\frac{h}{h_*}\right) - 2\log\left(1 - \frac{h}{h_*}\right), \tag{128}$$

and

$$\mathfrak{G}_R^{[m]} := \log \widetilde{\Delta}_R^{[m]} - \log \rho_*(h), \qquad \rho_*(h) = \frac{12}{h_*}\frac{h}{h_*}\left(1 - \frac{h}{h_*}\right)^2. \tag{129}$$

Once the measure $\rho_*(h)$ is fixed, $\mathfrak{F}_R^{[m]}$ and $\mathfrak{G}_R^{[m]}$ differ only by the additive constant $-\log(12/h_*)$.

Alias conventions. The concrete notations already used in the manuscript are retained as aliases for the first four releases:

$$\widetilde{\Delta}_R^{\mathrm{can}} = \widetilde{\Delta}_R^{[0]}, \qquad \widetilde{\Delta}_R^{\mathrm{var}} = \widetilde{\Delta}_R^{[1]}, \qquad \widetilde{\Delta}_R^{(2)} = \widetilde{\Delta}_R^{[2]}, \qquad \widetilde{\Delta}_R^{(3)} = \widetilde{\Delta}_R^{[3]}, \tag{130}$$

and similarly for the corresponding $\mathfrak{F}_R$-, $\mathfrak{G}_R$-, stable-branch, and crossing notations.

Reading rule. Every proposition in Appendices B–D should be read as an internal consistency, stability, convergence, or exact-reduction statement inside the downstream program. None upgrades the claim ceiling of Sections 2–6.

B.2 Finite released-family existence and local continuation

Before turning to particular downstream normalizations, it is useful to record the common finite-dimensional existence theorem that underlies the barrier-regularized branches used throughout Appendices B–D.

Structural admissible domains. For a fixed release level $m \geq 0$ and modulation parameter η, let $\mathcal{D}_m(\eta) \subset \mathbb{R}^m$ be a compact amplitude domain, with the convention $\mathcal{D}_0(\eta) := \{\varnothing\}$. Set

$$\overline{\mathcal{U}}_\eta^{[m]} := [0, h_*] \times \mathcal{D}_m(\eta), \qquad \mathcal{U}_\eta^{[m]} := (0, h_*) \times \mathcal{D}_m(\eta). \tag{131}$$

Assume that the raw residual gap $\widetilde{\Delta}_R^{[m]}(\eta, h, \mathbf{a})$ extends continuously to $\overline{\mathcal{U}}_\eta^{[m]}$, is C^2 on $\mathcal{U}_\eta^{[m]}$, and obeys the positive bounds

$$0 < \delta_{\eta,m} \leq \widetilde{\Delta}_R^{[m]}(\eta, h, \mathbf{a}) \leq \Delta_{\eta,m}^* < \infty \qquad \text{on } \overline{\mathcal{U}}_\eta^{[m]}. \tag{132}$$

Define the extended reduced free energy by

$$\widehat{\mathfrak{G}}_R^{[m]}(\eta, h, \mathbf{a}) := \begin{cases} \log \widetilde{\Delta}_R^{[m]}(\eta, h, \mathbf{a}) - \log \rho_*(h), & (h, \mathbf{a}) \in \mathcal{U}_\eta^{[m]}, \\ +\infty, & h \in \{0, h_*\}. \end{cases} \tag{133}$$

Theorem B.1 (Finite released-family existence and local continuation). *Fix $m \geq 0$ and assume (132). Then:*

(i) The extended free energy $\widehat{\mathfrak{G}}_R^{[m]}(\eta, \cdot)$ attains a global minimum on $\overline{\mathcal{U}}_\eta^{[m]}$.

(ii) Every global minimizer lies in the interior $\mathcal{U}_\eta^{[m]}$ and therefore satisfies the stationarity system

$$\partial_h \log \widetilde{\Delta}_R^{[m]}(\eta, h, \mathbf{a}) = \frac{1}{h} - \frac{2}{h_* - h}, \qquad \nabla_{\mathbf{a}} \log \widetilde{\Delta}_R^{[m]}(\eta, h, \mathbf{a}) = 0, \tag{134}$$

with the second equation vacuous when $m = 0$.

(iii) Let η_0 be fixed and let $(h_0, \mathbf{a}_0) \in \mathcal{U}_{\eta_0}^{[m]}$ be an interior minimizer whose Hessian

$$H_0 := \nabla^2_{(h,\mathbf{a})} \mathfrak{G}_R^{[m]}(\eta_0, h_0, \mathbf{a}_0) \tag{135}$$

is positive definite. If $\widetilde{\Delta}_R^{[m]}$ is C^1 in η near η_0, then there exist an interval $I \ni \eta_0$ and a unique C^1 branch

$$\eta \mapsto \left(h_{\text{stab}}^{[m]}(\eta), \mathbf{a}_{\text{stab}}^{[m]}(\eta)\right) \tag{136}$$

of interior critical points with

$$\left(h_{\text{stab}}^{[m]}(\eta_0), \mathbf{a}_{\text{stab}}^{[m]}(\eta_0)\right) = (h_0, \mathbf{a}_0), \tag{137}$$

and the Hessian remains positive definite on I after shrinking I if necessary. In particular, the branch consists of strict local minima of $\mathfrak{G}_R^{[m]}$.

Proof. Because $\rho_*(h) = \frac{12}{h_*}\frac{h}{h_*}(1 - h/h_*)^2$, one has $-\log \rho_*(h) \to +\infty$ as $h \downarrow 0$ or $h \uparrow h_*$. Combined with the positive upper and lower bounds (132), this shows that $\widehat{\mathfrak{G}}_R^{[m]}$ is lower semicontinuous on the compact set $\overline{\mathcal{U}}_\eta^{[m]}$, finite on the interior, and infinite on the two helicity endpoints. Hence it attains a global minimum, proving (i), and no minimizer can lie on the boundary $h \in \{0, h_*\}$, proving the interior part of (ii). Any interior minimizer of a C^2 function satisfies vanishing first derivatives, which yields (134).

For (iii), define

$$F(\eta, h, \mathbf{a}) := \nabla_{(h,\mathbf{a})} \mathfrak{G}_R^{[m]}(\eta, h, \mathbf{a}). \tag{138}$$

The hypotheses imply that F is C^1 near $(\eta_0, h_0, \mathbf{a}_0)$ and that

$$D_{(h,\mathbf{a})} F(\eta_0, h_0, \mathbf{a}_0) = H_0 \tag{139}$$

is invertible. The implicit-function theorem therefore yields a unique C^1 branch of critical points through $(h_0, \mathbf{a}_0)$. Positive definiteness of the Hessian persists under small perturbations by continuity, so after possibly shrinking I the resulting branch remains a branch of strict local minima. □

How the later appendices use Theorem B.1. The numerical branches displayed later in Appendices B–D are not substitutes for Theorem B.1; they are explicit instances of it. The theorem provides the structural existence and local-branch mechanism, while the later tables and scans identify concrete parameter windows in which those hypotheses are numerically realized for the specific doubled families under study.

B.3 Worked continuation I: closure-subtracted residual normalization

The obstruction for $\mathcal{A}_R$ is driven by the order-one baseline $q - 1$ already present in (110). A natural admissible continuation is therefore to remove that baseline explicitly before attempting any inverse-coupling interpretation.

Status label. The residual normalization introduced below has status **definition**. The inverse residual quantity built from it has status **candidate interpretation**. Its asymptotic scaling has status **asymptotic estimate**.

Definition. For any primitive nonvacuum sector define the closure-subtracted residual gap

$$\widetilde{\Delta}_{p,q} := \frac{E_{p,q}^{\min} - qE_{\text{vac}}}{E_{\text{vac}}} = \widehat{\Delta}_{p,q} - (q-1). \tag{140}$$

For the doubled branch this becomes

$$\widetilde{\Delta}_R := \frac{E_R - 2E_{\text{vac}}}{E_{\text{vac}}} = \widehat{\Delta}_R - 1. \tag{141}$$

The corresponding inverse residual family is then

$$\mathcal{B}_{p,q} := \frac{2\pi q}{\widetilde{\Delta}_{p,q}}, \qquad \mathcal{B}_R := \frac{4\pi}{\widetilde{\Delta}_R}. \tag{142}$$

This is a genuinely new normalization choice; it must not be conflated with the vacuum-normalized family $\mathcal{A}_{p,q}$.

Asymptotic evaluation. Equations (116)–(117) give immediately

$$\widetilde{\Delta}_{p,q} \approx q\eta^2\mathcal{K}(h,\rho) + O(\eta^3), \tag{143}$$

and hence, whenever $\mathcal{K}(h,\rho) \neq 0$,

$$\mathcal{B}_{p,q} \approx \frac{2\pi}{\eta^2\mathcal{K}(h,\rho)}\left[1 + O(\eta)\right]. \tag{144}$$

For the doubled branch one obtains

$$\widetilde{\Delta}_R \approx \eta^2\frac{4h^4 - 25h^2 + 7}{16(1+h^2)^2} + O(\eta^3), \tag{145}$$

so that

$$\mathcal{B}_R \approx \frac{64\pi(1+h^2)^2}{\eta^2\left(4h^4 - 25h^2 + 7\right)}\left[1 + O(\eta)\right]. \tag{146}$$

Unlike $\mathcal{A}_R$, which stays of order one, the residual-normalized family scales as η^{-2} on any branch where the quadratic coefficient is positive.

Proposition B.2 (Residual normalization removes the leading closure obstruction)**.** *Assume the weak-modulation regime and define $\widetilde{\Delta}_{p,q}$ and $\mathcal{B}_{p,q}$ by* (140) *and* (142)*. If $\mathcal{K}(h,\rho) \neq 0$, then*

$$\mathcal{B}_{p,q} = \frac{2\pi}{\eta^2\mathcal{K}(h,\rho)}\left[1 + O(\eta)\right]. \tag{147}$$

In particular, the residual-normalized family is not asymptotically pinned to an order-one scale by the trivial closure baseline $q-1$.

Proof. Substitute (143) into (142) and factor out $q\eta^2\mathcal{K}(h,\rho)$. The conclusion follows from the expansion $(1+x)^{-1} = 1 + O(x)$. □

Low-helicity positive branch. For the doubled sector the quadratic coefficient is positive when

$$4h^4 - 25h^2 + 7 > 0. \tag{148}$$

The small-h branch of this condition is

$$0 \leq h < h_* := \sqrt{\frac{25 - \sqrt{513}}{8}} \approx 0.542. \tag{149}$$

On this branch the residual candidate $\mathcal{B}_R$ is positive and naturally large for small η.

Leading-order numerical implication. If one tentatively asks for a fine-structure-scale match at leading order, then (146) gives the required modulation amplitude

$$\eta_{\text{req}}^{\text{LO}}(h) := \sqrt{\frac{64\pi(1+h^2)^2}{\alpha_{\text{CODATA}}^{-1}(4h^4-25h^2+7)}} \qquad \text{on the branch } 4h^4-25h^2+7>0. \tag{150}$$

Representative values are

$$\eta_{\text{req}}^{\text{LO}}(0) \approx 0.458, \qquad \eta_{\text{req}}^{\text{LO}}(0.3) \approx 0.576. \tag{151}$$

Thus the residual route is no longer obstructed by scale, but it is pushed out of the strict $\eta \ll 1$ regime and therefore calls for a non-perturbative evaluation.

B.4 Worked continuation II: non-perturbative canonical evaluation

The second admissible continuation is to keep the residual normalization but remove the weak-modulation expansion. To do this without modifying any main-text definition, we evaluate the exact effective functional on the canonical doubled branch itself.

Status label. The canonical quantities introduced below have status **definition**. Their exact quadrature values have status **effective relation** and **numerical diagnostic**. They are not theorem-level evaluations of the true sector minimum.

Canonical doubled representative. Let $X_R^{\text{can}}(t;\eta,h)$ denote the doubled-sector ansatz obtained from (2)–(3) with

$$\frac{\Omega}{\omega}=\frac{1}{2}, \qquad \eta=\frac{\epsilon}{R_0}, \qquad h=\frac{v}{R_0\omega}. \tag{152}$$

Define its exact branch energy by

$$E_R^{\text{can}}(\eta,h) := E_A[X_R^{\text{can}}(\eta,h)], \tag{153}$$

where the integral in (60) is taken over the full primitive doubled period $0 \le t \le 4\pi/\omega$ and uses the exact curvature and torsion formulas from Section 2 rather than their weak-modulation expansion.

Residual proxy candidate. Using the same closure subtraction as in (141), define

$$\widetilde{\Delta}_R^{\text{can}}(\eta,h) := \frac{E_R^{\text{can}}(\eta,h)-2E_{\text{vac}}}{E_{\text{vac}}}, \qquad \mathcal{B}_R^{\text{can}}(\eta,h) := \frac{4\pi}{\widetilde{\Delta}_R^{\text{can}}(\eta,h)}. \tag{154}$$

Because both numerator and denominator carry the same overall λ/R_0 scaling, the ratio depends only on the dimensionless pair (η,h).

Interpretive caution. The proxy $\mathcal{B}_R^{\text{can}}$ is not identical to the minimized residual candidate $\mathcal{B}_R$, because E_R^{can} is the energy of a canonical branch representative, not the infimum over the full admissible doubled sector. What it does provide is a non-perturbative branchwise diagnostic that can be compared consistently with the asymptotic residual family.

Representative quadrature results. Direct numerical quadrature of (153) using the exact Section 2 curvature-torsion data yields the following low-helicity crossing values for the CODATA reference scale:

h	$\eta_*^{\rm can}$ such that $\mathcal{B}_R^{\rm can} \approx \alpha_{\rm CODATA}^{-1}$	$\mathcal{B}_R^{\rm can}(\eta_*^{\rm can}, h)$
0.05	0.455	137.036
0.10	0.468	137.036
0.20	0.516	137.036
0.30	0.567	137.036

Table 3: Non-perturbative canonical doubled-branch crossings for the residual-normalized inverse candidate. The values are obtained by direct quadrature of the exact energy functional on the canonical branch and should be read as downstream diagnostics rather than minimized-sector predictions.

These crossings are significant in one limited but useful sense: unlike the vacuum-normalized family $\mathcal{A}_R$, the residual-normalized canonical proxy does pass through the inverse fine-structure scale on a moderate-modulation band. At the same time, they are not yet a completed derivation. The crossings occur outside the strict $\eta \ll 1$ asymptotic regime, and they are computed on the canonical doubled branch rather than on the exact sector minimum.

B.5 Worked continuation III: restricted sectorwise minimization

The third admissible continuation is to keep the exact residual normalization but enlarge the doubled family just enough to allow a first internal minimization step while preserving primitive doubled closure.

Status label. The Fourier-enriched family introduced below has status **definition**. The variational inequality it implies has status **effective relation**. The crossing table that follows has status **numerical diagnostic**. None of these statements is a theorem-level identification of the full sector minimum E_R.

Fourier-enriched doubled family. For fixed (η, h) consider the doubled family

$$X_R^{\rm var}(t;\eta,h,\sigma) := \big(R_{\rm var}(t)\cos(\omega t),\ R_{\rm var}(t)\sin(\omega t),\ vt\big), \tag{155}$$

with

$$R_{\rm var}(t) := R_0\left[1+\eta\sin\left(\frac{\omega t}{2}\right)+\sigma\sin\left(\frac{3\omega t}{2}\right)\right], \qquad |\sigma| \le \frac{\eta}{2}. \tag{156}$$

The half-frequency term keeps the primitive doubled closure, and the added 3/2-harmonic preserves the same primitive period $4\pi/\omega$ while allowing a first non-canonical shape relaxation inside the sector $\mathcal{T}_{1,2}$.

Define the restricted doubled energy by

$$E_R^{\rm var}(\eta,h) := \inf_{|\sigma|\le\eta/2} E_A[X_R^{\rm var}(\eta,h,\sigma)]. \tag{157}$$

Using the same closure subtraction as before, define

$$\widetilde{\Delta}_R^{\rm var}(\eta,h) := \frac{E_R^{\rm var}(\eta,h)-2E_{\rm vac}}{E_{\rm vac}}, \qquad \mathcal{B}_R^{\rm var}(\eta,h) := \frac{4\pi}{\widetilde{\Delta}_R^{\rm var}(\eta,h)}. \tag{158}$$

Proposition B.3 (Restricted variational dominance over the canonical branch). *For every admissible* (η, h)*,*

$$E_R^{\text{var}}(\eta, h) \le E_R^{\text{can}}(\eta, h). \tag{159}$$

Consequently, whenever $\widetilde{\Delta}_R^{\text{var}}(\eta, h) > 0$*, one has*

$$\mathcal{B}_R^{\text{var}}(\eta, h) \ge \mathcal{B}_R^{\text{can}}(\eta, h). \tag{160}$$

Proof. The canonical profile is recovered by the admissible choice $\sigma = 0$ in (156). Therefore the infimum in (157) cannot exceed the canonical branch energy. The second claim follows immediately from the common positive denominator E_{vac} and the monotonicity of $x \mapsto 4\pi/x$ on $x > 0$. □

Representative restricted-minimization scan. A bounded one-parameter minimization in σ, combined with exact quadrature of the Section 2 curvature-torsion functional, yields the following CODATA-scale crossings in the same low-helicity band used for Table 3:

h	η_*^{var}	σ_*^{var}	$\Delta\eta_* := \eta_*^{\text{var}} - \eta_*^{\text{can}}$
0.05	0.4559	−0.0077	0.0004
0.10	0.4676	0.0012	0.0000
0.20	0.5281	0.0186	0.0118
0.30	0.6650	0.0398	0.0980

Table 4: Restricted variational crossings for the residual-normalized inverse candidate in the Fourier-enriched doubled family. Each row satisfies $\mathcal{B}_R^{\text{var}} \approx \alpha_{\text{CODATA}}^{-1}$; the last column records the shift relative to the canonical crossings of Table 3.

Within this restricted minimization, the low-helicity crossings survive with essentially no displacement for $h = 0.05$ and $h = 0.10$, while the larger-helicity cases are pushed to higher η. This behavior is exactly what the variational inequality suggests: lowering the branch energy at fixed (η, h) reduces the residual gap and therefore enlarges the inverse residual quantity, so a fixed target value is reached only at slightly larger modulation. The important point is not the precise shift, which depends on the chosen enrichment family, but the persistence of a moderate-modulation crossing band under the first non-canonical relaxation.

Intermediate downstream conclusion. The first three continuation routes now point to a single numerically alive branch of downstream search. The vacuum-normalized family $\mathcal{A}_R$ is structurally obstructed at order 4π. By contrast, the residual-normalized family $\mathcal{B}_R$ removes the trivial doubled baseline, scales as η^{-2} on its positive branch, survives exact canonical quadrature well enough to produce fine-structure-scale crossings at moderate modulation, and retains those crossings under a first restricted variational improvement inside the doubled sector. The next admissible question is whether relaxing the low-helicity parameter h itself sharpens this route into an interior selector or instead drives it toward the edge of the admissible doubled branch. That reduced-envelope test is carried out next.

B.6 Worked continuation IV: ansatz-level reduced-envelope minimization

The phrase "progressively less restricted sectorwise minimization" hides an ambiguity that can already be probed within the frozen main-text architecture. Even without enlarging the admissible configuration class beyond the canonical doubled branch, one may minimize the residual proxy over the low-helicity parameter h while keeping the doubled closure ratio $\rho = 1/2$ fixed. This does *not* produce the full sector minimum $E_R = E_{1,2}^{\min}$, because it neither explores arbitrary doubled representatives nor derives an independent dynamical stationarity condition for h. It does, however, reveal whether the residual route develops an interior preferred scale or collapses onto the edge of its admissible low-helicity branch.

Status label. The envelope quantities below have status **definition**, **effective relation**, **asymptotic estimate**, and **numerical diagnostic**. They are not theorem-level evaluations of the true sector minimum.

Definition. On the low-helicity positive branch $0 < h < h_*$ from (149), define the canonical residual envelope

$$\widetilde{\Delta}_R^{\rm env}(\eta) := \inf_{0<h<h_*} \widetilde{\Delta}_R^{\rm can}(\eta, h), \qquad \mathcal{B}_R^{\rm env}(\eta) := \frac{4\pi}{\widetilde{\Delta}_R^{\rm env}(\eta)}, \tag{161}$$

whenever $\widetilde{\Delta}_R^{\rm env}(\eta) > 0$. By construction this is only a partial minimization inside the canonical doubled family.

Leading-order structure. Equation (145) can be written on the positive branch as

$$\widetilde{\Delta}_R^{\rm LO}(\eta, h) = \eta^2 C_R(h), \qquad C_R(h) := \frac{4h^4 - 25h^2 + 7}{16(1+h^2)^2}. \tag{162}$$

A direct derivative computation gives

$$C_R'(h) = \frac{33h^3 - 39h}{8(1+h^2)^3} = \frac{h(33h^2 - 39)}{8(1+h^2)^3}. \tag{163}$$

Since $h_*^2 = (25 - \sqrt{513})/8 < 39/33$, one has

$$C_R'(h) < 0 \qquad \text{for all } 0 < h < h_*. \tag{164}$$

Thus the leading residual coefficient decreases monotonically to zero as $h \uparrow h_*$.

Proposition B.4 (Asymptotic boundary dominance of the reduced envelope)**.** *On the low-helicity branch $0 < h < h_*$, the leading-order residual coefficient $C_R(h)$ is strictly decreasing and satisfies*

$$\inf_{0<h<h_*} C_R(h) = 0, \qquad \lim_{h\uparrow h_*} C_R(h) = 0. \tag{165}$$

Consequently, for every fixed $\eta > 0$,

$$\inf_{0<h<h_*} \widetilde{\Delta}_R^{\rm LO}(\eta, h) = 0, \qquad \sup_{0<h<h_*} \frac{4\pi}{\widetilde{\Delta}_R^{\rm LO}(\eta, h)} = +\infty. \tag{166}$$

Hence unrestricted low-helicity minimization does not pick an interior asymptotic scale; it drives the inverse residual candidate toward the boundary $h \uparrow h_*$.

Proof. Monotonicity follows from (163) and (164). The equation $4h^4 - 25h^2 + 7 = 0$ defining h_* implies $C_R(h_*) = 0$. Substituting $\widetilde{\Delta}_R^{\rm LO}(\eta, h) = \eta^2 C_R(h)$ completes the argument. □

Exact canonical envelope diagnostic. Direct numerical quadrature of the exact branch energy (153), followed by one-dimensional minimization of $\widetilde{\Delta}_R^{\rm can}(\eta, h)$ over $0 < h < h_*$, confirms the same pattern beyond the asymptotic regime.

For sufficiently small η, the function $h \mapsto \widetilde{\Delta}_R^{\rm can}(\eta, h)$ changes sign inside the interval, so the envelope becomes singular. Representative interior zero locations $h_0(\eta)$ solving

$$\widetilde{\Delta}_R^{\rm can}(\eta, h_0(\eta)) = 0 \tag{167}$$

are listed in Table 5.

η	interior zero $h_0(\eta)$ of $\tilde{\Delta}_R^{\mathrm{can}}(\eta, h)$
0.30	0.5303
0.35	0.5306
0.40	0.5365
0.415	0.5405

Table 5: Representative interior zero locations of the exact canonical residual gap on the low-helicity doubled branch. They show that below a critical envelope threshold the reduced envelope is singular rather than predictive.

Numerically, the sign-change regime ends at an envelope threshold

$$\eta_{\mathrm{crit}}^{\mathrm{env}} \approx 0.4195. \tag{168}$$

For $\eta > \eta_{\mathrm{crit}}^{\mathrm{env}}$, the envelope becomes positive but remains boundary dominated: throughout the scanned band, the minimizing helicity stays at

$$h_{\mathrm{env}}(\eta) \approx 0.5419, \tag{169}$$

which is numerically indistinguishable at the displayed precision from the upper edge $h_* \approx 0.5420$. Representative envelope values are shown in Table 6.

η	$h_{\mathrm{env}}(\eta)$	$\tilde{\Delta}_R^{\mathrm{env}}(\eta)$	$\mathcal{B}_R^{\mathrm{env}}(\eta)$
0.43	0.5419	5.33×10^{-4}	2.36×10^4
0.50	0.5419	1.01×10^{-2}	1.24×10^3
0.55	0.5419	3.11×10^{-2}	4.04×10^2
0.6058	0.5419	9.17×10^{-2}	137.036

Table 6: Reduced-envelope values obtained by exact canonical quadrature followed by minimization over the low-helicity interval $0 < h < h_*$. The last row shows where the envelope falls to the CODATA inverse fine-structure scale.

The last row shows that the CODATA scale is indeed reached again under this partial minimization. However, the match is no longer isolated in any robust sense: below $\eta_{\mathrm{crit}}^{\mathrm{env}}$ the envelope is singular because the residual gap crosses zero inside the interval, while above $\eta_{\mathrm{crit}}^{\mathrm{env}}$ the minimizing helicity remains pinned to the upper boundary. The envelope therefore behaves like a near-singular edge profile rather than like an interior optimized prediction.

Integrated downstream conclusion after the envelope test. The combined downstream picture is now more specific. Residual normalization plus non-perturbative evaluation is still the only route that remains numerically alive. The restricted Fourier relaxation of Section B.5 shows that the moderate-modulation crossing band is not an artifact of the strict canonical profile. But the first unrestricted ansatz-level minimization over h does *not* sharpen that route into a robust constant derivation; it over-relaxes the candidate and turns the result into a boundary-controlled envelope. The next coherent step is therefore

$$\begin{aligned} &\text{residual normalization} \\ &\quad + \text{non-perturbative evaluation} \\ &\quad + \text{shape relaxation inside } \mathcal{T}_{1,2} \\ &\quad + \text{an interior selector for } h. \end{aligned} \tag{170}$$

rather than further unconstrained minimization. Such a selector could be dynamical, variational, or regularity based, but it must fix the doubled branch away from the edge $h \uparrow h_*$ before any constant-level identification can become genuinely predictive.

B.7 Worked continuation V: a two-sided regularity selector for h

The envelope test shows that free minimization in h destroys predictivity by pushing the residual candidate toward the edge $h \uparrow h_*$. A minimal counter-move is therefore not to change the residual candidate itself, but to introduce a separate selector that chooses an interior helicity before the candidate is read off. Such a selector should suppress both endpoint degenerations: the planar limit $h \downarrow 0$, where the helix collapses toward a circle, and the upper edge $h \uparrow h_*$, where the positive low-helicity branch loses interior control.

Status label. The selector weight and selected branch below have status **definition**, **effective relation**, **asymptotic estimate**, and **numerical diagnostic**. They do not constitute a theorem-level identification of a physical constant.

Definition of a parameter-free two-sided selector. Define the dimensionless weight

$$W_*(h) := \frac{h}{h_*}\left(1 - \frac{h}{h_*}\right)^2, \qquad 0 < h < h_*. \tag{171}$$

The linear factor suppresses the planar endpoint $h = 0$, while the quadratic factor is the lowest-order algebraic barrier that vanishes faster than the asymptotic coefficient $C_R(h) \sim c_*(h_* - h)$ as $h \uparrow h_*$. In this sense, (171) is the weakest simple two-sided weight that removes the boundary-dominated envelope behavior identified in Section B.6 without introducing any new dimensionful input.

Using the exact canonical residual candidate itself for the constant comparison, define the selector score

$$\mathcal{Q}_R(\eta, h) := W_*(h)\, \mathcal{B}_R^{\mathrm{can}}(\eta, h) = \frac{4\pi W_*(h)}{\widetilde{\Delta}_R^{\mathrm{can}}(\eta, h)}, \tag{172}$$

whenever $\widetilde{\Delta}_R^{\mathrm{can}}(\eta, h) > 0$. Then define the selected helicity branch by

$$h_{\mathrm{sel}}(\eta) \in \underset{0<h<h_*}{\arg\max}\, \mathcal{Q}_R(\eta, h). \tag{173}$$

The physical candidate remains the unweighted residual quantity $\mathcal{B}_R^{\mathrm{can}}(\eta, h_{\mathrm{sel}}(\eta))$; the weighted score $\mathcal{Q}_R$ is used only to choose an interior value of h.

Proposition B.5 (Interior existence of the regularity-selected branch). *For every $\eta > \eta_{\mathrm{crit}}^{\mathrm{env}}$, the selector score $\mathcal{Q}_R(\eta, h)$ admits at least one maximizer $h_{\mathrm{sel}}(\eta)$ in the open interval $(0, h_*)$.*

Proof. For $\eta > \eta_{\mathrm{crit}}^{\mathrm{env}}$, the exact canonical residual gap is positive on $0 < h < h_*$ by the definition of the envelope threshold (168). Hence $\mathcal{B}_R^{\mathrm{can}}(\eta, h)$ is continuous on that interval. The weight $W_*(h)$ extends continuously to the closed interval $[0, h_*]$ and vanishes at both endpoints. Therefore $\mathcal{Q}_R(\eta, h)$ extends continuously to $[0, h_*]$ with endpoint values

$$\mathcal{Q}_R(\eta, 0) = \mathcal{Q}_R(\eta, h_*) = 0. \tag{174}$$

Since $\mathcal{Q}_R(\eta, h) > 0$ for interior points, its maximum on the compact interval $[0, h_*]$ cannot occur only at the endpoints. At least one maximizer therefore lies in $(0, h_*)$. □

Leading-order selector. At leading order one has

$$\mathcal{Q}_R^{\mathrm{LO}}(\eta, h) = \frac{4\pi}{\eta^2}\, G_*(h), \qquad G_*(h) := \frac{W_*(h)}{C_R(h)}. \tag{175}$$

A one-dimensional maximization of $G_*(h)$ gives a unique interior asymptotic selector at

$$h_{\mathrm{sel}}^{\mathrm{LO}} \approx 0.26525 \approx 0.48934\, h_*. \tag{176}$$

Thus the first regularity-based selector already picks a genuinely interior doubled branch, independent of η at leading order.

Exact canonical selector scan. Direct quadrature of $E_R^{\rm can}(\eta, h)$, combined with one-dimensional maximization of $\mathcal{Q}_R(\eta, h)$, yields the representative selected-branch values shown in Table 7. The selected helicity remains well inside the admissible interval and drifts gradually downward as the modulation strength increases.

η	$h_{\rm sel}(\eta)$	$W_*(h_{\rm sel})$	$\mathcal{B}_R^{\rm can}(\eta, h_{\rm sel})$	$\mathcal{Q}_R(\eta, h_{\rm sel})$
0.45	0.2885	0.1165	253.239	29.493
0.50	0.2745	0.1234	188.406	23.249
0.54595	0.2516	0.1333	137.036	18.261
0.60	0.2129	0.1448	89.484	12.960

Table 7: Interior-selected canonical branch determined by the two-sided regularity score $\mathcal{Q}_R$. The weighted score is used only to choose h; constant-level comparison is still made with the unweighted residual candidate $\mathcal{B}_R^{\rm can}$.

The third row shows that the same fixed reference value used throughout this appendix is attained on the selected branch at

$$\eta_\alpha^{\rm sel} \approx 0.54595, \qquad h_{\rm sel}\left(\eta_\alpha^{\rm sel}\right) \approx 0.25163. \tag{177}$$

This lies comfortably inside the admissible low-helicity interval, far from the boundary-pinned behavior of the reduced envelope.

Integrated downstream conclusion after the selector test. The first regularity selector does not by itself derive a named constant, but it removes the exact obstruction revealed by Section B.6: the residual route need no longer be read off from a boundary-controlled envelope. Instead, within the exact canonical doubled branch, one now obtains an interior-selected path through parameter space that reaches the same fine-structure scale at moderate modulation. The best current downstream route inside the frozen main-text architecture is therefore

$$\begin{aligned} &\text{residual normalization} \\ &\quad + \text{non-perturbative evaluation} \\ &\quad + \text{restricted shape relaxation inside } \mathcal{T}_{1,2} \\ &\quad + \text{a two-sided interior selector for } h. \end{aligned} \tag{178}$$

The next admissible refinement, carried out in the following subsection, is to combine this selector with the broader variational families already explored in Section B.5, rather than to return to unconstrained envelope minimization.

Remark B.6. Up to this point, the present appendix contains six distinct downstream messages. First, the original vacuum-normalized candidate is controlled and obstructed. Second, an admissible alternative normalization removes that obstruction at the asymptotic level. Third, non-perturbative canonical evaluation shows that this alternative route is numerically alive on a moderate-modulation band. Fourth, a first restricted variational improvement shows that the crossing band is not an artifact of the strict canonical profile, although its location becomes deformation-sensitive as h grows. Fifth, a reduced-envelope minimization over the admissible low-helicity branch reveals a new obstruction of a different kind: absent an interior selector, the residual candidate is boundary dominated rather than sharply minimized. Sixth, a first two-sided regularity selector restores an interior chosen canonical branch and produces a stable interior crossing point on that selected branch, but still only at the diagnostic downstream level. This is valuable route-selection information, but it is not yet a completed derivation of a named constant.

B.8 Worked continuation VI: selector-coupled restricted variational branch

The selector continuation of Section B.7 removed the boundary collapse of the canonical envelope by choosing an interior helicity before reading off the residual candidate. The next admissible test is to apply the same idea not on the strict canonical doubled branch, but on the first Fourier-enriched family from Section B.5. This is the smallest coupled continuation that simultaneously relaxes the shape through the parameter σ and chooses an interior helicity through the weight $W_*(h)$.

Status label. The selector-coupled branch defined below has status **definition, effective relation**, and **numerical diagnostic**. It is still not a theorem-level evaluation of the full constrained sector minimum.

Definition of the selector-coupled branch. Using (158) and the same two-sided weight (171), define

$$\mathcal{Q}_R^{\mathrm{var}}(\eta, h) := W_*(h)\, \mathcal{B}_R^{\mathrm{var}}(\eta, h) = \frac{4\pi W_*(h)}{\widetilde{\Delta}_R^{\mathrm{var}}(\eta, h)}, \tag{179}$$

whenever $\widetilde{\Delta}_R^{\mathrm{var}}(\eta, h) > 0$. Then define the selector-coupled helicity branch by

$$h_{\mathrm{sel}}^{\mathrm{var}}(\eta) \in \arg\max_{0<h<h_*} \mathcal{Q}_R^{\mathrm{var}}(\eta, h). \tag{180}$$

As before, the weighted score is used only to choose h; the constant-level comparison is made with the unweighted quantity

$$\mathcal{B}_R^{\mathrm{var}}\big(\eta, h_{\mathrm{sel}}^{\mathrm{var}}(\eta)\big). \tag{181}$$

Shifted positivity threshold. Because the Fourier-enriched relaxation lowers the branch energy, the positivity domain of the residual gap also shifts. A direct two-stage numerical scan, minimizing first in σ and then over $h \in (0, h_*)$, yields the threshold

$$\eta_{\mathrm{crit}}^{\mathrm{var}} := \inf\Big\{\eta > 0 : \widetilde{\Delta}_R^{\mathrm{var}}(\eta, h) > 0 \text{ for all } 0 < h < h_*\Big\} \approx 0.6186, \tag{182}$$

with the minimizing helicity still concentrated near the upper edge,

$$h_{\mathrm{crit}}^{\mathrm{var}} \approx 0.5419 \approx h_*. \tag{183}$$

Thus the selector-coupled restricted variational branch becomes finite only after a noticeably higher positivity threshold than in the exact canonical scan.

Representative selector-coupled scan. For $\eta > \eta_{\mathrm{crit}}^{\mathrm{var}}$, the selector-coupled branch is well-defined and remains interior. Representative values are shown in Table 8.

η	$h_{\mathrm{sel}}^{\mathrm{var}}(\eta)$	σ_*	$\mathcal{B}_R^{\mathrm{var}}$	$\mathcal{Q}_R^{\mathrm{var}}$
0.63	0.3311	0.0359	189.630	17.545
0.65	0.3192	0.0382	162.550	16.176
0.6714	0.3055	0.0407	137.035	14.707
0.70	0.2857	0.0442	108.193	12.754

Table 8: Selector-coupled restricted variational branch. The selected helicity stays interior, while the optimal Fourier relaxation grows gradually with η.

The third row shows that the same reference inverse scale used throughout this appendix is reached again on the selector-coupled branch at approximately

$$\eta_\alpha^{\mathrm{var,sel}} \approx 0.6714, \qquad h_{\mathrm{sel}}^{\mathrm{var}}\big(\eta_\alpha^{\mathrm{var,sel}}\big) \approx 0.3055, \qquad \sigma_*\big(\eta_\alpha^{\mathrm{var,sel}}\big) \approx 0.0407. \tag{184}$$

Relative to the canonical selected branch (177), the target crossing is therefore shifted upward by about

$$\eta_\alpha^{\text{var,sel}} - \eta_\alpha^{\text{sel}} \approx 0.1255. \tag{185}$$

This is the expected cost of coupling the selector to an energy-lowering variational relaxation: the residual inverse quantity is enhanced pointwise whenever the relaxed gap stays positive, so the same fixed reference scale is encountered only at a larger modulation depth.

Integrated downstream conclusion after the coupled test. The coupled selector-variational test adds a seventh downstream message. It confirms that the interior-selected fine-structure-scale crossing is not destroyed by the first non-canonical shape relaxation once the positivity threshold (182) has been crossed. But it also shows that the route is less rigid than the canonical selector alone suggested: the admissible branch becomes finite only deeper in the moderate-modulation regime, and the reference crossing migrates appreciably in η. The best current reading of the frozen main-text architecture is therefore not that a constant has been derived, but that a single downstream corridor remains open,

$$\begin{aligned}&\text{residual normalization}\\ &\quad + \text{non-perturbative evaluation}\\ &\quad + \text{restricted shape relaxation inside } \mathcal{T}_{1,2}\\ &\quad + \text{an explicit interior selector for } h\\ &\quad + \text{positivity control under selector-coupled relaxations.}\end{aligned} \tag{186}$$

The next admissible step would be to derive the selector itself from a genuine stability or variational principle, rather than to impose it externally while enlarging the shape family by hand.

B.9 Worked continuation VII: barrier-regularized joint selection principle

The selector-coupled continuation of Section B.8 leaves one conceptual defect: the helicity selector enters there as an externally written multiplicative weight. The next admissible step is not yet a microscopic derivation of that weight, but it is a first internalization. Inside the same Fourier-enriched doubled family, one can combine the residual gap and the endpoint barriers into a single barrier-regularized action and ask for its local minima in the joint variables (h, σ).

Status label. The joint barrier action below has status **definition**. The equivalence proposition and first-variation equations have status **effective relation**. The stability scan and Hessian test are **numerical diagnostics**. None of them upgrades the paper to a theorem-level derivation of a named constant or of the full doubled-sector minimum.

Barrier-regularized joint action. For the Fourier-enriched family (156), define the unreduced residual gap

$$\widetilde{\Delta}_R^{\text{raw}}(\eta, h, \sigma) := \frac{E_A[X_R^{\text{var}}(\eta, h, \sigma)] - 2E_{\text{vac}}}{E_{\text{vac}}}, \tag{187}$$

and the admissible positive-gap domain

$$\mathcal{D}_\eta := \left\{(h, \sigma) : 0 < h < h_*,\ |\sigma| \le \eta/2,\ \widetilde{\Delta}_R^{\text{raw}}(\eta, h, \sigma) > 0\right\}. \tag{188}$$

Whenever $\mathcal{D}_\eta \neq \varnothing$, define the barrier-regularized action

$$\mathfrak{F}_R(\eta, h, \sigma) := \log \widetilde{\Delta}_R^{\text{raw}}(\eta, h, \sigma) - \log\left(\frac{h}{h_*}\right) - 2\log\left(1 - \frac{h}{h_*}\right). \tag{189}$$

The logarithmic penalties are the additive form of the same two-sided endpoint suppression used earlier in (171); the difference is that they are now absorbed into a single action rather than written as an external multiplicative score.

Define the joint stability-selected branch by

$$\big(h_{\mathrm{stab}}(\eta), \sigma_{\mathrm{stab}}(\eta)\big) \in \arg\min_{(h,\sigma)\in\mathcal{D}_\eta} \mathfrak{F}_R(\eta, h, \sigma), \tag{190}$$

and read off the associated unweighted candidate as

$$\mathcal{B}_R^{\mathrm{stab}}(\eta) := \frac{4\pi}{\widetilde{\Delta}_R^{\mathrm{raw}}\big(\eta, h_{\mathrm{stab}}(\eta), \sigma_{\mathrm{stab}}(\eta)\big)}. \tag{191}$$

Proposition B.7 (Exact equivalence with the selector-coupled restricted variational branch). *For every $\eta > \eta_{\mathrm{crit}}^{\mathrm{var}}$, minimization of (189) over $\mathcal{D}_\eta$ is equivalent to the two-stage selector-coupled procedure of Section B.8. In particular,*

$$h_{\mathrm{stab}}(\eta) = h_{\mathrm{sel}}^{\mathrm{var}}(\eta), \qquad \sigma_{\mathrm{stab}}(\eta) = \sigma_*\big(\eta, h_{\mathrm{sel}}^{\mathrm{var}}(\eta)\big), \tag{192}$$

and therefore

$$\mathcal{B}_R^{\mathrm{stab}}(\eta) = \mathcal{B}_R^{\mathrm{var}}\big(\eta, h_{\mathrm{sel}}^{\mathrm{var}}(\eta)\big). \tag{193}$$

Proof. For fixed η and h, the barrier terms in (189) do not depend on σ. Hence minimizing $\mathfrak{F}_R$ in σ is equivalent to minimizing $\widetilde{\Delta}_R^{\mathrm{raw}}(\eta, h, \sigma)$, which gives the same restricted energy minimizer $\sigma_*(\eta, h)$ used in Section B.8. Substituting that minimizer back into (189) yields the reduced action

$$\mathfrak{F}_R^{\mathrm{red}}(\eta, h) := \log \widetilde{\Delta}_R^{\mathrm{var}}(\eta, h) - \log\left(\frac{h}{h_*}\right) - 2\log\left(1 - \frac{h}{h_*}\right) = \log\left(\frac{4\pi}{\mathcal{Q}_R^{\mathrm{var}}(\eta, h)}\right). \tag{194}$$

Since the logarithm is strictly increasing, minimizing $\mathfrak{F}_R^{\mathrm{red}}$ over $h \in (0, h_*)$ is equivalent to maximizing $\mathcal{Q}_R^{\mathrm{var}}$. This is exactly the definition of the selector-coupled branch in (180). The stated identities follow. ☐

First-variation equations. At any interior minimizer of (189), the stationarity equations take the form

$$\partial_\sigma \log \widetilde{\Delta}_R^{\mathrm{raw}}(\eta, h, \sigma) = 0, \tag{195}$$

$$\partial_h \log \widetilde{\Delta}_R^{\mathrm{raw}}(\eta, h, \sigma) = \frac{1}{h} - \frac{2}{h_* - h}. \tag{196}$$

Equation (195) is the usual restricted variational stationarity in the non-canonical mode σ, while (196) balances the logarithmic slope of the residual gap against the two endpoint barriers. In this sense, the selector is no longer merely appended to the candidate; it appears as part of a single first-variation condition.

Reduced-branch stiffness diagnostic. Along the stability-selected branch, define the reduced helicity stiffness by

$$\kappa_h^{\mathrm{stab}}(\eta) := \partial_h^2 \mathfrak{F}_R^{\mathrm{red}}\big(\eta, h_{\mathrm{stab}}(\eta)\big). \tag{197}$$

Finite-difference evaluation along the same representative branch used in Table 8 gives the values shown in Table 9. The positive values indicate that the selected helicity is a local minimum of the reduced barrier action rather than a flat or boundary-hugging artifact.

η	$h_{\text{stab}}(\eta)$	$\sigma_{\text{stab}}(\eta)$	$\kappa_h^{\text{stab}}(\eta)$
0.63	0.3310	0.0359	28.64
0.65	0.3193	0.0382	31.27
0.6714	0.3054	0.0407	33.65
0.70	0.2856	0.0442	36.28

Table 9: Reduced-action helicity stiffness along the barrier-regularized stability-selected branch. Positivity of κ_h^{stab} indicates local stability in the selected helicity direction.

Full two-variable stability at the reference crossing. At the selector-coupled reference crossing from (184), a direct finite-difference Hessian evaluation of the full joint action gives

$$\nabla^2_{(h,\sigma)}\mathfrak{F}_R\left(\eta_\alpha^{\text{var,sel}}, h_{\text{stab}}, \sigma_{\text{stab}}\right) \approx \begin{pmatrix} 34.08 & -27.67 \\ -27.67 & 1737.66 \end{pmatrix}, \tag{198}$$

with eigenvalues

$$\lambda_{\min} \approx 33.63, \qquad \lambda_{\max} \approx 1738.11. \tag{199}$$

Both are positive, so the reference crossing sits at a bona fide local minimum of the barrier-regularized action in both the helicity direction and the first non-canonical shape direction.

Integrated downstream conclusion after the barrier-action test. This continuation adds an eighth downstream message. It does not produce a new numerical crossing separate from Section B.8; by the equivalence (192), it reorganizes the same branch. But that reorganization matters conceptually: the previously external selector can now be absorbed into a single barrier-regularized variational principle, and the selected crossing is locally stiff rather than merely numerically visible. The best current reading of the frozen main-text architecture is therefore

$$\begin{aligned} &\text{residual normalization} \\ &\quad + \text{non-perturbative evaluation} \\ &\quad + \text{restricted shape relaxation inside } \mathcal{T}_{1,2} \\ &\quad + \text{an interior selector written as a joint barrier action} \\ &\quad + \text{local stability in } (h,\sigma) \text{ on the first enriched branch.} \end{aligned} \tag{200}$$

The remaining gap is equally clear: the barrier terms are still fixed by endpoint regularity rather than derived from a microscopic action, and the present local minimum has only been checked inside the first Fourier-enriched doubled family.

Remark B.8. The present appendix now carries eight distinct downstream messages. The new eighth message is also deliberately mixed. Positively, the previously external selector can be absorbed into a single barrier-regularized variational action, and at the first selector-coupled crossing that action has a bona fide local minimum in both the helicity and first non-canonical shape directions. Cautionarily, the barrier terms are still fixed by endpoint regularity orders rather than derived from a microscopic dynamics, so the result remains a downstream local-stability diagnostic rather than a completed derivation. This sharpens the route-selection picture once more without changing the theorem-level scope of the paper.

C Stability and Broader Released-Mode Tests

This appendix uses the standing conventions of Section B.1 and asks whether the selected doubled-sector branch survives progressively less restricted Fourier release. Its output is a sequence of local-stability and drift diagnostics inside broader doubled families, together with a first infinite-mode tail extrapolation.

C.1 Compactness of admissible released-mode ladders

Before examining particular higher-mode scans, it is useful to isolate the compactness mechanism that prevents the released-mode ladder from escaping merely because more odd half-modes are opened.

Common tail envelope. Fix $\eta > 0$ and let $\{b_n(\eta)\}_{n\geq 1}$ be positive numbers such that

$$\sum_{n=1}^{\infty}(2n+1)^3 b_n(\eta) < \infty. \tag{201}$$

For each $m \geq 0$, define the truncated coefficient box

$$\mathcal{K}_\eta^{[m]} := [0, h_*] \times \prod_{n=1}^{m}[-b_n(\eta), b_n(\eta)], \tag{202}$$

with the convention that the empty product is a point when $m = 0$. If $x_m = (h_m, a_1^{[m]}, \dots, a_m^{[m]}) \in \mathcal{K}_\eta^{[m]}$, pad it to an infinite coefficient vector by setting $a_n^{[m]} := 0$ for $n > m$.

Proposition C.1 (Compactness of admissible released-mode ladders). *Let $x_m \in \mathcal{K}_\eta^{[m]}$ be any sequence of padded released-mode points. Then there exist a subsequence, still denoted x_m, and a limit point*

$$x_\infty = (h_\infty, a_1^{(\infty)}, a_2^{(\infty)}, \dots) \in [0, h_*] \times \prod_{n=1}^{\infty}[-b_n(\eta), b_n(\eta)] \tag{203}$$

such that

$$h_m \to h_\infty, \qquad a_n^{[m]} \to a_n^{(\infty)} \quad \textit{for every fixed } n. \tag{204}$$

Moreover, the associated odd half-mode fields

$$\phi_m(u) := \sum_{n=1}^{\infty} a_n^{[m]} \sin\left(\frac{(2n+1)u}{2}\right) \tag{205}$$

converge, along the same subsequence, in $C^2([0, 4\pi])$ to

$$\phi_\infty(u) := \sum_{n=1}^{\infty} a_n^{(\infty)} \sin\left(\frac{(2n+1)u}{2}\right). \tag{206}$$

If one strengthens (201) *to $\sum_{n\geq 1}(2n+1)^4 b_n(\eta) < \infty$, then the same subsequence converges in $C^3([0, 4\pi])$.*

Proof. Each coordinate sequence $\{a_n^{[m]}\}_{m\geq 1}$ is bounded by $b_n(\eta)$. A diagonal extraction therefore yields a subsequence for which (204) holds simultaneously for every fixed n, together with $h_m \to h_\infty \in [0, h_*]$.

For $k = 0, 1, 2$, the kth derivative of the basis function $\sin\big((2n+1)u/2\big)$ is bounded in sup norm by $C_k(2n+1)^k$ for constants C_k independent of n. Hence (201) implies the summable majorant

$$\sum_{n=1}^{\infty} \sup_u \left|a_n^{[m]} \partial_u^k \sin\left(\frac{(2n+1)u}{2}\right)\right| \leq C_k \sum_{n=1}^{\infty}(2n+1)^k b_n(\eta) < \infty, \tag{207}$$

uniformly in m for $k \leq 2$. By the Weierstrass M-test the series for ϕ_m and their first two derivatives converge uniformly, and the limit may be passed through term by term along the extracted subsequence. This gives C^2 convergence to (206). The C^3 statement is identical under the stronger $(2n+1)^4$ summability. □

Use in the broader-family tests. Proposition C.1 does not determine which released-mode point is selected by the barrier-regularized dynamics. What it does guarantee is that, once a common tail envelope is imposed, a sequence of stable finite released-mode configurations cannot lose compactness before the infinite-mode analysis of Appendix D is even formulated.

C.2 Worked continuation VIII: a second higher-mode robustness test

The barrier-regularized branch of Section B.9 is locally stable on the first enriched doubled family, but that still leaves open whether the same branch survives once one more admissible odd half-mode is released. The next admissible continuation is therefore a controlled higher-mode robustness test: enlarge the doubled ansatz by a second non-canonical odd half-frequency, continue the interior stable branch into that larger family, and check whether the fine-structure-scale crossing remains a genuine local minimum.

Status label. The enlarged doubled family and its barrier action below have status **definition**. The stationarity system has status **effective relation**. The branch continuation, crossing value, and Hessian spectrum are **numerical diagnostics**. No part of this subsection claims a full doubled-sector minimization or a theorem-level derivation of a named constant.

Second enriched doubled family. Define

$$R_{\mathrm{var},2}(t) := R_0\left[1 + \eta \sin\left(\frac{\omega t}{2}\right) + \sigma \sin\left(\frac{3\omega t}{2}\right) + \chi \sin\left(\frac{5\omega t}{2}\right)\right], \tag{208}$$

with the subordinate amplitude bounds

$$|\sigma| \le \frac{\eta}{2}, \qquad |\chi| \le \frac{\eta}{3}. \tag{209}$$

These bounds are not presented as canonical; they simply keep the second released odd half-mode subordinate to the first and maintain the same doubled closure structure.

Define the corresponding raw residual gap by

$$\widetilde{\Delta}_R^{(2)}(\eta, h, \sigma, \chi) := \frac{E_A\left[X_R^{\mathrm{var},2}(\eta, h, \sigma, \chi)\right] - 2E_{\mathrm{vac}}}{E_{\mathrm{vac}}}, \tag{210}$$

and the associated barrier-regularized action by

$$\mathfrak{F}_R^{(2)}(\eta, h, \sigma, \chi) := \log \widetilde{\Delta}_R^{(2)}(\eta, h, \sigma, \chi) - \log\left(\frac{h}{h_*}\right) - 2\log\left(1 - \frac{h}{h_*}\right), \tag{211}$$

whenever the raw gap is positive.

Interior higher-mode branch. Rather than claim a global minimization over the enlarged four-parameter family, we continue the previously selected interior branch into the larger ansatz and track the resulting interior stationary point on the moderate-modulation window

$$0.67 \le \eta \le 0.75. \tag{212}$$

Write that continued branch as

$$\left(h_{\mathrm{stab}}^{(2)}(\eta), \sigma_{\mathrm{stab}}^{(2)}(\eta), \chi_{\mathrm{stab}}^{(2)}(\eta)\right). \tag{213}$$

At any interior point of this branch, the first-variation equations take the form

$$\partial_h \log \widetilde{\Delta}_R^{(2)} = \frac{1}{h} - \frac{2}{h_* - h}, \qquad \partial_\sigma \log \widetilde{\Delta}_R^{(2)} = 0, \qquad \partial_\chi \log \widetilde{\Delta}_R^{(2)} = 0. \tag{214}$$

The corresponding unweighted inverse residual candidate is read off as

$$\mathcal{B}_R^{(2)}(\eta) := \frac{4\pi}{\widetilde{\Delta}_R^{(2)}\left(\eta, h_{\mathrm{stab}}^{(2)}(\eta), \sigma_{\mathrm{stab}}^{(2)}(\eta), \chi_{\mathrm{stab}}^{(2)}(\eta)\right)}. \tag{215}$$

Representative branch samples. Numerical continuation on the window (212) gives the representative values in Table 10. Two features are immediate. First, the released second mode remains genuinely subordinate: along the branch one finds $|\chi_{\text{stab}}^{(2)}| \sim 10^{-2}$, well below the corresponding $\sigma_{\text{stab}}^{(2)}$. Second, the smallest Hessian eigenvalue stays positive on the displayed samples, so the continued branch remains locally stable in the enlarged variable set (h, σ, χ).

η	$h_{\text{stab}}^{(2)}$	$\sigma_{\text{stab}}^{(2)}$	$\chi_{\text{stab}}^{(2)}$	$\mathcal{B}_R^{(2)}$	$\lambda_{\min}$
0.6700	0.4009	0.0541	0.00656	373.10	20.40
0.7000	0.3630	0.0592	0.00721	226.93	34.91
0.7300	0.3368	0.0648	0.00793	163.52	40.38
0.7479	0.3223	0.0683	0.00839	137.04	42.49

Table 10: Representative points on the higher-mode relaxed interior branch. The last column gives the smallest eigenvalue of the full Hessian of $\mathfrak{F}_R^{(2)}$, showing local stability inside the enlarged variable set (h, σ, χ).

Crossing and full three-variable stability. On this continued higher-mode branch, the reference inverse scale is reached at approximately

$$\eta_\alpha^{(2)} \approx 0.7479, \qquad h_{\text{stab}}^{(2)}(\eta_\alpha^{(2)}) \approx 0.3223, \qquad \sigma_{\text{stab}}^{(2)}(\eta_\alpha^{(2)}) \approx 0.0683, \qquad \chi_{\text{stab}}^{(2)}(\eta_\alpha^{(2)}) \approx 0.00839, \tag{216}$$

with

$$\mathcal{B}_R^{(2)}(\eta_\alpha^{(2)}) \approx 137.037. \tag{217}$$

At that crossing, a finite-difference Hessian evaluation of the full action (211) gives

$$\nabla^2_{(h,\sigma,\chi)} \mathfrak{F}_R^{(2)}\left(\eta_\alpha^{(2)}, h_{\text{stab}}^{(2)}, \sigma_{\text{stab}}^{(2)}, \chi_{\text{stab}}^{(2)}\right) \approx \begin{pmatrix} 42.87 & -7.03 & -62.18 \\ -7.03 & 4674.03 & -8737.76 \\ -62.18 & -8737.76 & 31888.93 \end{pmatrix}, \tag{218}$$

with eigenvalues

$$\lambda_1 \approx 42.49, \qquad \lambda_2 \approx 2110.46, \qquad \lambda_3 \approx 34452.89. \tag{219}$$

All three are positive. Thus, within this larger doubled ansatz, the fine-structure-scale crossing is still a bona fide local minimum of the same barrier-regularized action, not merely a one-mode artifact.

Integrated downstream conclusion after the higher-mode test. This continuation adds a ninth downstream message. Positively, the locally stable selector-regularized crossing does survive one more admissible odd half-mode relaxation. Cautionarily, it does not remain numerically rigid: relative to the first enriched-branch crossing (184), the higher-mode relaxed branch reaches the same inverse scale only after an additional upward shift

$$\eta_\alpha^{(2)} - \eta_\alpha^{\text{var,sel}} \approx 0.0764. \tag{220}$$

The best current reading of the frozen main-text architecture is therefore

$$\begin{aligned} &\text{residual normalization} \\ &\quad + \text{non-perturbative evaluation} \\ &\quad + \text{restricted shape relaxation inside } \mathcal{T}_{1,2} \\ &\quad + \text{an interior selector written as a joint barrier action} \\ &\quad + \text{local stability on the first enriched branch} \\ &\quad + \text{persistence of that local minimum under one additional odd half-mode.} \end{aligned} \tag{221}$$

What remains open is equally clear: the crossing is now more robust than it was in Section B.9, but it is still only tracked across a narrow sequence of downstream ansätze rather than across the full admissible doubled sector.

Remark C.2. The present appendix now carries nine distinct downstream messages. The new ninth message is again mixed. Positively, the barrier-regularized crossing survives a second Fourier enrichment and remains a genuine local minimum in the enlarged variable set (h, σ, χ). Cautionarily, the crossing continues to drift in η as new shape freedom is released, so the route is stabilized locally but not yet rendered ansatz-independent. This pushes the downstream evidence further without changing the theorem-level scope of the paper.

C.3 Worked continuation IX: a broader doubled-family test

The second higher-mode test of Section C.2 still leaves open whether the same locally stable crossing survives when one more admissible odd half-mode is released. The next admissible continuation is therefore a broader doubled-family test: enlarge the radial profile by a third non-canonical odd half-frequency, continue the interior stable branch into that larger ansatz, and check whether the fine-structure-scale crossing persists as a bona fide local minimum in the enlarged variable set (h, σ, χ, ψ).

Status label. The broader doubled family and its barrier action below have status **definition**. The four-variable stationarity system has status **effective relation**. The branch continuation, crossing value, and Hessian spectrum are **numerical diagnostics**. No part of this subsection claims a global doubled-sector minimization or a theorem-level derivation of a named constant.

Third enriched doubled family. Define

$$R_{\mathrm{var},3}(t) := R_0\left[1 + \eta \sin\left(\frac{\omega t}{2}\right) + \sigma \sin\left(\frac{3\omega t}{2}\right) + \chi \sin\left(\frac{5\omega t}{2}\right) + \psi \sin\left(\frac{7\omega t}{2}\right)\right], \tag{222}$$

with subordinate amplitude bounds

$$|\sigma| \le \frac{\eta}{2}, \qquad |\chi| \le \frac{\eta}{3}, \qquad |\psi| \le \frac{\eta}{5}. \tag{223}$$

These bounds are again not presented as canonical. Their only role is to release one more odd half-mode while keeping the added shapes subordinate to the base doubled modulation.

Define the corresponding raw residual gap by

$$\widetilde{\Delta}_R^{(3)}(\eta, h, \sigma, \chi, \psi) := \frac{E_A\left[X_R^{\mathrm{var},3}(\eta, h, \sigma, \chi, \psi)\right] - 2E_{\mathrm{vac}}}{E_{\mathrm{vac}}}, \tag{224}$$

and the associated barrier-regularized action by

$$\mathfrak{F}_R^{(3)}(\eta, h, \sigma, \chi, \psi) := \log \widetilde{\Delta}_R^{(3)}(\eta, h, \sigma, \chi, \psi) - \log\left(\frac{h}{h_*}\right) - 2\log\left(1 - \frac{h}{h_*}\right), \tag{225}$$

whenever the raw gap is positive.

Interior four-variable branch. Rather than assert a global minimization on this larger family, we continue the previously stable branch into the broader ansatz and track the resulting interior stationary point on the window

$$0.76 \le \eta \le 0.81. \tag{226}$$

Write the continued branch as

$$\left(h_{\text{stab}}^{(3)}(\eta), \sigma_{\text{stab}}^{(3)}(\eta), \chi_{\text{stab}}^{(3)}(\eta), \psi_{\text{stab}}^{(3)}(\eta)\right). \tag{227}$$

At any interior point of this branch, the first-variation equations take the form

$$\partial_h \log \widetilde{\Delta}_R^{(3)} = \frac{1}{h} - \frac{2}{h_* - h}, \qquad \partial_\sigma \log \widetilde{\Delta}_R^{(3)} = 0, \qquad \partial_\chi \log \widetilde{\Delta}_R^{(3)} = 0, \qquad \partial_\psi \log \widetilde{\Delta}_R^{(3)} = 0. \tag{228}$$

The corresponding unweighted inverse residual candidate is

$$\mathcal{B}_R^{(3)}(\eta) := \frac{4\pi}{\widetilde{\Delta}_R^{(3)}\left(\eta, h_{\text{stab}}^{(3)}(\eta), \sigma_{\text{stab}}^{(3)}(\eta), \chi_{\text{stab}}^{(3)}(\eta), \psi_{\text{stab}}^{(3)}(\eta)\right)}. \tag{229}$$

Representative branch samples. Numerical continuation on the window (226) gives the representative values in Table 11. Three features are immediate. First, the newly released amplitude remains genuinely subordinate: along the displayed branch one finds $|\psi_{\text{stab}}^{(3)}| \sim 10^{-3}$, well below both $|\chi_{\text{stab}}^{(3)}|$ and $|\sigma_{\text{stab}}^{(3)}|$. Second, the smallest Hessian eigenvalue remains positive on the sample points, so the branch stays locally stable in the full variable set (h, σ, χ, ψ). Third, the crossing drift continues upward in η, but more mildly than in the previous higher-mode release.

η	$h_{\text{stab}}^{(3)}$	$\sigma_{\text{stab}}^{(3)}$	$\chi_{\text{stab}}^{(3)}$	$\psi_{\text{stab}}^{(3)}$	$\mathcal{B}_R^{(3)}$	$\lambda_{\min}$
0.7600	0.3563	0.0790	0.01366	0.00229	196.26	46.59
0.7800	0.3383	0.0839	0.01471	0.00245	156.35	49.26
0.7926	0.3277	0.0871	0.01541	0.00255	137.04	50.47
0.8040	0.3184	0.0900	0.01605	0.00264	122.24	51.33

Table 11: Representative points on the broader doubled-family branch. The last column gives the smallest eigenvalue of the full Hessian of $\mathfrak{F}_R^{(3)}$, showing local stability inside the enlarged variable set (h, σ, χ, ψ).

Crossing and full four-variable stability. On this broader branch, the reference inverse scale is reached at approximately

$$\eta_\alpha^{(3)} \approx 0.7926, \qquad h_{\text{stab}}^{(3)}(\eta_\alpha^{(3)}) \approx 0.3277, \qquad \sigma_{\text{stab}}^{(3)}(\eta_\alpha^{(3)}) \approx 0.0871, \tag{230}$$

$$\chi_{\text{stab}}^{(3)}(\eta_\alpha^{(3)}) \approx 0.01541, \qquad \psi_{\text{stab}}^{(3)}(\eta_\alpha^{(3)}) \approx 0.00255, \qquad \mathcal{B}_R^{(3)}(\eta_\alpha^{(3)}) \approx 137.036. \tag{231}$$

At that crossing, a finite-difference Hessian evaluation of the full action (225) gives

$$\nabla^2_{(h,\sigma,\chi,\psi)} \mathfrak{F}_R^{(3)}\left(\eta_\alpha^{(3)}, h_{\text{stab}}^{(3)}, \sigma_{\text{stab}}^{(3)}, \chi_{\text{stab}}^{(3)}, \psi_{\text{stab}}^{(3)}\right) \approx \begin{pmatrix} 50.75 & 6.55 & -30.78 & -113.26 \\ 6.55 & 6262.63 & -9960.09 & -772.74 \\ -30.78 & -9960.09 & 43970.62 & -55197.51 \\ -113.26 & -772.74 & -55197.51 & 203643.30 \end{pmatrix}, \tag{232}$$

with eigenvalues

$$\lambda_1 \approx 50.47, \qquad \lambda_2 \approx 2352.61, \qquad \lambda_3 \approx 30634.20, \qquad \lambda_4 \approx 220890.01. \tag{233}$$

All four are positive. Thus, within this broader doubled ansatz, the fine-structure-scale crossing remains a bona fide local minimum of the same barrier-regularized action, not merely an artifact of the first two released odd half-modes.

Integrated downstream conclusion after the broader-family test. This continuation adds a tenth downstream message. Positively, the locally stable selector-regularized crossing survives a third released odd half-mode, and the new mode remains numerically subordinate along the continued branch. Cautionarily, the branch still drifts in η: relative to the crossing (216) from Section C.2, the broader-family branch reaches the same inverse scale only after the additional shift

$$\eta_\alpha^{(3)} - \eta_\alpha^{(2)} \approx 0.0447. \tag{234}$$

The best current reading of the frozen main-text architecture is therefore

$$\begin{aligned} &\text{residual normalization} \\ &\quad + \text{non-perturbative evaluation} \\ &\quad + \text{restricted shape relaxation inside } \mathcal{T}_{1,2} \\ &\quad + \text{an interior selector written as a joint barrier action} \\ &\quad + \text{local stability on the first enriched branch} \\ &\quad + \text{persistence of that local minimum under two further odd half-modes.} \end{aligned} \tag{235}$$

What remains open is equally clear: the crossing is now locally robust across a broader sequence of downstream ansätze, but it is still being tracked along a controlled continuation corridor rather than obtained from a full minimization over the admissible doubled sector.

Remark C.3. The present appendix now carries ten distinct downstream messages. The new tenth message is again mixed. Positively, the barrier-regularized crossing survives a third released odd half-mode and remains a genuine local minimum in the enlarged variable set (h, σ, χ, ψ). Cautionarily, the crossing still drifts in η as new shape freedom is released, so the route is becoming broader-family robust only gradually rather than all at once. This pushes the downstream evidence further without changing the theorem-level scope of the paper.

C.4 Worked continuation X: a reduced-measure origin for the barrier terms

The broader-family test strengthens local robustness but also sharpens the remaining conceptual gap: the barrier action works, yet its endpoint terms still look externally written if one insists on a microscopic dynamics. The next admissible continuation is weaker than a microscopic derivation but still genuinely internal to the present route: derive the same barrier as the negative logarithm of the simplest normalized endpoint-stratified measure on the admissible helicity interval.

Status label. The endpoint-stratified density below has status **definition**. Its equivalence to the previously used barrier action has status **effective relation**. The resulting reinterpretation of the selected branches is a **candidate interpretation**. No part of this subsection claims a microscopic action or a theorem-level derivation of a named constant.

Endpoint-stratified reduced measure. Write

$$x := \frac{h}{h_*} \in (0, 1). \tag{236}$$

Within the present appendix route, the admissible helicity interval has two distinguished endpoint strata. The lower endpoint $x = 0$ corresponds to planar collapse of the doubled helix and removes one effective interior degree, so it contributes one power of distance to the reduced measure. The upper endpoint $x = 1$ is more singular inside this route: it is simultaneously the endpoint of admissibility and the point where the leading residual-gain coefficient $C_R(h)$ vanishes, so it contributes two powers of distance. The simplest normalized density carrying precisely those endpoint orders is therefore

$$\rho_*(h) := \frac{12}{h_*}\frac{h}{h_*}\left(1 - \frac{h}{h_*}\right)^2, \qquad \int_0^{h_*} \rho_*(h)\, dh = 1. \tag{237}$$

Equivalently, in the scaled variable x this is the Beta(2, 3) density

$$d\mu_*(x) = 12x(1-x)^2\, dx. \tag{238}$$

Measure-generated barrier action. For any downstream branch labeled by m — canonical, first variational, two-mode, or three-mode — define the measure-generated reduced action by

$$\mathfrak{G}_R^{(m)} := \log \widetilde{\Delta}_R^{(m)} - \log \rho_*(h), \tag{239}$$

whenever the raw residual gap is positive. By substituting (237) into (239), one obtains the exact identity

$$\mathfrak{G}_R^{(m)} = \mathfrak{F}_R^{(m)} - \log\left(\frac{12}{h_*}\right). \tag{240}$$

The additive constant is irrelevant to stationarity and Hessian signatures. Hence every local minimum of $\mathfrak{F}_R^{(m)}$ is equally a local minimum of $\mathfrak{G}_R^{(m)}$, and conversely.

The same statement may be written in multiplicative form. Define the reduced posterior weight

$$\Pi_R^{(m)} := \frac{\rho_*(h)}{\widetilde{\Delta}_R^{(m)}}. \tag{241}$$

Then minimizing $\mathfrak{G}_R^{(m)}$ is equivalent to maximizing $\Pi_R^{(m)}$. In this sense, the barrier action is no longer merely an appended selector: it is the negative log of a normalized endpoint-stratified measure divided by the residual gap.

Interpretive consequence for the selected branch. The density ρ_* by itself has a unique interior mode at

$$h_\mu = \frac{h_*}{3} \approx 0.1807. \tag{242}$$

By contrast, the broader-family crossing from Section C.3 occurs at

$$h_{\text{stab}}^{(3)}(\eta_\alpha^{(3)}) \approx 0.3277. \tag{243}$$

Thus the selected interior point is not imposed by the barrier alone. It results from competition between the endpoint-stratified measure and the residual gap. This is the strongest internal reading currently available in the frozen architecture: the barrier has an endogenous reduced-state-space origin, while the final selected helicity is still determined jointly with the energy proxy.

Integrated downstream conclusion after the reduced-measure reinterpretation. This continuation adds an eleventh downstream message. Positively, the barrier terms no longer need to be read as purely external at the appendix level: they can be derived from the simplest normalized endpoint-stratified measure on the admissible helicity interval, and the previously selected stable branches can be re-read as local maxima of the reduced posterior weight (241).

Cautionarily, this is still not a microscopic action. It is an endogenous origin only at the reduced state-space level. The best current reading of the frozen main-text architecture is therefore

$$\begin{aligned}&\text{residual normalization}\\&\quad+\text{non-perturbative evaluation}\\&\quad+\text{progressively broader doubled-family robustness}\\&\quad+\text{a selector absorbed into a joint barrier action}\\&\quad+\text{that same barrier rewritten as a normalized endpoint-stratified measure.}\end{aligned} \tag{244}$$

The remaining gap is now narrower but still genuine: the route has an internal barrier source, yet that source is effective and measure-theoretic rather than microscopic and dynamical.

Remark C.4. The present appendix now carries eleven distinct downstream messages. The new eleventh message is again mixed. Positively, the helicity barrier now has an endogenous reduced-measure origin inside the appendix route itself, so the interior selector is no longer only an external regularity device. Cautionarily, the origin is still reduced-state-space rather than microscopic. This advances the barrier question one layer further while keeping the theorem-level scope of the paper unchanged.

C.5 Worked continuation XI: mode-release stabilization of the broader-family crossing

The broader doubled-family test of Section C.3 strengthens local robustness, but it still leaves open whether the observed drift in the crossing location is runaway or stabilizing. The next admissible continuation is therefore not yet another mode release, but a numerical meta-diagnostic on the mode-release sequence already accumulated in this appendix: promote the crossing locations themselves to a downstream sequence and test whether their increments decay in a controlled fashion.

Status label. The mode-release sequence below has status **definition**. The geometric-tail ansatz and its fitted coefficients have status **numerical diagnostic**. The resulting infinite-release estimate is only a **candidate interpretation**. No part of this subsection claims an actual full-sector limit or an ansatz-independent theorem.

Mode-release crossing sequence. Let the index $m \geq 0$ count the number of released non-canonical odd half-modes beyond the base doubled branch. Using the crossing values already obtained in Sections B.7, B.8, C.2, and C.3, define

$$\eta_\alpha^{[0]} := \eta_\alpha^{\mathrm{sel}} \approx 0.54595, \qquad \eta_\alpha^{[1]} := \eta_\alpha^{\mathrm{var,sel}} \approx 0.6714, \tag{245}$$

$$\eta_\alpha^{[2]} := \eta_\alpha^{(2)} \approx 0.7479, \qquad \eta_\alpha^{[3]} := \eta_\alpha^{(3)} \approx 0.7926. \tag{246}$$

Define the successive release increments by

$$\delta_m := \eta_\alpha^{[m]} - \eta_\alpha^{[m-1]}, \qquad m \geq 1. \tag{247}$$

Numerically one finds

$$\delta_1 \approx 0.1255, \qquad \delta_2 \approx 0.0765, \qquad \delta_3 \approx 0.0447. \tag{248}$$

Hence the observed damping ratios are

$$\frac{\delta_2}{\delta_1} \approx 0.6098, \qquad \frac{\delta_3}{\delta_2} \approx 0.5843. \tag{249}$$

The upward drift is therefore still real, but its increments are already decaying rather than growing.

Geometric-tail fit. The smallest useful meta-model consistent with (249) is a geometric-tail ansatz

$$\eta_\alpha^{[m]} \approx \eta_\alpha^{[\infty]} - Ar^m, \qquad 0 < r < 1. \tag{250}$$

A least-squares fit to the four values (245)–(246) gives

$$\eta_\alpha^{[\infty]} \approx 0.8608, \qquad A \approx 0.3149, \qquad r \approx 0.6002. \tag{251}$$

The fitted sequence reproduces the observed crossings with a maximal absolute residual below

$$5.1 \times 10^{-4}. \tag{252}$$

Representative values are summarized in Table 12.

m	observed $\eta_\alpha^{[m]}$	fitted $\eta_\alpha^{[\infty]} - Ar^m$	residual
0	0.54595	0.54593	1.9×10^{-5}
1	0.67140	0.67184	-4.4×10^{-4}
2	0.74790	0.74740	5.0×10^{-4}
3	0.79260	0.79275	-1.5×10^{-4}

Table 12: Observed mode-release crossings versus the geometric-tail fit (250). The residuals are small compared with the total shift across the released-mode corridor.

Interpretive consequence. If the damping pattern encoded by (251) persists, then the current broader-family corridor is not drifting without bound. Instead it approaches a finite saturation scale,

$$\eta_\alpha^{[\infty]} - \eta_\alpha^{[3]} \approx 0.0682, \tag{253}$$

so the remaining tail beyond the third released odd half-mode is comparable to one more moderate shift rather than to a runaway instability. This is not yet a proof of ansatz independence, but it is stronger than mere survival of isolated crossings: the accumulated evidence now favors a *stabilizing* broader-family drift rather than an uncontrolled one.

Integrated downstream conclusion after the mode-release fit. This continuation adds a twelfth downstream message. Positively, the broader-family drift accumulated so far is numerically compatible with a finite saturation corridor near $\eta \approx 0.86$, rather than with runaway migration of the crossing. Cautionarily, the fitted limit (251) is still inferred from only four release levels and remains a meta-diagnostic rather than a true infinite-family minimization. The best current reading of the frozen main-text architecture is therefore

$$\begin{aligned}&\text{residual normalization}\\ &\quad + \text{non-perturbative evaluation}\\ &\quad + \text{progressively broader doubled-family robustness}\\ &\quad + \text{a local-stability corridor whose mode-release drift appears damped.}\end{aligned} \tag{254}$$

What remains open is equally clear: one now has evidence for stabilization of the corridor, but not yet an actual global minimization over the full admissible doubled family.

Remark C.5. The present appendix now carries twelve distinct downstream messages. The new twelfth message is again mixed. Positively, the sequence of broader-family crossings appears to organize itself into a damped release pattern with a finite extrapolated limit. Cautionarily, that limit is still inferred rather than derived. This strengthens the robustness side of the route without changing the theorem-level scope of the paper.

C.6 Worked continuation XII: a simplex-pushforward origin for the barrier measure

The reduced-measure reinterpretation of Section C.4 internalized the barrier at the level of the admissible helicity interval, but it still left the Beta(2, 3) density as the simplest effective choice rather than as the image of an explicit preimage space. The next admissible continuation is to supply exactly such a preimage: realize the same density as the pushforward of a flat finite-dimensional simplex measure under the grouped coordinate $x = h/h_*$.

Status label. The simplex construction below has status **definition.** Its pushforward to the Beta(2, 3) density has status **effective relation.** The resulting microcell interpretation of the barrier is a **candidate interpretation.** No part of this subsection claims a unique microscopic dynamics.

Flat five-cell precursor space. Let

$$\Sigma_4 := \Big\{(y_1, y_2, y_3, y_4, y_5) \in \mathbb{R}^5_{\geq 0} : y_1 + y_2 + y_3 + y_4 + y_5 = 1\Big\} \tag{255}$$

be the standard four-simplex equipped with its uniform Dirichlet(1, 1, 1, 1, 1) measure. Group the five nonnegative coordinates into a two-cell block and a three-cell block by defining

$$x := y_1 + y_2, \qquad 1 - x := y_3 + y_4 + y_5, \qquad 0 < x < 1. \tag{256}$$

The grouped coordinate x is then the sum of the first two coordinates of a uniform point on Σ_4.

Exact pushforward density. A standard Dirichlet marginal computation gives the exact pushforward law

$$p_\Sigma(x) = \frac{\Gamma(5)}{\Gamma(2)\Gamma(3)} x^{2-1}(1-x)^{3-1} = 12x(1-x)^2, \qquad 0 < x < 1. \tag{257}$$

Thus the grouped simplex coordinate is distributed precisely according to the Beta(2, 3) density already introduced in (238). Returning to the helicity variable through $h = h_* x$, one recovers

$$\rho_\Sigma(h) = \frac{1}{h_*} p_\Sigma\left(\frac{h}{h_*}\right) = \frac{12}{h_*}\frac{h}{h_*}\left(1 - \frac{h}{h_*}\right)^2 = \rho_*(h). \tag{258}$$

So the reduced-measure barrier from Section C.4 is not merely a convenient effective density: it is the exact image of a flat five-cell precursor measure under the grouped coordinate $x = h/h_*$.

Barrier as a grouped microcell counting term. Substituting (258) into (239) yields

$$\mathfrak{G}_R^{(m)} = \log \widetilde{\Delta}_R^{(m)} - \log \rho_\Sigma(h), \tag{259}$$

so the same barrier action may now be read as the negative logarithm of the grouped five-cell precursor density divided by the residual gap. In particular,

$$-\log \rho_\Sigma(h) = -\log\left(\frac{12}{h_*}\right) - \log\left(\frac{h}{h_*}\right) - 2\log\left(1 - \frac{h}{h_*}\right), \tag{260}$$

which reproduces exactly the barrier structure already used throughout Sections B.9–C.4. The lower-end exponent 1 and upper-end exponent 2 are therefore no longer inserted by hand: they arise from a concrete $2 + 3$ grouping of flat precursor coordinates.

Interpretive consequence. The simplex picture sharpens the internal source of the barrier in one precise sense. The previous subsection showed that the selected branches maximize a reduced posterior weight $\Pi_R^{(m)}$. The present subsection shows that the prior factor in that posterior can itself be obtained by coarse-graining a flat finite-dimensional precursor space. Equivalently, the selector can now be read as a grouped microcell counting term. This is still not a microscopic field theory, and the $2+3$ grouping is not claimed to be unique. But it is a stronger endogenous origin than the reduced-interval description alone, because it exhibits an explicit preimage whose pushforward is exactly the required Beta$(2, 3)$ law.

A useful check is that the simplex pushforward inherits the same interior mode already found in (242): since Beta$(2, 3)$ has mode $x_{\text{mode}} = 1/3$, one gets

$$h_{\Sigma,\text{mode}} = \frac{h_*}{3} \approx 0.1807, \tag{261}$$

in exact agreement with the reduced-measure route. The selected crossing helicities are therefore again not imposed by the prior alone; they remain the outcome of competition between grouped precursor counting and the residual gap.

Integrated downstream conclusion after the simplex lift. This continuation adds a thirteenth downstream message. Positively, the barrier now has an explicit finite-dimensional precursor: a flat five-cell simplex measure whose grouped coordinate $x = h/h_*$ pushes forward exactly to the Beta$(2, 3)$ density. Cautionarily, this is still a combinatorial precursor rather than a genuine microscopic dynamics, and the $2 + 3$ grouping is currently an admissible realization rather than a unique derivation. The best current reading of the frozen main-text architecture is therefore

$$\begin{aligned}
&\text{residual normalization} \\
&\quad + \text{non-perturbative evaluation} \\
&\quad + \text{progressively broader doubled-family robustness with damped drift} \\
&\quad + \text{a barrier having both a reduced-measure and a simplex-pushforward origin.}
\end{aligned} \tag{262}$$

The remaining conceptual gap is now narrower again: the route has not reached microscopic dynamics, but it does possess a concrete coarse-grained precursor space rather than only an effective interval weight.

Remark C.6. The present appendix now carries thirteen distinct downstream messages. The new thirteenth message is again mixed. Positively, the barrier density can now be lifted to an explicit flat precursor simplex whose grouped coordinate reproduces the exact Beta$(2, 3)$ law. Cautionarily, that lift is still combinatorial and non-unique rather than microscopic and dynamical. This pushes the barrier question another layer inward while keeping the theorem-level scope of the paper unchanged.

D Microscopic Lift, Infinite Released-Mode Limit, and Field Embedding

This appendix again uses the standing conventions of Section B.1 and converts the reduced-measure selector into explicit microscopic and field-theoretic constructions. Here the strongest statements are model-internal existence and exact coarse-graining statements at finite truncation, followed by projective and stochastic-field continuations whose rigor level is stated explicitly subsection by subsection.

D.1 Worked continuation XIII: a Gamma-microcell dynamical lift of the simplex barrier

The simplex-pushforward precursor of Section C.6 supplied an explicit static preimage of the Beta(2, 3) barrier. The next admissible continuation is to replace that static preimage by an explicit microscopic Markov dynamics whose stationary law reproduces the same grouped density exactly. A minimal choice is to lift the five precursor cells to five nonnegative stochastic intensities with exponential stationary marginals.

Status label. The stochastic process introduced below has status **definition**. Its stationary grouped law has status **effective relation**. Reading the five intensities as microscopic precursor cells is a **candidate interpretation**. No part of this subsection claims uniqueness of the microscopic lift.

Five nonnegative microcell intensities. For $i = 1, \ldots, 5$, let $S_i(\tau) \in \mathbb{R}_{\geq 0}$ solve

$$dS_i = (1 - S_i)\, d\tau + \sqrt{2S_i}\, dW_i, \qquad i = 1, \ldots, 5, \tag{263}$$

with independent standard Brownian motions W_i. For a single coordinate, the zero-current stationary equation

$$(1 - s)p(s) = \frac{d}{ds}\big(sp(s)\big) \tag{264}$$

is solved by $p(s) = e^{-s}\mathbf{1}_{s \geq 0}$. Hence the product density

$$\Pi_0(s_1, \ldots, s_5) = \exp\left(-\sum_{i=1}^{5} s_i\right) \prod_{i=1}^{5} \mathbf{1}_{s_i \geq 0} \tag{265}$$

is stationary for the uncoupled five-cell process.

Introduce the radial and normalized variables

$$r := \sum_{i=1}^{5} S_i, \qquad Y_i := \frac{S_i}{r}, \qquad X := Y_1 + Y_2, \qquad h := h_* X. \tag{266}$$

Whenever $r > 0$, one has $Y = (Y_1, \ldots, Y_5) \in \Sigma_4$ and $0 < X < 1$.

Proposition D.1 (Gamma microcells reproduce the barrier prior exactly). *For the stationary law* (265)*, the normalized vector Y is uniformly distributed on the simplex Σ_4, equivalently Dirichlet*$(1, 1, 1, 1, 1)$. *Consequently the grouped coordinate $X = Y_1 + Y_2$ has density*

$$p_X(x) = 12x(1 - x)^2, \qquad 0 < x < 1, \tag{267}$$

and the induced helicity density is exactly

$$\rho_{\text{micro}}(h) = \frac{1}{h_*} p_X\left(\frac{h}{h_*}\right) = \frac{12}{h_*}\frac{h}{h_*}\left(1 - \frac{h}{h_*}\right)^2 = \rho_*(h). \tag{268}$$

Proof. Write $s_i = r y_i$ with $r > 0$ and $y \in \Sigma_4$. The Jacobian is r^4, so the stationary density factorizes as

$$\Pi_0(s)\, ds \propto e^{-r} r^4\, dr\, d\nu_{\Sigma_4}(y), \tag{269}$$

where $d\nu_{\Sigma_4}$ is the flat measure on the simplex. After normalization, r and Y are independent, $r \sim \Gamma(5, 1)$, and Y is uniform on Σ_4. Grouping the first two simplex coordinates therefore gives the Beta(2, 3) marginal (267). Equation (268) follows from the linear change of variable $h = h_* x$. □

Interpretive consequence. Section C.6 exhibited the barrier as a static grouped-simplex counting term. Proposition D.1 strengthens that picture to a genuine microscopic dynamics: the same Beta(2, 3) law now arises as the stationary grouped distribution of five nonnegative stochastic intensities. In this sense the appendix route has crossed the threshold from a merely combinatorial precursor to a bona fide microscopic Markov process, even before the residual gap is coupled back in.

Integrated downstream conclusion after the Gamma lift. This continuation adds a fourteenth downstream message. Positively, the Beta(2, 3) barrier is now the exact grouped stationary law of an explicit five-cell stochastic microdynamics, not merely the pushforward of a static simplex measure. Cautionarily, the residual gap has not yet been coupled into that dynamics, so one has reached a microscopic precursor process but not yet the full microscopic doubled-sector dynamics. The next admissible step is therefore to bias the same microcell process by the residual gap and check whether the entire barrier-regularized branch becomes an exact coarse-grained free energy.

Remark D.2. The present appendix now carries fourteen distinct downstream messages. The new fourteenth message is again mixed. Positively, the grouped barrier prior has been upgraded from a static simplex picture to an explicit microscopic stochastic process. Cautionarily, that process is still only a precursor until the residual gap is coupled into it. This pushes the route to the threshold of microscopic dynamics while keeping the theorem-level scope of the paper unchanged.

D.2 Worked continuation XIV: exact coarse-graining of the microscopic doubled-sector dynamics

The Gamma-microcell lift of Section D.1 still leaves the residual factor $\widetilde{\Delta}_R^{(m)}$ outside the microscopic dynamics. The next admissible continuation is to couple that same five-cell process directly to any fixed finite doubled-family truncation and ask whether the previously used barrier-regularized action appears as an exact reduced free energy rather than as an appended effective rule.

Status label. The coupled stochastic process below has status **definition**. Its stationary density and grouped marginal have status **effective relation**. Reading the resulting free energy as a microscopic realization of the appendix branch is a **candidate interpretation**. Treating it as the unique microscopic physics of the present doubled-sector branch would still be a **conjectural identification** and is not claimed here.

Finite doubled-family truncation. Fix a finite release level $m \geq 0$. Let $\mathbf{a} \in \mathcal{D}_m(\eta)$ denote the non-canonical odd half-mode amplitudes for that truncation, where $\mathcal{D}_m(\eta) \subset \mathbb{R}^m$ is a compact admissible domain extending the one-, two-, and three-mode boxes already used in Sections B.8, C.2, and C.3. For $m = 0$ there is no $\mathbf{a}$-variable. Let

$$\widetilde{\Delta}_R^{(m)}(\eta, h, \mathbf{a}) > 0 \tag{270}$$

be the exact residual gap associated with that truncation on the positivity domain under consideration. Define

$$h(s) := h_* \frac{s_1 + s_2}{\sum_{j=1}^{5} s_j}. \tag{271}$$

Then, for $r = \sum_j s_j$ and $x = (s_1 + s_2)/r$,

$$\partial_{s_i} h(s) = \frac{h_*}{r} \begin{cases} 1 - x, & i = 1, 2, \\ -x, & i = 3, 4, 5. \end{cases} \tag{272}$$

So the residual-gap slope with respect to h couples the two-cell and three-cell microblocks with opposite signs.

Coupled microscopic dynamics. For $i = 1, \dots, 5$, define

$$dS_i = \Big[1 - S_i - S_i\, \partial_{s_i} \log \widetilde{\Delta}_R^{(m)}(\eta, h(S), \mathbf{a})\Big]\, d\tau + \sqrt{2S_i}\, dW_i, \tag{273}$$

and for each released amplitude coordinate a_k, define the reflected diffusion

$$da_k = -\partial_{a_k} \log \widetilde{\Delta}_R^{(m)}(\eta, h(S), \mathbf{a})\, d\tau + \sqrt{2}\, dB_k + dK_k, \qquad k = 1, \dots, m, \tag{274}$$

where the terms dK_k enforce reflection on the boundary of $\mathcal{D}_m(\eta)$. Equation (273) reduces to the uncoupled Gamma microcells of (263) when the residual gap is omitted. The drift correction simply inserts the logarithmic slope of the residual gap into the five-cell intensity dynamics.

Proposition D.3 (Exact microscopic lift of the barrier-regularized doubled branch)**.** *Assume* (270) *holds on* $\mathbb{R}^5_{>0} \times \mathcal{D}_m(\eta)$. *Then the coupled process* (273)–(274) *has reversible stationary density*

$$\Pi_m(s, \mathbf{a} \mid \eta) = \frac{1}{Z_m(\eta)} \exp\Big(-\sum_{i=1}^{5} s_i\Big)\, \widetilde{\Delta}_R^{(m)}(\eta, h(s), \mathbf{a})^{-1}\, \mathbf{1}_{\mathbf{a} \in \mathcal{D}_m(\eta)}. \tag{275}$$

Its grouped marginal in $(h, \mathbf{a})$ *is exactly*

$$\pi_m(h, \mathbf{a} \mid \eta) = \frac{1}{\mathcal{Z}_m(\eta)} \rho_*(h) \widetilde{\Delta}_R^{(m)}(\eta, h, \mathbf{a})^{-1} \mathbf{1}_{\mathbf{a} \in \mathcal{D}_m(\eta)}. \tag{276}$$

Consequently the negative logarithm of the reduced stationary density is

$$-\log \pi_m(h, \mathbf{a} \mid \eta) = \log \widetilde{\Delta}_R^{(m)}(\eta, h, \mathbf{a}) - \log \rho_*(h) + \mathit{const} = \mathfrak{G}_R^{(m)}(\eta, h, \mathbf{a}) + \mathit{const}. \tag{277}$$

Proof. For the microcell coordinates, the zero-current condition for a stationary density Π is

$$\Big(1 - s_i - s_i \partial_{s_i} \log \widetilde{\Delta}_R^{(m)}\Big)\Pi = \partial_{s_i}(s_i \Pi). \tag{278}$$

Substituting $\Pi = \Pi_m$ from (275) makes both sides equal identically. For the released amplitudes, the reflected additive diffusion has zero stationary current exactly when

$$-\partial_{a_k} \log \widetilde{\Delta}_R^{(m)}\, \Pi_m = \partial_{a_k} \Pi_m, \tag{279}$$

which again holds identically for (275). This proves reversibility of the full density.

To obtain the grouped marginal, write $s_i = r y_i$ as in Proposition D.1. Since $\widetilde{\Delta}_R^{(m)}$ depends on s only through the grouped coordinate $h = h_*(y_1 + y_2)$, the radial integral factors out, while the simplex integral over fibers of $x = y_1 + y_2$ gives exactly the Beta$(2, 3)$ density from (267). Therefore (276) follows, and (277) is immediate upon taking minus the logarithm. □

Microscopic meaning of the stable selected branches. Equation (277) closes the loop opened in Sections B.9–C.3. For every finite truncation m, the previously computed local minima of $\mathfrak{G}_R^{(m)}$ are exactly the local maxima of the reduced stationary density (276). In particular, the barrier-regularized stable crossings found for the one-, two-, and three-mode releases are not merely formal minima of an appendix action; within the present microscopic lift they are the most probable reduced configurations of an explicit Markov dynamics.

A compatible reduced overdamped description is therefore

$$\dot{h} = -M_h\, \partial_h \mathfrak{G}_R^{(m)}, \qquad \dot{\mathbf{a}} = -M_a\, \nabla_{\mathbf{a}} \mathfrak{G}_R^{(m)}, \tag{280}$$

with positive mobilities M_h and M_a, plus optional thermal noise. The stationary points identified in the previous continuations are linearly stable equilibria of (280) precisely when the Hessian of $\mathfrak{G}_R^{(m)}$ is positive definite, matching the stability diagnostics already computed in Sections B.9, C.2, and C.3.

Interpretive consequence. In this precise, model-internal sense, the appendix route now reaches a genuine microscopic dynamics. The barrier-regularized doubled-sector action is no longer merely an effective selector rule: it is the exact coarse-grained free energy of the coupled five-cell stochastic process (273)–(274). What remains open is not whether one can write down a microscopic dynamics at all, but whether this lift is unique, how it extends to genuinely broad or infinite doubled families, and whether it admits a deeper field-theoretic embedding.

Integrated downstream conclusion after the microscopic lift. This continuation adds a fifteenth downstream message. Positively, for every fixed finite doubled-family truncation there is now an explicit microscopic stochastic dynamics whose exact reduced stationary free energy is the barrier-regularized residual action already used to organize the appendix branches. Cautionarily, this microscopic lift is still model-dependent, finite-truncation based, and not yet a unique field-theoretic derivation. The best current reading of the frozen main-text architecture is therefore

$$
\begin{aligned}
&\text{residual normalization}\\
&\quad + \text{non-perturbative evaluation}\\
&\quad + \text{progressively broader doubled-family robustness with damped drift}\\
&\quad + \text{a barrier having reduced-measure, simplex, and Gamma-microcell origins}\\
&\quad + \text{an exact finite-truncation microscopic Markov lift whose reduction is } \mathfrak{G}_R^{(m)}.
\end{aligned}
\tag{281}
$$

This is the closest the present manuscript can come to microscopic completion without overstepping its frozen main-text scope: the downstream branch now has a bona fide microscopic dynamics, but not yet a unique untruncated one.

Remark D.4. The present appendix now carries fifteen distinct downstream messages. The new fifteenth message is the strongest so far. Positively, the barrier-regularized doubled-sector branch now arises as the exact coarse-grained free energy of an explicit microscopic stochastic process. Cautionarily, that process is still a model-dependent finite-truncation lift rather than a unique microscopic field theory. This reaches microscopic dynamics in a precise appendix-level sense while leaving the theorem-level scope of the paper unchanged.

D.3 Worked continuation XV: projective infinite released-mode closure

The finite-truncation microscopic lift of Section D.2 still leaves one structural step implicit: the released odd half-modes were handled one finite box at a time. The next admissible continuation is to pass from those finite boxes to a single tail-controlled infinite released family and ask whether the same microscopic branch closes to a genuine projective limit rather than only to a sequence of increasingly broad diagnostics.

Status label. The infinite released family and its projective stationary measure below have status **definition** and **effective relation**. Reading the resulting branch as the mode-independent completion of the finite releases is a **candidate interpretation**. A literal numerical identification of the infinite released-mode crossing with a specific named constant remains a **numerical diagnostic** and is not claimed here.

Tail-controlled infinite doubled family. Write the doubled angular variable as

$$
u := \omega t \in [0, 4\pi], \qquad e_0(u) := \sin\left(\frac{u}{2}\right), \qquad e_n(u) := \sin\left(\frac{(2n+1)u}{2}\right), \quad n \geq 1. \tag{282}
$$

Choose an admissible tail envelope $b_n(\eta) > 0$ such that

$$b_1(\eta) = \frac{\eta}{2}, \qquad b_2(\eta) = \frac{\eta}{3}, \qquad b_3(\eta) = \frac{\eta}{4}, \qquad \sum_{n=1}^{\infty}(2n+1)^3 b_n(\eta) < \infty. \tag{283}$$

The first three values reproduce the previously released one-, two-, and three-mode amplitude boxes, while the summability condition supplies a genuine infinite tail. Define the compact coefficient space

$$\mathcal{K}_\eta^{(\infty)} := \prod_{n=1}^{\infty}\left[-b_n(\eta), b_n(\eta)\right], \qquad P_m\mathbf{a} := (a_1, \ldots, a_m, 0, 0, \ldots). \tag{284}$$

For $\mathbf{a} = (a_1, a_2, \ldots) \in \mathcal{K}_\eta^{(\infty)}$, define the infinite released radial profile

$$R_\infty(t; \eta, \mathbf{a}) := R_0\left[1 + \eta e_0(\omega t) + \sum_{n=1}^{\infty} a_n e_n(\omega t)\right]. \tag{285}$$

Because each derivative of $e_n(\omega t)$ up to order three brings at most a factor $(2n+1)^3\omega^3/8$, the summability condition (283) implies that the series for R_∞, R'_∞, R''_∞, and R'''_∞ converge uniformly on $[0, 4\pi/\omega]$. Hence the corresponding curve

$$X_R^{(\infty)}(t; \eta, h, \mathbf{a}) := \big(R_\infty(t; \eta, \mathbf{a})\cos(\omega t),\ R_\infty(t; \eta, \mathbf{a})\sin(\omega t),\ vt\big) \tag{286}$$

is a well-defined C^3 doubled representative.

Infinite exact energy and residual gap. Assume one restricts attention to a compact admissible subset

$$\mathcal{U}_\eta \subset (0, h_*) \times \mathcal{K}_\eta^{(\infty)} \tag{287}$$

for which there exist positive constants $c_0(\eta)$ and $c_1(\eta)$ satisfying

$$|X_R^{(\infty)\prime}(t; \eta, h, \mathbf{a})| \geq c_0(\eta), \qquad |X_R^{(\infty)\prime}(t; \eta, h, \mathbf{a}) \times X_R^{(\infty)\prime\prime}(t; \eta, h, \mathbf{a})| \geq c_1(\eta) \tag{288}$$

for all $(h, \mathbf{a}) \in \mathcal{U}_\eta$ and all $t \in [0, 4\pi/\omega]$. Then the exact energy

$$E_R^{(\infty)}(\eta, h, \mathbf{a}) := E_A\left[X_R^{(\infty)}(\eta, h, \mathbf{a})\right] \tag{289}$$

is well defined, and the corresponding residual gap is

$$\widetilde{\Delta}_R^{(\infty)}(\eta, h, \mathbf{a}) := \frac{E_R^{(\infty)}(\eta, h, \mathbf{a}) - 2E_{\mathrm{vac}}}{E_{\mathrm{vac}}}. \tag{290}$$

Proposition D.5 (Uniform truncation convergence on admissible infinite families)**.** *Under* (283) *and* (288)*, one has*

$$\sup_{(h,\mathbf{a})\in\mathcal{U}_\eta}\left|E_R^{(m)}(\eta, h, P_m\mathbf{a}) - E_R^{(\infty)}(\eta, h, \mathbf{a})\right| \longrightarrow 0 \qquad (m \to \infty), \tag{291}$$

and therefore

$$\sup_{(h,\mathbf{a})\in\mathcal{U}_\eta}\left|\widetilde{\Delta}_R^{(m)}(\eta, h, P_m\mathbf{a}) - \widetilde{\Delta}_R^{(\infty)}(\eta, h, \mathbf{a})\right| \longrightarrow 0. \tag{292}$$

Proof. The tail bound (283) implies uniform convergence of R_∞ and of its first three derivatives, so

$$X_R^{(m)}(\eta, h, P_m\mathbf{a}) \longrightarrow X_R^{(\infty)}(\eta, h, \mathbf{a}) \quad \text{in } C^3 \tag{293}$$

uniformly on $\mathcal{U}_\eta$. On the admissible set (288), the curvature and torsion formulas are smooth functions of (X', X'', X'''). Hence the integrand $(\kappa^2 + \tau^2)|X'|$ in (60) converges uniformly on $[0, 4\pi/\omega] \times \mathcal{U}_\eta$. Integrating in t yields (291); the residual-gap convergence (292) follows by division by the fixed positive constant E_{vac}. □

Projective reduced stationary measures. Let

$$d\nu_\eta^{(\infty)}(\mathbf{a}) := \bigotimes_{n=1}^{\infty} \frac{1}{2b_n(\eta)} \mathbf{1}_{|a_n|\le b_n(\eta)}\, da_n \tag{294}$$

be the product probability measure on $\mathcal{K}_\eta^{(\infty)}$. For each finite release level m, define the reduced stationary measure

$$d\pi_m(h,\mathbf{a}\mid\eta) := \frac{1}{\mathcal{Z}_m(\eta)} \rho_*(h) \widetilde{\Delta}_R^{(m)}(\eta,h,P_m\mathbf{a})^{-1} \mathbf{1}_{(h,\mathbf{a})\in\mathcal{U}_\eta}\, dh\, d\nu_\eta^{(\infty)}(\mathbf{a}). \tag{295}$$

Assume in addition that there exist positive constants δ_η and Δ_η^* such that

$$0 < \delta_\eta \le \widetilde{\Delta}_R^{(\infty)}(\eta,h,\mathbf{a}) \le \Delta_\eta^* \qquad \text{on } \mathcal{U}_\eta. \tag{296}$$

Then (292) implies uniform convergence of the densities in (295).

Proposition D.6 (Projective closure of the finite-truncation microscopic branch). *Under* (288) *and* (296)*, the measures* $\pi_m(\cdot\mid\eta)$ *converge weakly to*

$$d\pi_\infty(h,\mathbf{a}\mid\eta) := \frac{1}{\mathcal{Z}_\infty(\eta)} \rho_*(h) \widetilde{\Delta}_R^{(\infty)}(\eta,h,\mathbf{a})^{-1} \mathbf{1}_{(h,\mathbf{a})\in\mathcal{U}_\eta}\, dh\, d\nu_\eta^{(\infty)}(\mathbf{a}). \tag{297}$$

Equivalently,

$$-\log \pi_\infty(h,\mathbf{a}\mid\eta) = \log \widetilde{\Delta}_R^{(\infty)}(\eta,h,\mathbf{a}) - \log \rho_*(h) + const = \mathfrak{G}_R^{(\infty)}(\eta,h,\mathbf{a}) + const, \tag{298}$$

where

$$\mathfrak{G}_R^{(\infty)}(\eta,h,\mathbf{a}) := \log \widetilde{\Delta}_R^{(\infty)}(\eta,h,\mathbf{a}) - \log \rho_*(h). \tag{299}$$

Proof. By (296) and Proposition D.5, the densities in (295) converge uniformly on $\mathcal{U}_\eta$ to the density in (297). Bounded convergence therefore gives convergence of all bounded cylinder test functions, hence weak convergence. Taking minus the logarithm of (297) yields (298). □

Corollary D.7 (Strict minima persist in the infinite released-mode limit). *Assume the hypotheses of Proposition D.6. Let $\mathcal{N}_\eta \subset \mathcal{U}_\eta$ be a compact neighborhood on which*

$$\mathfrak{G}_R^{(m)} \to \mathfrak{G}_R^{(\infty)} \qquad \text{in } C^2(\mathcal{N}_\eta), \tag{300}$$

and let $x_m \in \mathcal{N}_\eta$ be interior critical points such that

$$\nabla \mathfrak{G}_R^{(m)}(x_m) = 0, \qquad \nabla^2 \mathfrak{G}_R^{(m)}(x_m) \ge \lambda_0 I \quad \text{for some } \lambda_0 > 0 \text{ independent of } m. \tag{301}$$

If $x_m \to x_\infty \in \mathcal{N}_\eta$, then

$$\nabla \mathfrak{G}_R^{(\infty)}(x_\infty) = 0, \qquad \nabla^2 \mathfrak{G}_R^{(\infty)}(x_\infty) \ge \lambda_0 I. \tag{302}$$

In particular, x_∞ is a strict local minimum of the infinite released-mode free energy.

Proof. The C^2 convergence (300) implies convergence of gradients and Hessians uniformly on $\mathcal{N}_\eta$. Passing to the limit in (301) therefore gives (302). Positive definiteness of the limiting Hessian yields strict local minimality of x_∞. □

Continuation of the stable branch. The broader-family drift fit of Section C.5 already suggested a damped release sequence, with

$$\eta_\alpha^{[\infty]} \approx 0.8608 \tag{303}$$

as the minimal geometric-tail extrapolation from the first four releases. Proposition D.6 supplies the structural closure underlying that heuristic: there is now a single reduced free energy $\mathfrak{G}_R^{(\infty)}$ to which the finite-truncation free energies converge.

Corollary D.7 is the precise variational form of that continuation statement. Once a sequence of finite released-mode stable points enjoys uniform interior coercivity and converges inside a compact neighborhood, its accumulation point is automatically a strict local minimum of the infinite released-mode free energy. In this sense the stable selected branch passes from finite release to the infinite released-mode limit without changing its variational character.

Integrated downstream conclusion after projective closure. This continuation adds a sixteenth downstream message. Positively, the finite-truncation microscopic lift no longer stops at a sequence of ever-broader boxes: under explicit tail, admissibility, and positivity hypotheses it closes to a single infinite released-mode reduced measure whose free energy is exactly $\mathfrak{G}_R^{(\infty)}$. Cautionarily, the actual numerical location of the infinite released-mode crossing still depends on the chosen tail envelope and on a direct resolution of the full infinite reduced branch. The route is now structurally closed at the infinite released-mode level, but not yet numerically unique.

Remark D.8. The present appendix now carries sixteen distinct downstream messages. The new sixteenth message is the first one that directly closes the mode-release ladder itself: the broader-family drift is no longer only an extrapolation pattern but is attached to an explicit projective limit of reduced stationary measures. This is stronger than the previous finite-truncation microscopic lift, although it still depends on an admissible infinite tail prescription.

D.4 Worked continuation XVII: odd-sector field-theoretic embedding of the infinite lift

The projective limit of Section D.3 closes the released-mode sequence at the level of coefficient measures. The next admissible continuation is to rewrite that same limit as a stochastic field theory on the doubled parameter circle, so that the finite released families become Galerkin truncations of one reflected odd-sector field dynamics rather than a disconnected list of mode boxes.

Status label. The odd-sector field process and its stationary measure below have status **definition**. The identification of its Galerkin stationary laws with the finite released-mode microscopic lifts has status **effective relation**. Reading this process as the unique microscopic field theory of the present doubled-sector branch remains a **conjectural identification** and is not claimed here.

Odd half-mode field on the doubled circle. Let $\mathcal{H}_{\text{odd}}$ be the closed span of the basis (282) on the doubled circle $u \in [0, 4\pi]$. For $\mathbf{a} \in \mathcal{K}_\eta^{(\infty)}$, define the odd half-mode field

$$\phi_{\mathbf{a}}(u) := \sum_{n=1}^{\infty} a_n e_n(u), \qquad \phi_{\mathbf{a}}(u + 2\pi) = -\phi_{\mathbf{a}}(u). \tag{304}$$

By (283), this series converges in C^3. The antiperiodicity in (304) is the functional shadow of the interaction-type doubled-state condition (12): shifting by 2π keeps the same inertial plane background while passing to the mirror-symmetric projection-traversal partner and reversing the signed normal deviation with respect to that plane. The infinite released radial profile therefore becomes

$$R[u;\eta,\phi] := R_0\big[1+\eta e_0(u)+\phi(u)\big]. \tag{305}$$

The product coefficient law (294) therefore pushes forward to a probability measure $\mathbb{P}^{\phi}_{\eta}$ on odd-antiperiodic C^3 fields.

Microscopic field measure. Couple the five Gamma microcells to the field through the grouped helicity variable

$$h(s) = h_* \frac{s_1+s_2}{\sum_{j=1}^{5} s_j}. \tag{306}$$

Then define the microscopic field measure by

$$d\mathbb{P}^{\mathrm{AR}}_{\eta}(s,\phi) := \frac{1}{\mathbf{Z}_{\eta}} \exp\Big(-\sum_{i=1}^{5} s_i\Big)\, \widetilde{\Delta}^{(\infty)}_{R}\big(\eta, h(s), \phi\big)^{-1}\, ds\, d\mathbb{P}^{\phi}_{\eta}(\phi), \tag{307}$$

where $\widetilde{\Delta}^{(\infty)}_{R}(\eta,h,\phi)$ means (290) evaluated on the field profile (305). Integrating out the radial microcell variable exactly as in Proposition D.3 gives the reduced field measure

$$d\overline{\mathbb{P}}^{\mathrm{AR}}_{\eta}(h,\phi) := \frac{1}{\overline{\mathbf{Z}}_{\eta}} \rho_*(h) \widetilde{\Delta}^{(\infty)}_{R}(\eta,h,\phi)^{-1}\, dh\, d\mathbb{P}^{\phi}_{\eta}(\phi). \tag{308}$$

Hence the reduced free energy relative to the reference measure $dh\, d\mathbb{P}^{\phi}_{\eta}$ is again exactly $\mathfrak{G}^{(\infty)}_{R}$.

Reflected odd-sector stochastic field dynamics. In mode coordinates, the microscopic field dynamics is the countable reflected system

$$dS_i = \Big[1 - S_i - S_i\, \partial_{s_i} \log \widetilde{\Delta}^{(\infty)}_{R}\big(\eta, h(S), \mathbf{a}\big)\Big] d\tau + \sqrt{2S_i}\, dW_i, \qquad i=1,\dots,5, \tag{309}$$

$$da_n = -\partial_{a_n} \log \widetilde{\Delta}^{(\infty)}_{R}\big(\eta, h(S), \mathbf{a}\big)\, d\tau + \sqrt{2}\, dB_n + dK^{+}_{n} - dK^{-}_{n}, \qquad n \ge 1, \tag{310}$$

where the Skorokhod terms satisfy

$$|a_n(\tau)| \le b_n(\eta), \qquad \int_0^{\tau} \mathbf{1}_{a_n<b_n(\eta)}\, dK^{+}_{n} = 0, \qquad \int_0^{\tau} \mathbf{1}_{a_n>-b_n(\eta)}\, dK^{-}_{n} = 0. \tag{311}$$

Formally summing (310) against the basis functions e_n yields the reflected odd-sector SPDE

$$\partial_\tau \phi(u,\tau) = -P_{\mathrm{odd}} \frac{\delta}{\delta\phi(u)} \log \widetilde{\Delta}^{(\infty)}_{R}\big(\eta, h(S), \phi\big) + \sqrt{2}\, \xi_{\mathrm{odd}}(u,\tau) + \zeta_{\mathrm{ref}}(u,\tau), \tag{312}$$

where P_{odd} is the projector onto the odd half-mode sector, ξ_{odd} is cylindrical white noise in that sector, and ζ_{ref} is the reflected constraint force enforcing $|a_n| \le b_n(\eta)$ mode by mode.

Proposition D.9 (Galerkin recovery of the finite microscopic lifts). *For each $M \ge 0$, project (312) onto the first M released odd half-modes. The resulting coupled diffusion has stationary law*

$$d\mathbb{P}^{\mathrm{AR}}_{\eta,M}(s,\mathbf{a}_{1:M}) = \frac{1}{\mathbf{Z}_{\eta,M}} \exp\Big(-\sum_{i=1}^{5} s_i\Big)\, \widetilde{\Delta}^{(M)}_{R}\big(\eta, h(s), \mathbf{a}_{1:M}\big)^{-1} \prod_{n=1}^{M} \frac{1}{2b_n(\eta)} \mathbf{1}_{|a_n|\le b_n(\eta)}\, da_n\, ds. \tag{313}$$

Its reduced $(h,\mathbf{a}_{1:M})$-marginal is exactly the finite-truncation barrier-regularized measure of Section D.2 written with the same amplitude boxes. Under the hypotheses of Proposition D.6, these Galerkin stationary laws converge weakly to (307) and (308) as $M \to \infty$.

Proof. At fixed M, the coefficient system (309)–(310) is exactly the finite-dimensional reflected diffusion already used in Section D.2, now written in the uniform product-box coordinates $a_1, \dots, a_M$. The zero-current computation is therefore the same as in Proposition D.3, giving (313). The reduction to $(h, \mathbf{a}_{1:M})$ again uses the Beta(2, 3) grouped microcell factor. Weak convergence of the Galerkin laws to the infinite field measures follows from Proposition D.6. □

Theorem D.10 (Microscopic-to-field cylinder consistency of the infinite lift). *Assume the hypotheses of Proposition D.6. Let F be a bounded cylinder functional on the full microscopic-field state space of the form*

$$F(s, \phi) := f\big(s_1, \dots, s_5, \langle \phi, e_1 \rangle, \dots, \langle \phi, e_N \rangle\big), \tag{314}$$

with f bounded and continuous and $N < \infty$. Then, for every $M \geq N$,

$$\lim_{M \to \infty} \int F \, d\mathbb{P}^{\mathrm{AR}}_{\eta,M} = \int F \, d\mathbb{P}^{\mathrm{AR}}_{\eta}. \tag{315}$$

Likewise, if G is a bounded reduced cylinder functional of the form

$$G(h, \phi) := g\big(h, \langle \phi, e_1 \rangle, \dots, \langle \phi, e_N \rangle\big), \tag{316}$$

with g bounded and continuous, then

$$\lim_{M \to \infty} \int G \, d\overline{\mathbb{P}}^{\mathrm{AR}}_{\eta,M} = \int G \, d\overline{\mathbb{P}}^{\mathrm{AR}}_{\eta}. \tag{317}$$

In particular, the finite microscopic lifts and the infinite odd-sector field embedding define the same model-internal branch when tested by finitely resolved observables.

Proof. For $M \geq N$, the functional F depends only on the microcell variables and on the first N released coefficients. Under the identification $\langle \phi, e_n \rangle = a_n$ for the odd half-mode basis, F becomes a bounded continuous cylinder test function on coefficient space. Proposition D.9 identifies $\mathbb{P}^{\mathrm{AR}}_{\eta,M}$ with the M-mode microscopic lift, while Proposition D.6 gives weak convergence of the corresponding coefficient-space laws to the infinite measure. Equation (315) follows immediately. The reduced statement (317) is identical after integrating out the radial microcell variable and using the reduced laws from (308). □

Interpretive consequence. Proposition D.9 completes the alignment that was only partial in Section D.2. The finite one-, two-, and three-mode microscopic lifts are no longer isolated model choices: they are the first Galerkin sectors of a single reflected stochastic field theory on the doubled parameter circle. The projective infinite released-mode measure and the field embedding therefore describe the same microscopic branch in two equivalent languages, one coefficient-based and one field-based. In particular, the odd-sector field is not an arbitrary Fourier container: by Proposition 2.5 it is exactly the function-space realization of the interaction-type doubled normal form singled out geometrically in the minimal doubled sector.

Integrated downstream conclusion after the field embedding. This continuation adds a seventeenth downstream message. Positively, the appendix route is now connected all the way from finite released-mode diagnostics to a single infinite odd-sector field dynamics whose Galerkin truncations reproduce the earlier microscopic lifts and whose reduced free energy remains exactly $\mathfrak{G}_R^{(\infty)}$. Cautionarily, the resulting field theory is still reflected, nonlocal through $\widetilde{\Delta}_R^{(\infty)}$, and tied to the doubled ansatz rather than derived from a unique local microscopic action. But the missing interface itself has now been supplied: the finite-truncation microscopic lift, the infinite released-mode limit, and the field-theoretic embedding are fully aligned inside one model-internal construction.

Remark D.11. The present appendix now carries seventeen distinct downstream messages. The new seventeenth message is the strongest microscopic completion available within the frozen scope of the manuscript: the appendix branch is no longer merely a finite sequence of stochastic lifts, but the Galerkin shadow of a single reflected odd-sector field theory on the doubled circle. What remains genuinely open after this point is uniqueness, locality, renormalized continuum control, and independence from the present doubled ansatz, not the existence of a coherent microscopic-to-field continuation itself.

D.5 Worked continuation XVIII: reduced reversible generators and projective cylinder martingale consistency

The field embedding of Section D.4 aligns the stationary measures of the finite microscopic lifts and the infinite odd-sector field. The next admissible strengthening is not yet a full pathwise theory of the reflected SPDE, but it does make two further steps rigorous inside the manuscript's own model class: first, one can isolate the unique reversible reduced generator carried by each finite Galerkin level; second, one can pass that generator-level stationarity to the infinite field at cylinder resolution.

Status label. The reduced finite-Galerkin generator theorem below has status **effective relation**. The projective cylinder martingale statement for the infinite field has status **effective relation**. A complete pathwise well-posedness or uniqueness theorem for the full infinite reflected SPDE remains a **conjectural continuation** and is not claimed here.

Reduced finite-Galerkin operator. Fix $M \geq 0$ and let

$$\Omega^{\mathrm{red}}_{\eta,M} := (0, h_*) \times \mathcal{D}_M(\eta). \tag{318}$$

Let $\overline{\mathbb{P}}^{\mathrm{AR}}_{\eta,M}$ denote the reduced $(h, \mathbf{a}_{1:M})$-law with density $\pi_M(h, \mathbf{a} \mid \eta)$ from (276), and write

$$A(h) := \frac{h(h_* - h)}{h_*}. \tag{319}$$

Then

$$A'(h) + A(h)\,\partial_h \log \rho_*(h) = 2 - \frac{5h}{h_*}. \tag{320}$$

For smooth test functions F on $\Omega^{\mathrm{red}}_{\eta,M}$ satisfying the Neumann reflection conditions $\partial_{a_n} F = 0$ on $a_n = \pm b_n(\eta)$, define

$$\mathcal{L}^{\mathrm{red}}_{\eta,M} F := \frac{1}{\pi_M}\partial_h\big(A(h)\pi_M\,\partial_h F\big) + \sum_{n=1}^{M} \frac{1}{\pi_M}\partial_{a_n}\big(\pi_M\,\partial_{a_n} F\big). \tag{321}$$

Expanding (321) gives

$$\begin{aligned}\mathcal{L}^{\mathrm{red}}_{\eta,M} F = {} & A(h)\,\partial_h^2 F + \Big(2 - \frac{5h}{h_*} - A(h)\partial_h \log \widetilde{\Delta}^{(M)}_R\Big)\partial_h F \\ & + \sum_{n=1}^{M}\Big(\partial_{a_n}^2 F - \partial_{a_n} \log \widetilde{\Delta}^{(M)}_R\,\partial_{a_n} F\Big),\end{aligned} \tag{322}$$

with the convention that the amplitude sum is absent when $M = 0$.

Theorem D.12 (Finite-Galerkin reversible generator and ergodic class). *Assume $\widetilde{\Delta}^{(M)}_R(\eta, h, \mathbf{a})$ extends to a strictly positive C^2 function on a neighborhood of $[0, h_*] \times \mathcal{D}_M(\eta)$. Then the following hold.*

(i) $\mathcal{L}^{\mathrm{red}}_{\eta,M}$ *is the unique local second-order operator on* $\Omega^{\mathrm{red}}_{\eta,M}$ *with diffusion coefficients* $A(h)$ *in the helicity coordinate and* 1 *in each released amplitude, Neumann reflection on* $\partial\mathcal{D}_M(\eta)$*, and zero stationary current with respect to* π_M.

(ii) The bilinear form

$$\mathcal{E}_{\eta,M}(F,G) := \int_{\Omega^{\mathrm{red}}_{\eta,M}} \Big(A(h)\,\partial_h F\,\partial_h G + \sum_{n=1}^{M} \partial_{a_n} F\,\partial_{a_n} G\Big)\, d\overline{\mathbb{P}}^{\mathrm{AR}}_{\eta,M} \tag{323}$$

satisfies

$$\int_{\Omega^{\mathrm{red}}_{\eta,M}} G\,\mathcal{L}^{\mathrm{red}}_{\eta,M} F\, d\overline{\mathbb{P}}^{\mathrm{AR}}_{\eta,M} = -\mathcal{E}_{\eta,M}(F,G) \tag{324}$$

for all such test functions F, G.

(iii) If F *is a smooth invariant of the generator,* $\mathcal{L}^{\mathrm{red}}_{\eta,M}F = 0$*, then* F *is constant* $\overline{\mathbb{P}}^{\mathrm{AR}}_{\eta,M}$*-almost everywhere. Hence the reduced finite-Galerkin branch is ergodic within the manuscript's reversible zero-current class.*

Proof. For a local second-order operator with diffusion matrix $\mathrm{diag}(A(h), 1, \ldots, 1)$, the zero-current condition with respect to the density π_M determines the drift uniquely in divergence form. This is exactly (321), and expanding it with (320) gives (322). This proves item (i).

For item (ii), integrate (321) by parts. The boundary terms in the amplitude directions vanish because of the Neumann reflection conditions on $\partial\mathcal{D}_M(\eta)$; the boundary terms in the helicity direction vanish because $A(0) = A(h_*) = 0$. The resulting identity is precisely (324).

Finally, if $\mathcal{L}^{\mathrm{red}}_{\eta,M}F = 0$, choose $G = F$ in (324). Then $\mathcal{E}_{\eta,M}(F,F) = 0$, so $A(h)|\partial_h F|^2 + \sum_{n=1}^{M} |\partial_{a_n} F|^2 = 0$ almost everywhere with respect to $\overline{\mathbb{P}}^{\mathrm{AR}}_{\eta,M}$. Since $A(h) > 0$ on $(0, h_*)$ and $\Omega^{\mathrm{red}}_{\eta,M}$ is connected, F is constant almost everywhere. This proves item (iii). □

Proposition D.13 (Projective cylinder martingale consistency of the infinite field). *Let*

$$F(h,\phi) := f\big(h, \langle\phi, e_1\rangle, \ldots, \langle\phi, e_N\rangle\big) \tag{325}$$

be a bounded smooth cylinder functional depending on the first N *odd half-modes. Define its projective cylinder generator by*

$$(\mathcal{L}^{\mathrm{cyl}}_{\eta,\infty}F)(h,\phi) := \big(\mathcal{L}^{\mathrm{red}}_{\eta,N} f\big)\big(h, \langle\phi, e_1\rangle, \ldots, \langle\phi, e_N\rangle\big). \tag{326}$$

Then

$$\int \mathcal{L}^{\mathrm{cyl}}_{\eta,\infty} F\, d\overline{\mathbb{P}}^{\mathrm{AR}}_{\eta} = 0. \tag{327}$$

If, in addition, $\mathcal{L}^{\mathrm{cyl}}_{\eta,\infty}F = 0$*, then* F *is constant* $\overline{\mathbb{P}}^{\mathrm{AR}}_{\eta}$*-almost everywhere on cylinder sets. Thus the infinite lifted field measure is stationary and cylinder-ergodic under the projective generator family induced from the finite Galerkin levels.*

Proof. For each $M \geq N$, the cylinder functional $\mathcal{L}^{\mathrm{cyl}}_{\eta,\infty}F$ depends only on h and on the first N released coefficients. By Theorem D.12,

$$\int \mathcal{L}^{\mathrm{red}}_{\eta,N} f\, d\overline{\mathbb{P}}^{\mathrm{AR}}_{\eta,N} = 0. \tag{328}$$

The cylinder-level convergence statement of Theorem D.10 then passes (328) to the infinite law, giving (327). If $\mathcal{L}^{\mathrm{cyl}}_{\eta,\infty}F = 0$, then $\mathcal{L}^{\mathrm{red}}_{\eta,N} f = 0$, so Theorem D.12 implies that f is constant $\overline{\mathbb{P}}^{\mathrm{AR}}_{\eta,N}$-almost everywhere. Applying Theorem D.10 once more yields the constancy of F under $\overline{\mathbb{P}}^{\mathrm{AR}}_{\eta}$. □

Interpretive consequence. The new point is not that a pathwise infinite SPDE theory has been completed; it has not. The new point is that the manuscript now controls the reversible generator class exactly at every finite released level and transfers its stationarity, and its ergodicity on cylinder observables, to the projective infinite field law. In this precise sense the microscopic-to-field route is no longer merely a measure-level coincidence but a generator-level continuation on all finitely resolved observables.

Integrated downstream conclusion after the reversible-generator continuation. This continuation adds an eighteenth downstream message. Positively, each finite reduced Galerkin branch now carries a uniquely determined reversible zero-current generator, and the infinite odd-sector field law is stationary and cylinder-ergodic for the induced projective generator family. Cautionarily, this is still weaker than a full pathwise well-posedness, uniqueness, or ergodicity theorem for the complete reflected SPDE. The manuscript has therefore closed the finite reversible class and the cylinder-level infinite continuation, while leaving the genuinely infinite stochastic dynamics as the next mathematical frontier.

Remark D.14. The present appendix now carries eighteen distinct downstream messages. The new eighteenth message is a controlled strengthening of the reflected odd-sector field layer: finite reversible generators are now fixed uniquely by the reduced stationary law, and the infinite field inherits stationary cylinder dynamics by projective consistency. What remains open after this point is not finite-mode reversibility, but full infinite-dimensional pathwise control, locality, renormalized continuum analysis, and independence from the present doubled ansatz.

D.6 Worked continuation XIX: finite reflected martingale problems and cylinder-level infinite continuation

The reversible-generator package of Section D.5 fixes the stationary divergence-form operator at each finite released level, but it still leaves open whether that operator is realized by a genuine reflected diffusion and whether the resulting finite martingale problems are compatible in the projective limit. The present continuation resolves exactly those two points inside the manuscript's finite-mode scope.

Status label. The finite reflected-diffusion theorem below has status **effective relation**. The projective cylinder martingale continuation for the infinite field also has status **effective relation**. A full pathwise theorem for the complete infinite reflected odd-sector SPDE remains a **conjectural continuation** and is not claimed here.

Theorem D.15 (Finite reflected diffusion realizing the reduced reversible generator)**.** *Assume the hypotheses of Theorem D.12. Fix $M \geq 0$. Then the reduced operator $\mathcal{L}^{\mathrm{red}}_{\eta,M}$ is realized by a reflected diffusion on the compact closure $\overline{\Omega^{\mathrm{red}}_{\eta,M}}$ with state variables $(h, \mathbf{a}_{1:M})$ solving*

$$dh_\tau = \Big(2 - \frac{5h_\tau}{h_*} - A(h_\tau)\,\partial_h \log \widetilde{\Delta}_R^{(M)}(\eta, h_\tau, \mathbf{a}_\tau)\Big)d\tau + \sqrt{2A(h_\tau)}\,d\beta_\tau, \tag{329}$$

$$da_{n,\tau} = -\partial_{a_n} \log \widetilde{\Delta}_R^{(M)}(\eta, h_\tau, \mathbf{a}_\tau)\,d\tau + \sqrt{2}\,dB_{n,\tau} + dK^+_{n,\tau} - dK^-_{n,\tau}, \qquad n = 1, \ldots, M, \tag{330}$$

where the Skorokhod terms enforce $|a_{n,\tau}| \leq b_n(\eta)$. More precisely:

(i) for every initial condition in $\overline{\Omega^{\mathrm{red}}_{\eta,M}}$, the reflected martingale problem for $\mathcal{L}^{\mathrm{red}}_{\eta,M}$ is well posed;

(ii) the resulting Markov family is Feller and has unique invariant law $\overline{\mathbb{P}}^{\mathrm{AR}}_{\eta,M}$;

(iii) for every bounded measurable F,

$$\lim_{T\to\infty} \mathbb{E}_x\big[F(h_T, \mathbf{a}_T)\big] = \int F\, d\overline{\mathbb{P}}^{\mathrm{AR}}_{\eta,M}, \tag{331}$$

uniformly for initial data x in compact subsets of $\overline{\Omega^{\mathrm{red}}_{\eta,M}}$.

Proof. Because $\widetilde{\Delta}^{(M)}_R$ is strictly positive and C^2 on a neighborhood of the compact box $[0, h_*] \times \mathcal{D}_M(\eta)$, all drift coefficients in (329)–(330) are continuous and locally Lipschitz on the interior. The amplitude coordinates evolve in a compact product of intervals with normal reflection, so the standard Skorokhod problem on a rectangular domain gives existence and uniqueness for the reflected amplitude subsystem. In the helicity coordinate, the diffusion coefficient vanishes only at the endpoints while the drift term $2 - 5h/h_*$ points inward there, so the interval $[0, h_*]$ is invariant and the one-dimensional martingale problem is well posed. Combining the helicity coordinate with the reflected amplitude block yields well posedness of the full reflected martingale problem, proving item (i).

The divergence form (321) and Theorem D.12 show that $\overline{\mathbb{P}}^{\mathrm{AR}}_{\eta,M}$ is invariant and reversible. Since item (iii) of Theorem D.12 rules out nonconstant harmonic functions, the invariant law is unique. Feller continuity follows from continuous dependence of reflected diffusions on initial data in compact domains. This proves item (ii).

Finally, on a compact state space a Feller Markov process with unique invariant probability converges to that invariant law along Cesaro averages, and in the present reversible ergodic class the absence of nonconstant invariants upgrades that convergence to ordinary long-time convergence on bounded observables. This yields (331) and proves item (iii). □

Proposition D.16 (Cylinder martingale continuation of the infinite odd-sector field). *Let $F(h, \phi) = f(h, \langle \phi, e_1\rangle, \dots, \langle \phi, e_N\rangle)$ be a bounded smooth cylinder functional, and define $\mathcal{L}^{\mathrm{cyl}}_{\eta,\infty}F$ as in (326). Then there exists a projective cylinder process (h_T, ϕ_T) under the infinite reduced law $\overline{\mathbb{P}}^{\mathrm{AR}}_{\eta}$ such that*

$$M^F_T := F(h_T, \phi_T) - F(h_0, \phi_0) - \int_0^T (\mathcal{L}^{\mathrm{cyl}}_{\eta,\infty}F)(h_s, \phi_s)\, ds \tag{332}$$

is a martingale for every such cylinder observable F. Moreover, every finite-mode projection of this cylinder process agrees in law with the reflected diffusion of Theorem D.15.

Proof. For each $M \geq N$, Theorem D.15 gives a well-posed reflected martingale problem with generator $\mathcal{L}^{\mathrm{red}}_{\eta,M}$. Applied to the cylinder test function determined by f, this yields a martingale

$$M^{F,M}_T := F(h^{(M)}_T, \phi^{(M)}_T) - F(h^{(M)}_0, \phi^{(M)}_0) - \int_0^T (\mathcal{L}^{\mathrm{cyl}}_{\eta,\infty}F)(h^{(M)}_s, \phi^{(M)}_s)\, ds, \tag{333}$$

because the first N coordinates of $\mathcal{L}^{\mathrm{red}}_{\eta,M}$ agree with $\mathcal{L}^{\mathrm{cyl}}_{\eta,\infty}$ on F. The projective consistency from Proposition D.13 and the cylinder-measure convergence from Theorem D.10 allow passage to a projective limit along the finite-dimensional distributions. The limit process therefore satisfies (332), and its finite-mode marginals coincide with the corresponding reflected diffusions. □

Interpretive consequence. The manuscript still does not claim a pathwise solution theory for the complete infinite reflected odd-sector SPDE. What it now does claim is more precise: every finite released sector is carried by a bona fide reflected diffusion with unique invariant law, and the infinite field inherits a consistent cylinder martingale structure whose finite projections are exactly those reflected diffusions. The microscopic-to-field continuation is therefore no longer only measure-theoretic or generator-theoretic, but martingale-theoretic on all finitely resolved observables.

Integrated downstream conclusion after the reflected-martingale continuation. This continuation adds a nineteenth downstream message. Positively, the reduced finite released sectors now possess well-posed reflected dynamics with unique invariant laws, and the infinite interaction-type odd-sector field inherits their projective cylinder martingale structure. Cautionarily, this still falls short of a full pathwise well-posedness, uniqueness, or strong mixing theorem for the complete infinite reflected field dynamics. The manuscript has therefore reached the natural boundary of what can be proved inside its current finite-to-projective framework without introducing a new infinite-dimensional analytic package.

Remark D.17. The present appendix now carries nineteen distinct downstream messages. The new nineteenth message is the strongest stochastic-control statement presently justified inside the manuscript's frozen scope before any path-space compactness argument is introduced: finite reflected dynamics are well posed and uniquely ergodic, and the infinite interaction-type odd-sector field is controlled as their projective cylinder martingale continuation. What remains open after this point is to pass from that cylinder-level control to bona fide infinite weak processes on path space.

D.7 Worked continuation XX: path-space tightness and subsequential weak interaction-field solutions

The reflected-martingale continuation of Section D.6 controls every finite released level and its cylinder projections, but it does not yet produce an actual infinite stochastic process on path space. The present continuation closes exactly that gap at the weak level: it proves tightness of the finite reflected interaction-type dynamics on every finite observation window and extracts subsequential infinite weak solutions whose cylinder martingales are generated by the interaction-type odd-sector field operator.

Status label. The tightness theorem and weak-solution extraction below have status **effective relation**. They do *not* claim pathwise uniqueness, strong existence, or full ergodicity for the complete infinite reflected field dynamics. Those stronger statements remain outside the present scope.

For a fixed horizon $T > 0$, let $\mathbb{P}^{\text{path}}_{\eta,M}$ denote the law on

$$C([0,T],[0,h_*]) \times C([0,T],[-b_1(\eta),b_1(\eta)]) \times \cdots \times C([0,T],[-b_M(\eta),b_M(\eta)])$$

of the finite reflected diffusion of Theorem D.15 started from its invariant law $\overline{\mathbb{P}}^{\text{AR}}_{\eta,M}$. For each $N \le M$, write Π_N for projection onto $(h, a_1, \ldots, a_N)$.

Theorem D.18 (Path-space tightness of the finite reflected interaction branches)**.** *Fix $T > 0$ and assume the standing hypotheses of Appendices B–D together with the tail bound* (283)*. Then for every fixed $N \ge 0$, the projected path laws*

$$\{(\Pi_N)_\# \mathbb{P}^{\text{path}}_{\eta,M} : M \ge N\}$$

are tight on

$$C\Big([0,T],[0,h_*] \times \prod_{n=1}^{N} [-b_n(\eta), b_n(\eta)]\Big).$$

Moreover, for each coordinate $Y \in \{h, a_1, \ldots, a_N\}$ there exists a constant $C_{N,T,\eta}$ independent of $M \ge N$ such that

$$\mathbb{E}\Big[|Y_t^{(M)} - Y_s^{(M)}|^4\Big] \le C_{N,T,\eta}|t-s|^2, \qquad 0 \le s \le t \le T. \tag{334}$$

Proof. For each fixed N, the first N amplitude coordinates of the M-level reflected diffusion solve SDEs with diffusion coefficient $\sqrt{2}$ and drift $-\partial_{a_n} \log \widetilde{\Delta}_R^{(M)}$. Because the state space is compact and $\widetilde{\Delta}_R^{(M)}$ is strictly positive and C^2 on a neighborhood of the admissible box, all first derivatives of $\log \widetilde{\Delta}_R^{(M)}$ in the first N coordinates are uniformly bounded in $M \geq N$. The same holds for the helicity drift in (329), since $A(h)$ is bounded on $[0, h_*]$. Standard Burkholder–Davis–Gundy estimates for the martingale parts, together with the bounded-variation character of the reflection terms and the uniform boundedness of the drifts, give the increment estimate (334). Kolmogorov's continuity criterion therefore yields tightness of each coordinate family on $C([0, T])$, and the finite product tightness follows. □

Proposition D.19 (Subsequential weak interaction-field continuation). *Fix $T > 0$. There exists a sequence $M_k \to \infty$ and a probability law*

$$\mathbb{P}_{\eta,\infty}^{\text{weak},T}$$

on the path space of continuous trajectories $\tau \mapsto (h_\tau, \phi_\tau)$ with values in the interaction-type cylinder state space such that, for every finite N, the projected laws of $(h, \langle \phi, e_1 \rangle, \ldots, \langle \phi, e_N \rangle)$ are limits of $(\Pi_N)_\# \mathbb{P}_{\eta,M_k}^{\text{path}}$. Under $\mathbb{P}_{\eta,\infty}^{\text{weak},T}$, every bounded smooth cylinder functional F satisfies the martingale identity

$$F(h_t, \phi_t) - F(h_0, \phi_0) - \int_0^t (\mathcal{L}_{\eta,\infty}^{\text{cyl}} F)(h_s, \phi_s)\, ds \quad \textit{is a martingale on } [0, T]. \tag{335}$$

The one-time marginal at every $t \in [0, T]$ equals $\overline{\mathbb{P}}_\eta^{\text{AR}}$ on cylinder observables.

Proof. By Theorem D.18, for each finite N the family of projected path laws is tight. A diagonal extraction over $N = 0, 1, 2, \ldots$ gives a sequence $M_k \to \infty$ such that every finite projection converges weakly on path space. Since the finite projected laws are projectively consistent by Proposition D.16, these limits define a cylinder-consistent probability law $\mathbb{P}_{\eta,\infty}^{\text{weak},T}$.

For a bounded smooth cylinder functional F depending on the first N released amplitudes, the finite processes satisfy the martingale relation (333) for all $M_k \geq N$. The convergence of finite-dimensional path laws, together with the boundedness and continuity of F and of $\mathcal{L}_{\eta,\infty}^{\text{cyl}} F$ on the relevant compact cylinder boxes, passes this martingale identity to the limit and yields (335). Finally, each finite process was started from its invariant law, so every one-time finite projection is stationary; passing to the subsequential limit and using Theorem D.10 gives the claimed one-time marginal on cylinder observables. □

Interpretive consequence. The manuscript still does not claim a full pathwise well-posedness theorem for the complete infinite reflected interaction-type field. What it now claims is strictly stronger than cylinder generator consistency alone: on every finite time window there exist subsequential weak interaction-field processes whose finite observations are limits of the bona fide reflected finite-mode diffusions and whose cylinder observables solve the infinite martingale problem. In this precise sense the manuscript has crossed from stationary infinite laws to actual weak stochastic trajectories.

Integrated downstream conclusion after the path-space-tightness continuation. This continuation adds a twentieth downstream message. Positively, the finite reflected interaction branches are now tight on path space at every finite observation level, and the infinite interaction-type odd-sector field admits subsequential weak solutions carrying the correct stationary cylinder marginals and martingale relations. Cautionarily, this still leaves open uniqueness of the limiting path law, pathwise well posedness, and full infinite-dimensional ergodicity. The manuscript has therefore advanced from cylinder-level stochastic consistency to genuine weak-process continuation, while stopping short of a complete strong theory.

Remark D.20. The present appendix now carries twenty distinct downstream messages. The new twentieth message is the first bona fide path-space result at the infinite interaction-field level: although strong uniqueness is still absent, the manuscript now produces subsequential weak stochastic trajectories whose finite observations converge from the reflected released-mode diffusions and whose cylinder martingale structure matches the infinite reduced field generator. What remains open after this point is the upgrade from weak existence to a fully well posed infinite reflected field theory.

D.8 Worked continuation XXI: stationary cylinder-Markov selections on the infinite interaction field

The weak path-space package of Section D.7 still leaves open whether one can select an infinite interaction-type continuation that is not merely a weak subsequential limit, but also carries a genuine Markov property at the cylinder level on all finite time horizons simultaneously. The present continuation supplies exactly that intermediate upgrade. It does not claim full uniqueness of the infinite reflected field dynamics; instead, it proves the existence of at least one stationary cylinder-Markov selection obtained from the reflected finite-mode diffusions by a single diagonal extraction over times and cylinder levels.

Let

$$\mathcal{X}_{\mathrm{cyl}} := [0, h_*] \times \prod_{n\geq 1} [-b_n(\eta), b_n(\eta)]$$

be the interaction-type cylinder state space endowed with the product topology, and write $\Pi_N(h, \mathbf{a}) = (h, a_1, \ldots, a_N)$ for the finite coordinate projections. For each finite released level M, let $\mathbb{P}^{\mathrm{path}}_{\eta,M}$ denote the stationary reflected path law from Section D.6. The preceding continuation produced, on every finite time window $[0, T]$, subsequential weak path laws whose cylinder observables solve the infinite martingale problem. We now make that extraction simultaneous in T and in the cylinder rank.

Theorem D.21 (Global stationary weak continuation and cylinder-Markov selection). *Fix $\eta > \eta^{\mathrm{var}}_{\mathrm{crit}}$. There exists a subsequence $M_k \to \infty$ and a probability law*

$$\mathbb{P}^{\mathrm{sel}}_{\eta,\infty}$$

on the path space $C([0,\infty), \mathcal{X}_{\mathrm{cyl}})$ such that:

1. *for every $T > 0$ and every finite N, the projected laws of $\Pi_N X^{(M_k)}|_{[0,T]}$ converge weakly to the corresponding projection of $\mathbb{P}^{\mathrm{sel}}_{\eta,\infty}$;*

2. *the one-time marginal of $\mathbb{P}^{\mathrm{sel}}_{\eta,\infty}$ equals $\overline{\mathbb{P}}^{\mathrm{AR}}_{\eta}$ on all cylinder observables;*

3. *for every bounded smooth cylinder functional F depending on the first N released amplitudes and every $t \geq 0$, there exists a measurable bounded function $P^{(N)}_t F$ on $[0, h_*] \times \prod_{n=1}^{N} [-b_n(\eta), b_n(\eta)]$ such that under $\mathbb{P}^{\mathrm{sel}}_{\eta,\infty}$,*

$$\mathbb{E}[F(X_{s+t}) \mid \sigma(\Pi_N X_r : 0 \leq r \leq s)] = (P^{(N)}_t F)(\Pi_N X_s) \quad \textit{for a.e. } s \geq 0; \tag{336}$$

4. *for every such cylinder functional F and every $s, t \geq 0$,*

$$P^{(N)}_{s+t} F = P^{(N)}_s \big(P^{(N)}_t F\big) \qquad \textit{a.e. with respect to } (\Pi_N)_\# \overline{\mathbb{P}}^{\mathrm{AR}}_{\eta}, \tag{337}$$

   and $\overline{\mathbb{P}}^{\mathrm{AR}}_{\eta}$ is invariant under this cylinder semigroup.

Proof. Apply Theorem D.18 simultaneously on the integer windows $[0, m]$, $m = 1, 2, \ldots$, and for the finite projections $N = 0, 1, 2, \ldots$. A Cantor diagonal extraction yields a single subsequence $M_k \to \infty$ for which every finite window and every finite projection converge weakly. By restriction consistency of the finite reflected path laws, these limiting finite-window laws patch together into a probability law $\mathbb{P}^{\text{sel}}_{\eta,\infty}$ on $C([0,\infty), \mathcal{X}_{\text{cyl}})$. Statement (1) is built into the construction, while statement (2) follows because each finite process is stationary with one-time marginal $\overline{\mathbb{P}}^{\text{AR}}_{\eta}$ on cylinder observables.

For fixed N and $t \geq 0$, the finite N-projected reflected diffusions are time-homogeneous Feller-Markov processes, so there are bounded measurable operators $P^{(N)}_{t,M}$ satisfying

$$\mathbb{E}\Big[F(\Pi_N X^{(M)}_{s+t}) \mid \sigma(\Pi_N X^{(M)}_r : 0 \leq r \leq s)\Big] = (P^{(N)}_{t,M} F)(\Pi_N X^{(M)}_s).$$

Because the state space for Π_N is compact and F is bounded, the family $\{P^{(N)}_{t,M} F\}_{M \geq N}$ is uniformly bounded in L^∞. Extracting further along a countable dense set of times and cylinder test functions and using stationarity, one obtains measurable limits $P^{(N)}_t F$ on the N-cylinder state space. Passing to the weak path-law limit gives (336). The semigroup law (337) is inherited from the finite identities $P^{(N)}_{s+t,M} = P^{(N)}_{s,M} P^{(N)}_{t,M}$ by the same limiting argument, and invariance of $\overline{\mathbb{P}}^{\text{AR}}_{\eta}$ follows from stationarity of the selected path law. □

Remark D.22 (What the selection theorem does and does not prove)**.** Theorem D.21 upgrades the infinite interaction-type field from a merely subsequential weak process on finite windows to a globally defined stationary weak process that is Markov when tested through every finite cylinder. This is a genuine strengthening of the stochastic package. At the same time it still stops short of full uniqueness of the infinite path law, pathwise strong well posedness, or a theorem asserting that two different subsequences must produce the same cylinder-Markov family.

Integrated downstream conclusion after the cylinder-Markov-selection continuation. This continuation adds a twenty-first downstream message. Positively, the infinite interaction-type field now admits at least one globally defined stationary weak realization whose finite observations form invariant cylinder semigroups inherited from the reflected finite-mode diffusions. Cautionarily, the manuscript still does not prove uniqueness of that selection, nor does it prove a strong infinite-dimensional Markov process in the usual pathwise sense. The manuscript has therefore advanced from finite-window weak continuation to a global stationary cylinder-Markov package, while leaving the genuinely unique strong theory open.

Remark D.23. The present appendix now carries twenty-one distinct downstream messages. The new twenty-first message is the strongest stochastic-control statement presently justified within the manuscript's frozen scope: not only do subsequential infinite weak trajectories exist, but at least one such trajectory can be selected globally in time so that every finite cylinder sees a stationary Markov semigroup with the correct invariant law. What remains open after this point is whether this cylinder-Markov selection is unique, whether it upgrades to a bona fide strong infinite reflected field dynamics, and whether the resulting dynamics enjoy genuinely infinite-dimensional ergodicity.

D.9 Worked continuation XXII: canonical cylinder resolvents and mean-ergodic stabilization

The global cylinder-Markov package of Section D.8 still leaves open whether different stationary selections might generate different cylinder evolutions on the same finite observable algebra. A full answer would require uniqueness of the infinite reflected interaction-field dynamics, which remains outside the frozen scope. The present continuation isolates the strongest canonically selection-independent object that can already be justified: the finite-cylinder resolvent family and the associated mean-ergodic stabilization.

For each fixed $N \geq 0$, let

$$\mu_{\eta,N} := (\Pi_N)_\# \overline{\mathbb{P}}_\eta^{\mathrm{AR}}$$

be the stationary N-cylinder marginal on

$$\mathcal{X}_N := [0, h_*] \times \prod_{n=1}^{N} [-b_n(\eta), b_n(\eta)].$$

Write $\mathcal{C}_N$ for the algebra of bounded smooth cylinder test functions depending only on $(h, a_1, \ldots, a_N)$.

Theorem D.24 (Canonical cylinder resolvent and mean-ergodic stabilization). *Fix $\eta > \eta_{\mathrm{crit}}^{\mathrm{var}}$ and $N \geq 0$. There exists a family of bounded operators*

$$R_\lambda^{(N)} : L^2(\mu_{\eta,N}) \to L^2(\mu_{\eta,N}), \qquad \lambda > 0,$$

with the following properties:

1. *for every stationary cylinder-Markov selection from Theorem D.21, every $F \in \mathcal{C}_N$, and every $\lambda > 0$,*

$$R_\lambda^{(N)} F = \int_0^\infty e^{-\lambda t} P_t^{(N)} F \, dt \qquad \text{in } L^2(\mu_{\eta,N}); \tag{338}$$

2. *the operator $R_\lambda^{(N)}$ is independent of the chosen stationary cylinder-Markov selection on $\mathcal{C}_N$;*
3. *for every $F, G \in \mathcal{C}_N$,*

$$\langle G, (\lambda - \mathcal{L}_{\eta,\infty}^{\mathrm{cyl},N}) R_\lambda^{(N)} F \rangle_{L^2(\mu_{\eta,N})} = \langle G, F \rangle_{L^2(\mu_{\eta,N})}, \tag{339}$$

so $R_\lambda^{(N)}$ is the canonical weak resolvent of the N-cylinder limit generator;

4. *the* Cesaro *averages of the selected cylinder semigroup satisfy*

$$\frac{1}{T} \int_0^T P_t^{(N)} F \, dt \longrightarrow \int_{\mathcal{X}_N} F \, d\mu_{\eta,N} \qquad \text{in } L^2(\mu_{\eta,N}) \tag{340}$$

for every $F \in \overline{\mathcal{C}_N}^{L^2(\mu_{\eta,N})}$.

Proof. Fix N. For every finite released level $M \geq N$, the N-projected reflected diffusion is a reversible Feller-Markov process on the compact state space $\mathcal{X}_N$ with invariant measure $\mu_{\eta,N}^{(M)} = (\Pi_N)_\# \pi_M$. Its semigroup $P_{t,M}^{(N)}$ is therefore a self-adjoint contraction semigroup on $L^2(\mu_{\eta,N}^{(M)})$, and its resolvent

$$R_{\lambda,M}^{(N)} F := \int_0^\infty e^{-\lambda t} P_{t,M}^{(N)} F \, dt$$

solves the finite weak identity

$$\langle G, (\lambda - \mathcal{L}_{\eta,M}^{\mathrm{red},N}) R_{\lambda,M}^{(N)} F \rangle_{L^2(\mu_{\eta,N}^{(M)})} = \langle G, F \rangle_{L^2(\mu_{\eta,N}^{(M)})}$$

for $F, G \in \mathcal{C}_N$. Because $\mathcal{X}_N$ is compact, the coefficients of the projected generators converge uniformly on $\mathcal{X}_N$, and the invariant measures converge weakly to $\mu_{\eta,N}$. Passing to the diagonal subsequence selected in Theorem D.21 yields the limit formula (338). Since the finite resolvents depend only on the finite projected dynamics and not on any later selection, the limit on the common core $\mathcal{C}_N$ is canonical, which proves selection independence.

The weak identity (339) follows by passing to the limit in the finite resolvent equation using the uniform convergence of coefficients and the weak convergence of invariant measures. For the mean-ergodic statement, note that each selected $P_t^{(N)}$ is a contraction semigroup on $L^2(\mu_{\eta,N})$ and is symmetric on the core inherited from the finite reversible semigroups. The finite projected processes are ergodic with unique invariant measure, so the only L^2 functions annihilated by the weak generator on the core are constants. Von Neumann's mean-ergodic theorem therefore gives convergence of Cesaro averages to the orthogonal projection onto constants, which is exactly the mean under $\mu_{\eta,N}$. □

Remark D.25 (What becomes canonical at this stage)**.** Theorem D.24 does not yet prove that all infinite interaction-field selections coincide as path laws. What it does prove is more precise than mere existence of one stationary selection: once one restricts attention to any fixed finite observable cylinder, the weak resolvent and the long-time averaged response are already canonical. In this precise sense the manuscript now reaches selection-independent reduced dynamics at every finite observation depth, even though full infinite-dimensional uniqueness remains open.

Integrated downstream conclusion after the canonical-resolvent continuation. This continuation adds a twenty-second downstream message. Positively, the global stationary cylinder-Markov package now carries a canonical finite-cylinder resolvent and a mean-ergodic stabilization principle that are independent of the chosen stationary selection. Cautionarily, this still does not settle uniqueness of full path laws or strong well posedness of the complete infinite reflected interaction field. The manuscript has therefore advanced from existence of a stationary cylinder-Markov selection to selection-independent reduced evolution on every finite observable cylinder.

Remark D.26. The present appendix now carries twenty-two distinct downstream messages. The new twenty-second message is the strongest uniqueness surrogate presently justified within the manuscript's frozen scope: even before full infinite-dimensional uniqueness is available, every fixed finite observable cylinder already sees a canonical weak resolvent and a canonical mean-ergodic limit. What remains open after this point is whether these cylinderwise canonical objects lift to a unique global infinite reflected interaction-field dynamics.

D.10 Worked continuation XXIII: uniqueness of the admissible reduced measure

The canonical-resolvent continuation of the previous subsection made the finite observable dynamics selection-independent, but the first item in the open-problem list was still purely static: whether the Beta$(2,3)$ barrier prior is merely an admissible example or the unique reduced measure compatible with the manuscript's own endpoint geometry and reduced reversible balance law. The enlarged arc-helicon picture sharpened in the present revision suggests exactly the missing intrinsic input. The local doubled branch occupies a closed interior field while the full light and magnetic arcs define an open exterior relation; at the light pole the two fields meet in a unique geometric tangency, whereas at the magnetic side the exterior continuation separates into two transverse open directions. At the reduced helicity level this singles out one lower-end tangency stratum and two upper-end separation strata. The next admissible continuation is therefore to turn those strata, together with the minimal reversible reduced balance law already encoded in the finite reflected generators, into a uniqueness theorem.

Status label. The intrinsic reduced class introduced below has status **definition**. Its uniqueness theorem has status **theorem**. The translation from light-pole tangency / magnetic-side separation to endpoint orders is a **candidate interpretation** of the reduced geometry. No part of this subsection enlarges the theorem-level scope of Sections 2–7.

Definition D.27 (Intrinsic admissible reduced class). Let $x = h/h_* \in (0,1)$. An *intrinsic admissible reduced measure* is a probability measure $d\mu(x) = w(x)\,dx$ on $(0,1)$ whose density satisfies all of the following:

(I1) $w \in C^1(0,1)$ and $w(x) > 0$ for every $x \in (0,1)$;

(I2) the endpoint strata are exactly one lower tangency stratum and two upper separation strata, in the sense that

$$x^{-1}w(x) \to c_0 \in (0,\infty), \qquad (1-x)^{-2}w(x) \to c_1 \in (0,\infty) \qquad (x \downarrow 0,\ x \uparrow 1); \tag{341}$$

(I3) μ is stationary for the reversible zero-current reduced operator whose scaled diffusion coefficient is

$$\widetilde{A}(x) = x(1-x) \tag{342}$$

and whose reduced drift is the minimal affine balance law

$$\widetilde{B}(x) = 2 - 5x. \tag{343}$$

Theorem D.28 (Uniqueness of the intrinsic admissible reduced measure). *Definition D.27 determines a unique probability density on* $(0,1)$, *namely*

$$w_*(x) = 12x(1-x)^2. \tag{344}$$

Equivalently, the unique intrinsic admissible reduced measure on $(0, h_*)$ *is*

$$\rho_*(h) = \frac{12}{h_*}\frac{h}{h_*}\left(1 - \frac{h}{h_*}\right)^2. \tag{345}$$

In particular, the reduced-measure density introduced in (237), *the simplex-pushforward density of Section C.6, and the grouped Gamma-microcell stationary density of Section D.1 all coincide with the unique member of the manuscript's intrinsic reduced class.*

Proof. Let w satisfy Definition D.27. Reversibility with zero current for the one-dimensional scaled reduced operator means precisely that

$$\widetilde{B}(x) = \widetilde{A}'(x) + \widetilde{A}(x)\,\partial_x \log w(x) \qquad (0 < x < 1). \tag{346}$$

Substituting (342) and (343) gives

$$2 - 5x = (1-2x) + x(1-x)\partial_x \log w(x),$$

so

$$\partial_x \log w(x) = \frac{1-3x}{x(1-x)} = \frac{1}{x} - \frac{2}{1-x}. \tag{347}$$

Integrating on $(0,1)$ yields

$$w(x) = C\,x(1-x)^2 \tag{348}$$

for some constant $C > 0$. This form already matches the endpoint orders (341); indeed the lower endpoint vanishes linearly and the upper endpoint vanishes quadratically. Imposing normalization gives

$$1 = \int_0^1 w(x)\,dx = C\int_0^1 x(1-x)^2\,dx = \frac{C}{12},$$

so $C = 12$, which proves (344). The change of variable $h = h_* x$ then gives (345). Since Sections C.4, C.6, and D.1 already constructed the same density through reduced, simplex, and Gamma-microcell routes, those realizations are all identified with the unique measure of the intrinsic class. □

Remark D.29 (Why the uniqueness theorem is genuinely internal)**.** Theorem D.28 is not an external identification imposed from outside the manuscript. Its three inputs are all already present internally: the endpoint orders come from the interaction-type doubled geometry and its light-end / magnetic-end stratification, the reversible zero-current structure comes from the finite reflected reduced generators, and the affine balance law (343) is exactly the drift already encoded in those generators after scaling. The theorem therefore closes the first former open problem *within the manuscript's own intrinsic reduced class.* What it does not claim is uniqueness across arbitrary unrelated priors or across external field-theoretic dictionaries.

Integrated downstream conclusion after the reduced-measure uniqueness theorem. This continuation adds a twenty-third downstream message. Positively, the reduced barrier prior is no longer merely admissible: within the manuscript's intrinsic reduced class it is uniquely forced, and the reduced, simplex, and Gamma-microcell constructions are now seen to be three realizations of the same unique object. Cautionarily, the theorem settles uniqueness only at the reduced-measure level; it does not yet solve ansatz independence, projective-to-continuum uniqueness, or strong well posedness of the full infinite reflected interaction field.

Remark D.30. The present appendix now carries twenty-four distinct downstream messages. The new twenty-fourth message turns the former attainment problem into an intrinsic theorem: the preferred branch is attained in the infinite released class itself, and its finite-cylinder content is schedule-independent inside that class. The remaining open problems now concern extension beyond the intrinsic class, sharper projective-to-continuum uniqueness, and stronger infinite-dimensional dynamical control.

D.11 Worked continuation XXIV: intrinsic attainment and cylinder-level ansatz independence

The previous continuations have already supplied all of the ingredients needed for a first genuine answer to the second former open problem, provided one states the admissible class intrinsically and does not overclaim beyond it. The aim of the present continuation is therefore not to assert ansatz independence across arbitrary external parametrizations, but to prove it within the manuscript's own intrinsic released family: common tail envelope, interaction-type doubled geometry, the unique reduced prior of Theorem D.28, and the barrier-regularized free energies $\mathfrak{G}_R^{[m]}$ and $\mathfrak{G}_R^{(\infty)}$.

Definition D.31 (Intrinsic admissible released ladder)**.** Fix $\eta > 0$ and a tail envelope $\{b_n(\eta)\}_{n\geq 1}$ satisfying (201). An *intrinsic admissible released ladder* is a sequence

$$x_m = (h_m, a_1^{[m]}, \ldots, a_m^{[m]}) \in \mathcal{U}_\eta^{[m]} \cap \mathcal{K}_\eta^{[m]} \qquad (m \geq 0)$$

of interior points such that:

(L1) each x_m is a global minimizer of the extended free energy $\widehat{\mathfrak{G}}_R^{[m]}(\eta, \cdot)$ on $\overline{\mathcal{U}}_\eta^{[m]} \cap \mathcal{K}_\eta^{[m]}$;

(L2) there exist $\varepsilon_0, \lambda_0 > 0$, independent of m, such that

$$\varepsilon_0 \leq h_m \leq h_* - \varepsilon_0, \qquad \nabla^2 \mathfrak{G}_R^{[m]}(x_m) \geq \lambda_0 I; \tag{349}$$

(L3) after padding by zeros for $n > m$, the ladder lies in the common compact product box of Proposition C.1.

Proposition D.32 (Attainment in the intrinsic infinite released class). *Fix $\eta > 0$ and assume:*

(A1) a tail envelope satisfying (201) *and the stronger C^2-control hypotheses of Proposition D.5;*

(A2) positivity and continuity hypotheses sufficient for Proposition D.6;

(A3) the existence of at least one intrinsic admissible released ladder in the sense of Definition D.31.

Then the infinite released free energy $\mathfrak{G}_R^{(\infty)}(\eta, \cdot)$ attains a minimum at some point

$$x_\infty = (h_\infty, a_1^{(\infty)}, a_2^{(\infty)}, \dots) \in \mathcal{U}_\eta \cap \mathcal{K}_\eta^{(\infty)}.$$

Moreover, every intrinsic admissible released ladder possesses a subsequence converging coordinatewise and in C^2-field topology to such an attainer, and every such attainer is a strict local minimum of $\mathfrak{G}_R^{(\infty)}$.

Proof. Let x_m be an intrinsic admissible released ladder. By Proposition C.1, after extraction there is a limit point $x_\infty \in [0, h_*] \times \prod_{n\geq 1}[-b_n(\eta), b_n(\eta)]$ with coordinatewise convergence and C^2-convergence of the associated odd half-mode fields. The uniform interior bound in (349) excludes the boundary $h = 0, h_*$, hence $x_\infty \in \mathcal{U}_\eta$.

By Proposition D.5, $\mathfrak{G}_R^{[m]} \to \mathfrak{G}_R^{(\infty)}$ uniformly on a compact neighborhood $\mathcal{N}_\eta$ containing the ladder and its limit. Since each x_m minimizes $\widehat{\mathfrak{G}}_R^{[m]}$ on the truncated box, for every fixed comparison point $y \in \mathcal{U}_\eta \cap \mathcal{K}_\eta^{(\infty)}$ and every sufficiently large m, its projection $y^{[m]}$ lies in the truncated box and satisfies

$$\mathfrak{G}_R^{[m]}(x_m) \leq \mathfrak{G}_R^{[m]}(y^{[m]}).$$

Passing to the limit along the convergent subsequence gives

$$\mathfrak{G}_R^{(\infty)}(x_\infty) \leq \mathfrak{G}_R^{(\infty)}(y).$$

Hence x_∞ is a global minimizer of $\mathfrak{G}_R^{(\infty)}$ on the intrinsic infinite released class. The strict local minimality follows from Corollary D.7, using the uniform Hessian lower bound from (349). $\square$

Theorem D.33 (Cylinder-level ansatz independence inside the intrinsic released class). *Fix $\eta > 0$ and assume the hypotheses of Proposition D.32. Let $\mathcal{B}_\eta \subset \mathcal{U}_\eta \cap \mathcal{K}_\eta^{(\infty)}$ be a compact basin on which:*

(B1) $\mathfrak{G}_R^{(\infty)}$ has a unique global minimizer $x_\eta^\star$ in $\mathcal{B}_\eta$;

(B2) every intrinsic admissible released ladder is eventually contained in $\mathcal{B}_\eta$.

Then the preferred branch is ansatz independent at every finite cylinder depth within the intrinsic released class*: for every $N \geq 1$, every intrinsic admissible released ladder x_m satisfies*

$$\Pi_N x_m \longrightarrow \Pi_N x_\eta^\star, \tag{350}$$

as $m \to \infty$, where Π_N denotes projection onto $(h, a_1, \dots, a_N)$. Equivalently, every bounded continuous cylinder observable $F = F(h, a_1, \dots, a_N)$ obeys

$$F(x_m) \longrightarrow F(x_\eta^\star). \tag{351}$$

In particular, the infinite preferred branch is independent of the particular released-mode schedule used to approach it, provided that schedule remains intrinsic in the sense of Definition D.31.

Proof. Take any intrinsic admissible released ladder x_m. By Proposition D.32, every subsequence has a further subsequence converging to an infinite attainer of $\mathfrak{G}_R^{(\infty)}$. Assumption (B2) places all sufficiently large terms inside $\mathcal{B}_\eta$, while (B1) says that the only attainer in $\mathcal{B}_\eta$ is $x_\eta^\star$. Therefore every convergent subsubsequence has the same limit $x_\eta^\star$. In a metrizable compact product this forces the full sequence to converge to $x_\eta^\star$, and therefore every finite projection $\Pi_N x_m$ converges to $\Pi_N x_\eta^\star$. The cylinder-observable statement follows by continuity. □

Remark D.34 (What has and has not been solved)**.** Proposition D.32 closes the attainment part of the second former open problem *inside the manuscript's intrinsic released class.* Theorem D.33 goes further: once the infinite free energy has a unique minimizer in the compact intrinsic basin actually selected by the released ladders, all finite-observable predictions become independent of the particular released-mode schedule. What is *not* claimed is ansatz independence across arbitrary external parametrizations or across unrelated non-doubled families. The remaining difficulty is therefore no longer mere attainment, but extension from the intrinsic doubled released class to broader external ansatz universes.

Integrated downstream conclusion after intrinsic attainment. This continuation adds a twenty-fourth downstream message. Positively, the preferred branch is no longer merely observed along successive released-mode scans: within the intrinsic released class it is now attained by the infinite free energy itself, and, under a unique-basin hypothesis, its finite-observable content is independent of the particular released-mode schedule used to approach it. Cautionarily, this still falls short of absolute ansatz independence across all conceivable parametrizations. What has been closed is the intrinsic attainment problem and the cylinder-level release-schedule ambiguity; what remains open is the extension of that uniqueness beyond the manuscript's own doubled released universe.

D.12 Worked continuation XXV: intrinsic continuum support and projective-to-continuum uniqueness

The intrinsic attainment theorem of Section D.11 fixes a preferred infinite released branch inside the manuscript's doubled universe, but the third former open problem still asks whether the projective infinite released-mode closure really determines a unique continuum law rather than only a cylinder shadow. The present continuation resolves that question at the intrinsic continuum-support level: one identifies a canonical compact support class in C^3, shows that the projective law lives on that class with uniform tail regularity, and proves that its cylinder marginals determine a unique Borel probability measure there.

Status label. The continuum support class below has status **definition**. The compactness and tail estimates have status **effective relation**. The uniqueness theorem for the continuum law on that support class has status **theorem-level downstream result**. Extension beyond this intrinsic support class to broader renormalized continuum universes remains a **conjectural continuation** and is not claimed here.

Definition D.35 (Intrinsic continuum support class)**.** Fix $\eta > 0$ and an admissible tail envelope satisfying (283). Let

$$\ell^1_{(3)} := \left\{ \mathbf{a} = (a_n)_{n\geq 1} : \ \|\mathbf{a}\|_{(3)} := \sum_{n=1}^{\infty} (2n+1)^3 |a_n| < \infty \right\}.$$

For $(h, \mathbf{a}) \in \mathcal{U}_\eta$, define

$$\Gamma_\eta(h, \mathbf{a}) := (h, \phi_{\mathbf{a}}), \qquad \phi_{\mathbf{a}}(u) = \sum_{n=1}^{\infty} a_n e_n(u).$$

The *intrinsic continuum support class* is the compact image

$$\mathfrak{S}_\eta^{(3)} := \Gamma_\eta(\mathcal{U}_\eta) \subset (0, h_*) \times C^3_{\mathrm{odd}}([0, 4\pi]).$$

For each finite release level m, write

$$\widehat{\pi}_m := (\Gamma_\eta)_\# \pi_m, \qquad \widehat{\pi}_\infty := (\Gamma_\eta)_\# \pi_\infty.$$

Proposition D.36 (Canonical intrinsic continuum support and tail regularity). *Assume the hypotheses of Proposition D.6. Then:*

(C1) $\mathcal{K}_\eta^{(\infty)} \subset \ell^1_{(3)}$*, and this set is compact in the* $\ell^1_{(3)}$ *norm;*

(C2) $\Gamma_\eta : \mathcal{U}_\eta \to \mathfrak{S}_\eta^{(3)}$ *is a homeomorphism onto a compact subset of* $(0, h_*) \times C^3_{\mathrm{odd}}([0, 4\pi])$*;*

(C3) there exist constants $C_{j,\eta} < \infty$ *for* $j = 0, 1, 2, 3$ *such that every* $(h, \phi) \in \mathfrak{S}_\eta^{(3)}$ *satisfies*

$$\|\partial_u^j \phi\|_{L^\infty([0,4\pi])} \leq C_{j,\eta};$$

in particular, $\widehat{\pi}_\infty$ *is supported on a compact* C^3*-regular odd-antiperiodic continuum class.*

If, in addition, $\sum_{n\geq 1}(2n+1)^{3+r} b_n(\eta) < \infty$ *for some integer* $r \geq 1$*, then the same support class is compact in* C^{3+r}*.*

Proof. For $\mathbf{a} \in \mathcal{K}_\eta^{(\infty)}$,

$$\|\mathbf{a}\|_{(3)} \leq \sum_{n=1}^{\infty} (2n+1)^3 b_n(\eta) < \infty,$$

so $\mathcal{K}_\eta^{(\infty)} \subset \ell^1_{(3)}$. If $\mathbf{a}^{(m)} \to \mathbf{a}$ coordinatewise inside the product box, dominated convergence against the summable majorant $(2n+1)^3 b_n(\eta)$ gives $\|\mathbf{a}^{(m)} - \mathbf{a}\|_{(3)} \to 0$. Tychonoff compactness of the product box therefore upgrades to compactness in $\ell^1_{(3)}$.

The map $\mathbf{a} \mapsto \phi_{\mathbf{a}}$ is continuous from $\ell^1_{(3)}$ to C^3_{odd} because the j-th derivative of e_n is bounded by $c_j(2n+1)^j$, so for $0 \leq j \leq 3$,

$$\|\partial_u^j(\phi_{\mathbf{a}} - \phi_{\mathbf{b}})\|_\infty \leq c_j \sum_{n\geq 1} (2n+1)^j |a_n - b_n| \leq c_j \|\mathbf{a} - \mathbf{b}\|_{(3)}.$$

Since $\mathcal{U}_\eta$ is compact and Γ_η is injective, Γ_η is a homeomorphism onto its compact image $\mathfrak{S}_\eta^{(3)}$. The derivative bounds follow by the same estimate with $\mathbf{b} = 0$. The C^{3+r} statement is identical under the stronger summability hypothesis. □

Theorem D.37 (Intrinsic projective-to-continuum uniqueness). *Assume the hypotheses of Proposition D.6. Then:*

(P1) the pushforward measures $\widehat{\pi}_m$ *converge weakly to* $\widehat{\pi}_\infty$ *as Borel probability measures on the compact metric space* $\mathfrak{S}_\eta^{(3)}$*;*

(P2) $\widehat{\pi}_\infty$ *is the unique Borel probability measure on* $\mathfrak{S}_\eta^{(3)}$ *whose integrals against every bounded continuous cylinder functional*

$$F(h, \phi) = f\big(h, \langle \phi, e_1 \rangle, \ldots, \langle \phi, e_N \rangle\big)$$

agree with the projective cylinder limits generated by π_m*.*

Equivalently, the manuscript's projective infinite released-mode closure determines a unique continuum stationary law on the intrinsic C^3 *odd-sector support class.*

Proof. Weak convergence of $\widehat{\pi}_m$ to $\widehat{\pi}_\infty$ follows from Proposition D.6 and the continuity of Γ_η: if $F \in C(\mathfrak{S}_\eta^{(3)})$, then $F \circ \Gamma_\eta$ is bounded and continuous on $\mathcal{U}_\eta$, so

$$\int F\, d\widehat{\pi}_m = \int (F \circ \Gamma_\eta)\, d\pi_m \longrightarrow \int (F \circ \Gamma_\eta)\, d\pi_\infty = \int F\, d\widehat{\pi}_\infty.$$

For uniqueness, let μ be another Borel probability measure on $\mathfrak{S}_\eta^{(3)}$ agreeing with $\widehat{\pi}_\infty$ on all bounded continuous cylinder functionals. Those cylinder functionals form a unital subalgebra of $C(\mathfrak{S}_\eta^{(3)})$. Because Γ_η is injective and the coordinates $h, a_1, a_2, \ldots$ separate points on $\mathcal{U}_\eta$, the cylinder functionals separate points of $\mathfrak{S}_\eta^{(3)}$. By the Stone-Weierstrass theorem, this algebra is uniformly dense in $C(\mathfrak{S}_\eta^{(3)})$. Hence μ and $\widehat{\pi}_\infty$ agree on all continuous functions, so $\mu = \widehat{\pi}_\infty$. □

Corollary D.38 (Selection-independent continuum law on finite observables). *Under the hypotheses of Theorem D.37, any two global stationary weak path selections from Section D.8 having the same stationary cylinder marginals induce the same time-t continuum law on $\mathfrak{S}_\eta^{(3)}$ for every fixed $t \geq 0$. In particular, every bounded continuous continuum observable depending on finitely many released modes has a selection-independent stationary expectation.*

Proof. At each fixed time t, the stationary law of any such selection is a Borel probability measure on $\mathfrak{S}_\eta^{(3)}$ with the same cylinder marginals as $\widehat{\pi}_\infty$. Theorem D.37 then forces equality of these laws on $\mathfrak{S}_\eta^{(3)}$. The statement for finite-observable expectations is immediate. □

Integrated downstream conclusion after intrinsic continuum closure. This continuation adds a twenty-fifth downstream message. Positively, the projective infinite released-mode law is no longer merely a coherent cylinder object: within the manuscript's intrinsic tail-controlled class it now lives on a canonical compact C^3 odd-sector support, its finite cylinder data determine a unique continuum Borel measure there, and the projective-to-continuum passage becomes selection-independent at the level of stationary finite observables. Cautionarily, the theorem still does not cover broader non-product tail priors, renormalized continuum limits, or external non-doubled ansatz classes. What has been closed is the manuscript-intrinsic projective-to-continuum problem; what remains is how far that control extends beyond the present support universe.

Remark D.39. The present appendix now carries twenty-five distinct downstream messages. The new twenty-fifth message closes the former projective-to-continuum problem inside the manuscript's intrinsic support class: support, tail regularity, and continuum-law uniqueness are now fixed by the same tail envelope and odd-sector geometry already used in the projective closure. What remains open is no longer whether the intrinsic projective limit determines a continuum law, but whether that law is still canonical beyond the present tail-controlled doubled universe and whether the full infinite reflected dynamics is strongly well posed.

D.13 Worked continuation XXVI: canonical cylinder semigroups and uniqueness in law of the stationary interaction field

The previous continuations have already fixed every static ingredient of the manuscript-intrinsic infinite interaction field: the reduced measure is unique, the preferred branch is attained inside the intrinsic released class, and the projective law determines a unique continuum stationary measure on the compact odd-sector support $\mathfrak{S}_\eta^{(3)}$. What still remains from the third open problem is therefore genuinely dynamical. Could two global stationary cylinder-Markov selections still differ as path laws after sharing the same stationary marginals and the same canonical finite-cylinder resolvents? The present continuation shows that they cannot. Within the manuscript's intrinsic stationary cylinder-Markov class, the interaction-type reflected field now carries a unique global stationary weak path law.

Status label. The semigroup and path-law statements below have status **theorem**. They close uniqueness in law inside the intrinsic stationary cylinder-Markov class. They do *not* claim pathwise strong existence, pathwise uniqueness, or full ergodicity for genuinely non-cylinder observables.

Proposition D.40 (Canonical finite-cylinder semigroups). *Fix $\eta > \eta_{\text{crit}}^{\text{var}}$ and a cylinder depth $N \geq 0$. The canonical weak resolvent family*

$$R_\lambda^{(N)} : L^2(\mu_{\eta,N}) \to L^2(\mu_{\eta,N}), \qquad \lambda > 0,$$

of Theorem D.24 is self-adjoint, positivity preserving, and satisfies the resolvent identity

$$R_\lambda^{(N)} - R_\mu^{(N)} = (\mu - \lambda) R_\lambda^{(N)} R_\mu^{(N)} \qquad (\lambda, \mu > 0) \tag{352}$$

on $L^2(\mu_{\eta,N})$. Consequently there exists a unique nonpositive self-adjoint operator $A_{\eta,N}$ and a unique strongly continuous self-adjoint contraction semigroup $(S_t^{(N)})_{t \geq 0}$ on $L^2(\mu_{\eta,N})$ such that

$$R_\lambda^{(N)} = (\lambda - A_{\eta,N})^{-1}, \qquad S_t^{(N)} = e^{tA_{\eta,N}}. \tag{353}$$

For every global stationary cylinder-Markov selection from Theorem D.21, every bounded smooth cylinder observable $F \in \mathcal{C}_N$, and every $t \geq 0$, the selected cylinder evolution agrees with the canonical semigroup in the sense that

$$P_t^{(N)} F = S_t^{(N)} F \qquad \text{in } L^2(\mu_{\eta,N}). \tag{354}$$

Proof. For every finite released level $M \geq N$, the projected reflected diffusion on $\mathcal{X}_N$ has a reversible self-adjoint contraction semigroup on $L^2(\mu_{\eta,N}^{(M)})$, and its finite resolvent satisfies the usual resolvent identity. Passing to the intrinsic limit exactly as in the proof of Theorem D.24, and using the weak convergence of the invariant measures together with the uniform convergence of the projected coefficients on the compact set $\mathcal{X}_N$, yields (352) on the common core $\mathcal{C}_N$. Density of $\mathcal{C}_N$ in $L^2(\mu_{\eta,N})$ extends the identity and self-adjointness to the full space. The Hille–Yosida theorem together with the spectral theorem for self-adjoint resolvents then yields the unique operator-semigroup pair $(A_{\eta,N}, S_t^{(N)})$ in (353).

Now fix any stationary cylinder-Markov selection and write $P_t^{(N)}$ for its cylinder semigroup from Theorem D.21. For bounded cylinder observables $F, G \in \mathcal{C}_N$, define

$$u_{F,G}(t) := \langle G, P_t^{(N)} F \rangle_{L^2(\mu_{\eta,N})}, \qquad \sigma_{F,G}(t) := \langle G, S_t^{(N)} F \rangle_{L^2(\mu_{\eta,N})}.$$

Because the selected paths are continuous and F, G are bounded and continuous on the compact cylinder box, both functions of t are continuous. Their Laplace transforms coincide by (338) and (353):

$$\int_0^\infty e^{-\lambda t} u_{F,G}(t)\, dt = \langle G, R_\lambda^{(N)} F \rangle = \int_0^\infty e^{-\lambda t} \sigma_{F,G}(t)\, dt.$$

Uniqueness of Laplace transforms for continuous exponentially bounded functions gives $u_{F,G}(t) = \sigma_{F,G}(t)$ for all $t \geq 0$. Since this holds for every $G \in \mathcal{C}_N$ and the latter is dense in $L^2(\mu_{\eta,N})$, equation (354) follows. □

Theorem D.41 (Uniqueness in law of the intrinsic stationary interaction field). *Fix $\eta > \eta_{\text{crit}}^{\text{var}}$. Let*

$$\mathbb{P}_{\eta,\infty}^{(1)}, \ \mathbb{P}_{\eta,\infty}^{(2)}$$

be two global stationary cylinder-Markov selections on $C([0,\infty), \mathcal{X}_{\text{cyl}})$ satisfying Theorem D.21. Then

$$\mathbb{P}_{\eta,\infty}^{(1)} = \mathbb{P}_{\eta,\infty}^{(2)} \qquad \text{as Borel probability laws on } C([0,\infty), \mathcal{X}_{\text{cyl}}). \tag{355}$$

Equivalently, there exists a unique global stationary weak path law in the manuscript's intrinsic stationary cylinder-Markov class; we denote it by

$$\mathbb{P}^{\mathrm{stat}}_{\eta,\infty}.$$

Moreover, for every finite cylinder depth N, every times $0 \le t_0 < t_1 < \cdots < t_m$, and every bounded cylinder observables $F_0, \ldots, F_m \in \mathcal{C}_N$, one has the canonical finite-dimensional formula

$$\begin{aligned} &\mathbb{E}_{\mathbb{P}^{\mathrm{stat}}_{\eta,\infty}}\Big[\prod_{j=0}^{m} F_j(\Pi_N X_{t_j})\Big] \\ &\quad = \int_{\mathcal{X}_N} F_0(x) S^{(N)}_{t_1-t_0}\Big(F_1\, S^{(N)}_{t_2-t_1}\big(\cdots F_{m-1}\, S^{(N)}_{t_m-t_{m-1}} F_m\big)\Big)(x)\, d\mu_{\eta,N}(x). \end{aligned} \tag{356}$$

Proof. Fix a depth N and times $0 \le t_0 < t_1 < \cdots < t_m$. Let $i \in \{1,2\}$. By stationarity of $\mathbb{P}^{(i)}_{\eta,\infty}$, the law of $\Pi_N X_{t_0}$ is $\mu_{\eta,N}$. Repeatedly applying the cylinder-Markov property (336) and then replacing the selected cylinder operators by the canonical semigroup from Proposition D.40 yields

$$\mathbb{E}_{\mathbb{P}^{(i)}_{\eta,\infty}}\Big[\prod_{j=0}^{m} F_j(\Pi_N X_{t_j})\Big] = \int_{\mathcal{X}_N} F_0(x) S^{(N)}_{t_1-t_0}\Big(F_1\, S^{(N)}_{t_2-t_1}\big(\cdots F_{m-1}\, S^{(N)}_{t_m-t_{m-1}} F_m\big)\Big)(x)\, d\mu_{\eta,N}(x),$$

which is independent of i. Hence the two laws have the same finite-dimensional distributions on every finite cylinder level.

Now let $\mathcal{A}_{\mathrm{rat}}$ be the algebra generated by the maps

$$\omega \longmapsto \big(\Pi_N\omega(q_1), \ldots, \Pi_N\omega(q_r)\big), \qquad N \in \mathbb{N}, \quad q_1, \ldots, q_r \in \mathbb{Q}_+.$$

Because $\mathcal{X}_{\mathrm{cyl}}$ is a compact metric countable product and the path space $C([0,\infty), \mathcal{X}_{\mathrm{cyl}})$ carries the standard compact-open topology, its Borel σ-algebra is generated by rational-time coordinate maps. The algebra $\mathcal{A}_{\mathrm{rat}}$ therefore separates path points and generates the full Borel σ-field. Since the two candidate laws agree on all finite-dimensional cylinder distributions, they agree on $\mathcal{A}_{\mathrm{rat}}$, hence on the generated Borel σ-field. This proves (355). Formula (356) is exactly the recursive computation above, now with the unique law. □

Remark D.42 (What the third open problem now reduces to)**.** Theorem D.41 closes the *uniqueness in law* part of the third former open problem inside the manuscript's intrinsic stationary cylinder-Markov class. What still remains open is pathwise strong existence and uniqueness for the full infinite reflected interaction field, together with ergodicity beyond the cylinder observable algebra and beyond the presently frozen intrinsic tail class.

Integrated downstream conclusion after stationary path-law uniqueness. This continuation adds a twenty-sixth downstream message. Positively, the infinite interaction-type field is no longer represented only by an existence theorem and canonical finite-cylinder resolvents: within the manuscript's intrinsic stationary cylinder-Markov class it now has a unique global stationary weak path law, and all finite-dimensional cylinder distributions are fixed by the canonical semigroups and the stationary continuum marginal. Cautionarily, the theorem still does not produce a pathwise strong reflected SPDE, nor does it settle ergodicity for genuinely non-cylinder observables. The manuscript has therefore upgraded the stochastic package from selection-independent finite-cylinder evolution to full uniqueness in law of the stationary weak interaction field.

Remark D.43. The present appendix now carries twenty-six distinct downstream messages. The new twenty-sixth message is the strongest stochastic uniqueness statement presently justified within the manuscript's frozen scope: once one stays inside the intrinsic stationary cylinder-Markov class, there is only one global stationary weak interaction-field law. What remains open after this point is no longer uniqueness in law, but the stronger pathwise and beyond-cylinder ergodic theory.

D.14 Worked continuation XXVII: canonical strong cylinder lifts and spectral-gap stabilization

The stationary path-law uniqueness theorem of Section D.13 closes uniqueness in law inside the manuscript's intrinsic stationary cylinder-Markov class, but it still leaves open whether any genuinely pathwise statement can be recovered before one solves the full infinite reflected interaction-field problem. The present continuation shows that there is indeed one more intrinsic upgrade available. Every *finite observable cylinder* already admits a canonical pathwise strong realization, these strong realizations are compatible across cylinder depth, and their canonical semigroups have a spectral gap. Thus the remaining difficulty is no longer pathwise control on finite cylinders or cylinder-level ergodicity, but the full continuum upgrade beyond finitely observed coordinates.

Status label. The strong-realization and spectral-gap statements below have status **theorem.** They close pathwise strong well-posedness and exponential ergodicity on every *finite* observable cylinder. They do *not* claim a pathwise strong reflected process on the full intrinsic continuum support, nor ergodicity for genuinely non-cylinder observables.

Theorem D.44 (Canonical strong realization on every finite observable cylinder). *Fix $\eta > \eta_{\mathrm{crit}}^{\mathrm{var}}$ and a cylinder depth $N \geq 0$. Then:*

(i) the reduced finite-cylinder operator $\mathcal{L}_{\eta,N}^{\mathrm{red}}$ is realized by a pathwise unique strong Markov process $X^{[N]} = (H^{[N]}, A_1^{[N]}, \dots, A_N^{[N]})$ on $\mathcal{X}_N$;

(ii) one may choose a filtered probability space carrying independent Brownian motions

$$\beta,\ B_1, B_2, \dots,$$

such that for every $M \geq N$, if the M-cylinder process $X^{[M]}$ and the N-cylinder process $X^{[N]}$ are started from initial data satisfying

$$\Pi_N X_0^{[M]} = X_0^{[N]} \qquad a.s.,$$

and are driven by the same first $N+1$ Brownian coordinates $(\beta, B_1, \dots, B_N)$, then

$$\Pi_N X_t^{[M]} = X_t^{[N]} \qquad a.s.\ for\ all\ t > 0; \tag{357}$$

(iii) if X_0 is distributed according to the intrinsic stationary continuum law $\widehat{\pi}_\infty$, then the induced cylindrical strong family

$$X^{\mathrm{str}} = (X^{[N]})_{N \geq 0}$$

has, on every finite cylinder level, semigroup $S_t^{(N)}$ from Proposition D.40; consequently its finite-cylinder distributions agree with those of the unique stationary weak path law $\mathbb{P}_{\eta,\infty}^{\mathrm{stat}}$ from Theorem D.41.

Proof. For fixed N, Theorem D.15 already gives well posedness of the reflected martingale problem for $\mathcal{L}^{\mathrm{red}}_{\eta,N}$. To upgrade this to a strong solution, note first that the helicity coordinate satisfies the one-dimensional SDE (329) with diffusion coefficient

$$\sigma(h) = \sqrt{2A(h)} = \sqrt{\frac{2h(h_* - h)}{h_*}},$$

which is of Jacobi/Wright–Fisher type on the compact interval $[0, h_*]$. Such coefficients satisfy pathwise uniqueness in one dimension; one may check this either directly by the standard Yamada–Watanabe criterion for square-root diffusions or after a Lamperti transform to a Lipschitz-drift equation on a compact interval. For the amplitude block, the drifts $-\partial_{a_n} \log \widetilde{\Delta}^{(N)}_R$ are Lipschitz on the compact box $\prod_{n=1}^N [-b_n(\eta), b_n(\eta)]$, and the normal-reflection Skorokhod problem on a rectangular domain is pathwise well posed. Combining the one-dimensional helicity equation with the reflected amplitude block yields pathwise uniqueness for the coupled $(N+1)$-dimensional system, and Yamada–Watanabe upgrades the weak solution from Theorem D.15 to a strong Markov solution. This proves item (i).

For item (ii), realize simultaneously all finite-cylinder systems on one probability space carrying the Brownian family $(\beta, B_1, B_2, \dots)$. If $M \geq N$, then by definition of the reduced generators the first N coordinates of $\mathcal{L}^{\mathrm{red}}_{\eta,M}$ act on cylinder test functions exactly as $\mathcal{L}^{\mathrm{red}}_{\eta,N}$. Hence the first N coordinates of $X^{[M]}$, when driven by $(\beta, B_1, \dots, B_N)$, solve the same reflected SDE system as $X^{[N]}$ with the same initial condition. Pathwise uniqueness from item (i) therefore forces (357).

For item (iii), start from a random initial condition with law $\widehat{\pi}_\infty$ and define $X_0^{[N]} := \Pi_N X_0$. By item (ii), the resulting family is projectively compatible and therefore defines a cylindrical strong process X^{str}. On each finite cylinder level its semigroup is exactly the one generated by $\mathcal{L}^{\mathrm{red}}_{\eta,N}$, which by Proposition D.40 is the canonical semigroup $S_t^{(N)}$. The canonical finite-dimensional formula (356) from Theorem D.41 then shows that the finite-cylinder distributions of X^{str} coincide with those of $\mathbb{P}^{\mathrm{stat}}_{\eta,\infty}$. □

Proposition D.45 (Compact resolvent and exponential cylinder ergodicity). *Fix $\eta > \eta^{\mathrm{var}}_{\mathrm{crit}}$ and a cylinder depth $N \geq 0$. Then the canonical self-adjoint generator $A_{\eta,N}$ of Proposition D.40 has compact resolvent on $L^2(\mu_{\eta,N})$. Consequently there exists a spectral gap*

$$\lambda_{\eta,N} > 0$$

such that for every $F \in L^2(\mu_{\eta,N})$ with $\int F \, d\mu_{\eta,N} = 0$,

$$\|S_t^{(N)} F\|_{L^2(\mu_{\eta,N})} \leq e^{-\lambda_{\eta,N} t} \|F\|_{L^2(\mu_{\eta,N})}, \qquad t \geq 0. \tag{358}$$

Equivalently, if $X^{[N]}$ is the canonical strong cylinder diffusion started from stationarity, then for every $F, G \in L^2(\mu_{\eta,N})$,

$$\left|\mathrm{Cov}(F(X_t^{[N]}), G(X_0^{[N]}))\right| \leq e^{-\lambda_{\eta,N} t} \|F - \mu_{\eta,N}(F)\|_{L^2(\mu_{\eta,N})} \|G - \mu_{\eta,N}(G)\|_{L^2(\mu_{\eta,N})}. \tag{359}$$

Hence every finite observable cylinder is exponentially mixing.

Proof. The Dirichlet form from Theorem D.12,

$$\mathcal{E}_{\eta,N}(f, f) = \int_{\mathcal{X}_N} \left(A(h) |\partial_h f|^2 + \sum_{n=1}^N |\partial_{a_n} f|^2 \right) d\mu_{\eta,N},$$

closes on a weighted Sobolev space over the compact cylinder box $\mathcal{X}_N$. Because the amplitude directions are compact intervals and the helicity coordinate carries the Jacobi-type weight

$\rho_*(h) \sim h(h_* - h)^2$, the corresponding weighted Rellich lemma gives a compact embedding of the form domain into $L^2(\mu_{\eta,N})$. Therefore the self-adjoint resolvent $(\lambda - A_{\eta,N})^{-1}$ is compact for every $\lambda > 0$, so the spectrum of $-A_{\eta,N}$ is discrete with finite-multiplicity eigenvalues accumulating only at $+\infty$.

The kernel of $-A_{\eta,N}$ consists exactly of constants by item (iii) of Theorem D.12; hence the first positive eigenvalue $\lambda_{\eta,N}$ is strictly positive. The spectral theorem for self-adjoint contraction semigroups then yields (358) on the mean-zero subspace. Formula (359) is the usual stationary covariance estimate obtained by applying (358) to $F - \mu_{\eta,N}(F)$ and pairing with $G - \mu_{\eta,N}(G)$ in $L^2(\mu_{\eta,N})$. □

What the third open problem now reduces to. The two results above close everything that can be asked purely at finite observation depth: each finite cylinder now has a pathwise unique strong reflected realization, those realizations are compatible across cylinder rank, and their canonical semigroups converge exponentially to equilibrium. What remains open is therefore genuinely infinite-dimensional: a pathwise strong reflected process on the full intrinsic continuum support, together with ergodicity for observables not captured by any finite cylinder algebra.

Integrated downstream conclusion after the strong-cylinder continuation. This continuation adds a twenty-seventh downstream message. Positively, the stochastic package now reaches pathwise strong well-posedness on every finite observable cylinder, compatibility of those strong realizations across cylinder depth, and exponential cylinder mixing through a spectral-gap estimate. Cautionarily, this still does not yield a pathwise strong reflected process on the full continuum support, nor a beyond-cylinder ergodic theorem for genuinely infinite observables. The manuscript has therefore reduced the third former open problem to its irreducibly full-continuum part.

Remark D.46. The present appendix now carries twenty-seven distinct downstream messages. The new twenty-seventh message is the strongest pathwise statement presently justified without leaving the manuscript's frozen scope: finite observable cylinders are now strongly and uniquely realized, and they mix exponentially to equilibrium. What remains open after this point is not finite-cylinder pathwise or ergodic control, but the genuinely infinite reflected interaction-field theory and its non-cylinder long-time behavior.

D.15 Worked continuation XXVIII: intrinsic continuum Hunt realization and full L^2-ergodicity

The previous continuation closed pathwise strong control and exponential mixing on every *finite observable cylinder*, but it still left open whether the full intrinsic continuum support itself carries a canonical Markov process and whether ergodicity can be stated for observables not belonging to any finite cylinder algebra. The present continuation shows that both questions admit a manuscript-intrinsic answer. The canonical finite-cylinder semigroups glue to a symmetric Feller semigroup on the compact continuum support $\mathfrak{S}^{(3)}_{//}$; this semigroup yields a conservative Hunt process with invariant law $\widehat{\pi}_\infty$; and the uniqueness of that invariant continuum law forces L^2-ergodicity, hence stationary time-average ergodicity, for genuinely non-cylinder observables. What remains open after this point is therefore no longer mere existence of a continuum Markov process or ergodicity at the measure-theoretic level, but a pathwise strong reflected SDE/SPDE realization and sharper quantitative continuum mixing estimates.

Status label. The semigroup, Hunt-process, and full-L^2-ergodicity statements below have status **theorem**. They close full-continuum strong Markov existence and non-cylinder ergodicity inside the manuscript's intrinsic compact support class. They do *not* claim a pathwise strong SDE/SPDE realization driven by a fixed infinite Brownian family, nor quantitative exponential mixing for the full continuum semigroup, nor any extension beyond the intrinsic tail-controlled support universe.

Write $\Pi_N(h,\phi) := (h, \langle\phi, e_1\rangle, \dots, \langle\phi, e_N\rangle)$, and let

$$\mathcal{C}_\infty := \bigcup_{N\geq 0} \left\{ F : \mathfrak{S}_\eta^{(3)} \to \mathbb{R} : \; F = f \circ \Pi_N \text{ for some bounded smooth } f \text{ on } \mathcal{X}_N \right\}$$

be the algebra of bounded smooth cylinder observables on the intrinsic continuum support.

Theorem D.47 (Canonical intrinsic continuum semigroup and Hunt realization). *Fix $\eta > \eta_{\mathrm{crit}}^{\mathrm{var}}$. For $F = f \circ \Pi_N \in \mathcal{C}_\infty$, define*

$$(T_t^\infty F)(h,\phi) := \big(S_t^{(N)} f\big)\big(\Pi_N(h,\phi)\big), \qquad t \geq 0. \tag{360}$$

Then the following hold.

(i) the definition (360) *is well defined on $\mathcal{C}_\infty$, positivity preserving, Markov, and contractive both in $\|\cdot\|_\infty$ and in $L^2(\widehat{\pi}_\infty)$;*

(ii) $(T_t^\infty)_{t\geq 0}$ extends uniquely to a conservative Feller semigroup on $C(\mathfrak{S}_\eta^{(3)})$ and to a strongly continuous self-adjoint Markov semigroup on $L^2(\widehat{\pi}_\infty)$;

(iii) on the cylinder core $\mathcal{C}_\infty$, the bilinear form

$$\mathcal{E}_{\eta,\infty}(F,G) := \int_{\mathfrak{S}_\eta^{(3)}} \left(A(h)\, \partial_h F\, \partial_h G + \sum_{n\geq 1} \partial_{a_n} F\, \partial_{a_n} G \right) d\widehat{\pi}_\infty \tag{361}$$

is well defined (the sum being finite for cylinder functions), closable, and its closure is a regular symmetric Dirichlet form on the compact metric space $\mathfrak{S}_\eta^{(3)}$;

(iv) consequently there exists a unique conservative $\widehat{\pi}_\infty$-symmetric Hunt process

$$Y^\infty = (H^\infty, \Phi^\infty)$$

on $\mathfrak{S}_\eta^{(3)}$ associated with that regular Dirichlet form;

(v) for every finite cylinder depth N, the projected process $\Pi_N Y^\infty$ has semigroup $S_t^{(N)}$, so the full continuum Hunt process recovers all canonical finite-cylinder dynamics.

Proof. Suppose first that a cylinder observable admits two representations, $F = f \circ \Pi_N = g \circ \Pi_M$ with $M \geq N$. Then g is obtained from f by ignoring the last $M - N$ coordinates. The strong compatibility statement (357) together with Proposition D.40 therefore gives

$$\big(S_t^{(M)} g\big) \circ \Pi_M = \big(S_t^{(N)} f\big) \circ \Pi_N,$$

so (360) is well defined. Positivity preservation, the Markov property, and both sup- and L^2-contractivity are inherited from the finite-cylinder semigroups $S_t^{(N)}$.

The cylinder algebra $\mathcal{C}_\infty$ separates points of $\mathfrak{S}_\eta^{(3)}$ and is uniformly dense in $C(\mathfrak{S}_\eta^{(3)})$ by the same Stone–Weierstrass argument used in Theorem D.37. For each fixed cylinder observable, strong continuity at $t = 0$ follows from the strong continuity of the corresponding finite-cylinder semigroup. Because $\mathfrak{S}_\eta^{(3)}$ is compact and the operators are sup-norm contractions on a dense subalgebra, $(T_t^\infty)_{t\geq 0}$ extends uniquely to a conservative Feller semigroup on $C(\mathfrak{S}_\eta^{(3)})$. The finite-cylinder semigroups are symmetric with respect to their invariant measures, and the continuum stationary law has those same cylinder marginals; hence T_t^∞ is symmetric on $\mathcal{C}_\infty$, and by density it extends to a strongly continuous self-adjoint Markov semigroup on $L^2(\widehat{\pi}_\infty)$.

For cylinder observables $F = f \circ \Pi_N$ and $G = g \circ \Pi_N$, formula (361) coincides exactly with the finite-cylinder form (323) under the identification of $\widehat{\pi}_\infty$-cylinder marginals with $\mu_{\eta,N}$. Thus the pre-form is nonnegative, symmetric, and Markovian on $\mathcal{C}_\infty$. Since it is represented by a symmetric Feller semigroup on the compact metric state space and $\mathcal{C}_\infty$ is dense in both $C(\mathfrak{S}_\eta^{(3)})$ and $L^2(\widehat{\pi}_\infty)$, its closure is a regular symmetric Dirichlet form. Standard regular Dirichlet-form theory on compact metric spaces then yields a unique conservative $\widehat{\pi}_\infty$-symmetric Hunt process Y^∞. Finally, item (v) is immediate from the defining identity (360): the finite-cylinder semigroup seen by Y^∞ is exactly $S_t^{(N)}$ at every depth N. □

Theorem D.48 (Unique invariant continuum law and full non-cylinder ergodicity)**.** *Assume the hypotheses of Theorem D.47. Then:*

(i) $\widehat{\pi}_\infty$ is the unique invariant Borel probability measure for the continuum semigroup $(T_t^\infty)_{t\geq 0}$;

(ii) if $A_{\eta,\infty}$ denotes the self-adjoint generator of $(T_t^\infty)_{t\geq 0}$ on $L^2(\widehat{\pi}_\infty)$, then

$$\ker A_{\eta,\infty} = \{\textit{constants}\};$$

(iii) consequently, for every $F \in L^2(\widehat{\pi}_\infty)$,

$$\|T_t^\infty F - \widehat{\pi}_\infty(F)\|_{L^2(\widehat{\pi}_\infty)} \longrightarrow 0 \qquad (t \to \infty); \tag{362}$$

(iv) if the Hunt process Y^∞ is started from stationarity, then for every bounded Borel observable F on $\mathfrak{S}_\eta^{(3)}$,

$$\frac{1}{T}\int_0^T F(Y_t^\infty)\,dt \longrightarrow \int_{\mathfrak{S}_\eta^{(3)}} F\,d\widehat{\pi}_\infty \qquad \textit{a.s. as } T \to \infty. \tag{363}$$

In particular, the manuscript's intrinsic continuum interaction field is ergodic for genuinely non-cylinder observables.

Proof. Let μ be an invariant Borel probability measure for $(T_t^\infty)_{t\geq 0}$. Fix N and a bounded continuous test function f on $\mathcal{X}_N$. Then

$$\int f\,d(\Pi_N)_\#\mu = \int (f \circ \Pi_N)\,d\mu = \int T_t^\infty (f \circ \Pi_N)\,d\mu = \int (S_t^{(N)} f)\,d(\Pi_N)_\#\mu.$$

Hence $(\Pi_N)_\#\mu$ is invariant for the finite-cylinder semigroup $S_t^{(N)}$. Proposition D.45 gives uniqueness of the finite-cylinder invariant law, so $(\Pi_N)_\#\mu = \mu_{\eta,N}$ for every N. Theorem D.37 then forces $\mu = \widehat{\pi}_\infty$, proving item (i).

For item (ii), let $u \in D(A_{\eta,\infty})$ satisfy $A_{\eta,\infty}u = 0$. Equivalently, u lies in the kernel of the closed Dirichlet form $\mathcal{E}_{\eta,\infty}$. If u were nonconstant, choose a level $c \in \mathbb{R}$ with $0 < \widehat{\pi}_\infty(\{u > c\}) < 1$. By the Markov property of the Dirichlet form, the truncated functions

$$\Phi_k(u) := \big(0 \vee k(u-c)\big) \wedge 1$$

also have zero energy. Letting $k \to \infty$ and using closure of the form yields zero energy for the indicator $\mathbf{1}_{\{u>c\}}$, which therefore defines an invariant set for the associated Hunt process. Normalizing the restrictions of $\widehat{\pi}_\infty$ to $\{u > c\}$ and its complement produces two distinct invariant probability measures, contradicting item (i). Thus the kernel consists only of constants.

Item (iii) is now the spectral theorem for the self-adjoint contraction semigroup $(T_t^\infty)_{t\geq 0}$: because the fixed-point space is exactly the constants, the strong limit of T_t^∞ as $t \to \infty$ is the orthogonal projection onto constants, namely $F \mapsto \widehat{\pi}_\infty(F)$. Finally, item (iv) is the standard stationary ergodic theorem for the conservative Hunt process Y^∞ under the now-ergodic invariant measure $\widehat{\pi}_\infty$. □

What the third open problem now reduces to. The two theorems above close everything that can be asked purely at the level of the manuscript's intrinsic continuum support and its measure-theoretic long-time behavior: the full intrinsic continuum class now carries a canonical conservative symmetric Hunt process, and that process is ergodic on all bounded Borel observables. What remains open is therefore no longer non-cylinder ergodicity itself, but a *pathwise strong* reflected SDE/SPDE realization on the full continuum support and sharper quantitative control of the full continuum semigroup (for example a genuinely infinite-dimensional spectral gap, log-Sobolev inequality, or mixing estimate uniform beyond finite cylinders).

Integrated downstream conclusion after the continuum-Hunt continuation. This continuation adds a twenty-eighth downstream message. Positively, the stochastic package now reaches a canonical full-continuum Markov object inside the intrinsic tail-controlled support class: the finite-cylinder semigroups glue to a conservative symmetric Hunt process, the stationary continuum law is uniquely invariant for that process, and genuinely non-cylinder observables obey both L^2-ergodicity and the stationary time-average ergodic theorem. Cautionarily, this still does not yield a pathwise strong reflected SDE/SPDE driven by a fixed infinite Brownian family, nor a quantitative exponential-mixing theorem on the full continuum support, nor any extension beyond the intrinsic support universe. The manuscript has therefore reduced the third former open problem to its irreducibly pathwise-and-quantitative full-continuum part.

Remark D.49. The present appendix now carries twenty-eight distinct downstream messages. The new twenty-eighth message is the strongest continuum-dynamical statement presently justified within the manuscript's frozen scope: the intrinsic interaction field now has a canonical conservative symmetric Hunt realization on the full compact support, and that realization is ergodic for genuinely non-cylinder observables. What remains open after this point is not existence or ergodicity of the intrinsic continuum process, but the stronger reflected SDE/SPDE realization and quantitative full-continuum mixing theory.

D.16 Worked continuation XXIX: coordinate Fukushima decompositions and conditional quantitative continuum mixing

The previous continuation produced a canonical conservative symmetric Hunt process on the full intrinsic continuum support and closed ergodicity at the measure-theoretic level. Two natural questions remained. First, does the continuum interaction field admit an explicit coordinate-level stochastic calculus compatible with the finite reflected cylinders? Second, can the finite-cylinder spectral-gap and entropy estimates be lifted to genuinely continuum observables? The present continuation answers the first question unconditionally inside the manuscript's intrinsic support class and the second question conditionally under a transparent uniform-coercivity hypothesis. The result is a canonical coordinate Fukushima decomposition for the continuum interaction field together with a full-continuum Poincare/log-Sobolev package whenever the finite observable cylinders admit uniform lower bounds on their coercivity constants.

Status label. The coordinate-decomposition theorem below has status **theorem**. The quantitative full-continuum mixing theorem is also a **theorem**, but it is explicitly *conditional*: it requires uniform cylinder-level spectral-gap and (optionally) log-Sobolev bounds. These results do *not* claim a pathwise strong reflected SDE/SPDE driven by a fixed infinite Brownian family, and they do *not* prove the required uniform coercivity bounds intrinsically. What they do show is that once those finite-cylinder bounds are available, the full intrinsic continuum semigroup inherits them exactly.

For the continuum state $(h, \phi) \in \mathfrak{S}_\eta^{(3)}$, write

$$H(h, \phi) := h, \qquad A_n(h, \phi) := \langle \phi, e_n \rangle, \qquad n \geq 1,$$

for the helicity coordinate and the released odd half-mode coordinates.

Theorem D.50 (Coordinate Fukushima decomposition on the intrinsic continuum support). *Fix $\eta > \eta_{\rm crit}^{\rm var}$, and let*

$$Y^\infty = (H^\infty, \Phi^\infty)$$

be the canonical conservative symmetric Hunt process from Theorem D.47. Then for each $n \geq 1$ the coordinate maps H and A_n belong to the local Dirichlet-form domain of $\mathcal{E}_{\eta,\infty}$, and there exist continuous martingale additive functionals M^H, $M^{(n)}$ and continuous zero-energy additive functionals N^H, $N^{(n)}$ such that, for quasi-every starting point and all $t \geq 0$,

$$H_t^\infty - H_0^\infty = M_t^H + N_t^H, \tag{364}$$

$$A_n(Y_t^\infty) - A_n(Y_0^\infty) = M_t^{(n)} + N_t^{(n)}. \tag{365}$$

Moreover, their predictable quadratic covariations are given by

$$\langle M^H \rangle_t = 2 \int_0^t A(H_s^\infty)\, ds, \tag{366}$$

$$\langle M^H, M^{(n)} \rangle_t = 0, \tag{367}$$

$$\langle M^{(n)}, M^{(m)} \rangle_t = 2t\, \delta_{nm}. \tag{368}$$

Consequently, for every fixed N, the block process

$$\big(H^\infty, A_1(\Phi^\infty), \ldots, A_N(\Phi^\infty)\big)$$

is a continuous semimartingale whose finite-dimensional martingale problem coincides with the canonical finite-cylinder generator $A_{\eta,N}$.

Proof. The coordinate map H is a smooth cylinder observable of depth 0, while each A_n is a smooth cylinder observable of depth n. Hence all of them lie in the cylinder core $\mathcal{C}_\infty \subset D(\mathcal{E}_{\eta,\infty})$. Because $\mathcal{E}_{\eta,\infty}$ is a regular symmetric Dirichlet form on the compact metric state space $\mathfrak{S}_\eta^{(3)}$, the Fukushima decomposition applies to every function in the local Dirichlet-form domain. This gives decompositions (364)–(365) with continuous martingale and zero-energy parts.

It remains to compute the carré du champ on the cylinder core. For cylinder observables the form is exactly (361), so

$$\Gamma(H,H) = A(h), \qquad \Gamma(H, A_n) = 0, \qquad \Gamma(A_n, A_m) = \delta_{nm}$$

on $\mathcal{C}_\infty$. The standard identification between predictable quadratic covariations of Fukushima martingale parts and the carré du champ therefore yields (366)–(368). Finally, fix N and a bounded smooth test function f on $\mathcal{X}_N$. The observable $f \circ \Pi_N$ belongs to $\mathcal{C}_\infty$, and Theorem D.47 says that its semigroup is exactly $S_t^{(N)}$. Therefore the projected block process solves the martingale problem for the same generator that governs the canonical strong finite-cylinder realization, namely $A_{\eta,N}$. □

Theorem D.51 (Conditional full-continuum spectral gap and log-Sobolev lift). *Fix $\eta > \eta_{\rm crit}^{\rm var}$, and assume that the finite observable cylinders admit uniform coercivity constants in the following sense:*

$$\lambda_* := \inf_{N \geq 1} \lambda_{\eta,N} > 0, \tag{369}$$

$$\alpha_* := \inf_{N \geq 1} \alpha_{\eta,N} > 0, \tag{370}$$

where $\lambda_{\eta,N}$ is the spectral gap from Proposition D.45, and $\alpha_{\eta,N}$ is any log-Sobolev constant satisfying

$$\mathrm{Ent}_{\mu_{\eta,N}}(f^2) \leq \frac{2}{\alpha_{\eta,N}}\, \mathcal{E}_{\eta,N}(f,f) \tag{371}$$

for bounded smooth f on $\mathcal{X}_N$. Then the following hold.

(i) for every $F \in D(\mathcal{E}_{\eta,\infty})$,

$$\|F - \widehat{\pi}_\infty(F)\|^2_{L^2(\widehat{\pi}_\infty)} \le \frac{1}{\lambda_*}\,\mathcal{E}_{\eta,\infty}(F,F); \tag{372}$$

(ii) consequently, for every $F \in L^2(\widehat{\pi}_\infty)$,

$$\|T_t^\infty F - \widehat{\pi}_\infty(F)\|_{L^2(\widehat{\pi}_\infty)} \le e^{-\lambda_* t}\|F - \widehat{\pi}_\infty(F)\|_{L^2(\widehat{\pi}_\infty)}, \qquad t \ge 0; \tag{373}$$

(iii) if the log-Sobolev lower bound (370) also holds, then every $F \in D(\mathcal{E}_{\eta,\infty})$ satisfies

$$\operatorname{Ent}_{\widehat{\pi}_\infty}(F^2) \le \frac{2}{\alpha_*}\,\mathcal{E}_{\eta,\infty}(F,F), \tag{374}$$

and, for bounded nonnegative F,

$$\operatorname{Ent}_{\widehat{\pi}_\infty}\big((T_t^\infty F)^2\big) \le e^{-2\alpha_* t}\operatorname{Ent}_{\widehat{\pi}_\infty}(F^2), \qquad t \ge 0. \tag{375}$$

In particular, uniform finite-cylinder coercivity propagates to the full intrinsic continuum semigroup without loss.

Proof. It is enough to prove the claims first on the cylinder core $\mathcal{C}_\infty$. Let $F = f \circ \Pi_N \in \mathcal{C}_\infty$. Then F has continuum law $\widehat{\pi}_\infty$ exactly when f has finite-cylinder law $\mu_{\eta,N}$, and $\mathcal{E}_{\eta,\infty}(F,F) = \mathcal{E}_{\eta,N}(f,f)$ by definition of the continuum Dirichlet form. Therefore the finite-cylinder Poincare inequality with constant $\lambda_{\eta,N}$ yields

$$\|F - \widehat{\pi}_\infty(F)\|^2_{L^2(\widehat{\pi}_\infty)} = \|f - \mu_{\eta,N}(f)\|^2_{L^2(\mu_{\eta,N})} \le \frac{1}{\lambda_{\eta,N}}\,\mathcal{E}_{\eta,N}(f,f) \le \frac{1}{\lambda_*}\,\mathcal{E}_{\eta,\infty}(F,F).$$

Because $\mathcal{C}_\infty$ is dense in $D(\mathcal{E}_{\eta,\infty})$ for the form norm and both sides are lower semicontinuous under that norm, (372) extends to the full Dirichlet-form domain.

The exponential estimate (373) is now the standard spectral-gap consequence for the symmetric semigroup $(T_t^\infty)_{t\ge 0}$: inequality (372) says precisely that the bottom of the spectrum of $-A_{\eta,\infty}$ on the mean-zero subspace is bounded below by λ_*.

For the log-Sobolev statement, again start with $F = f \circ \Pi_N \in \mathcal{C}_\infty$. The finite-cylinder inequality (371) gives

$$\operatorname{Ent}_{\widehat{\pi}_\infty}(F^2) = \operatorname{Ent}_{\mu_{\eta,N}}(f^2) \le \frac{2}{\alpha_{\eta,N}}\,\mathcal{E}_{\eta,N}(f,f) \le \frac{2}{\alpha_*}\,\mathcal{E}_{\eta,\infty}(F,F).$$

Density of $\mathcal{C}_\infty$ in the form domain plus lower semicontinuity of entropy along bounded truncations extends (374) to the natural domain. Finally, the entropy decay (375) is the standard semigroup consequence of the log-Sobolev inequality for symmetric Markov semigroups. □

What the third open problem now reduces to. The two theorems above identify the strongest remaining gap with precision. On the one hand, the intrinsic continuum interaction field now possesses a coordinate-wise stochastic calculus through the Fukushima decomposition, so the full continuum process is no longer a black-box Hunt object. On the other hand, any uniform spectral-gap or log-Sobolev control available on finite observable cylinders lifts automatically to the full intrinsic continuum semigroup. What still remains open is therefore the genuinely pathwise strong reflected SDE/SPDE realization driven by a fixed infinite Brownian family, together with an intrinsic proof — or a sharp obstruction — for the uniform coercivity bounds required to make the continuum mixing package unconditional.

Integrated downstream conclusion after the coordinate-and-coercivity continuation. This continuation adds a twenty-ninth downstream message. Positively, the full intrinsic interaction field now has a canonical coordinate-wise Fukushima decomposition, and its full-continuum quantitative mixing theory is reduced to a transparent finite-cylinder coercivity question: once uniform cylinder spectral-gap or log-Sobolev bounds are available, the continuum semigroup inherits them exactly. Cautionarily, the manuscript still does not produce a pathwise strong reflected SDE/SPDE driven by a fixed infinite Brownian family, nor an intrinsic derivation of the required uniform coercivity constants, nor any extension beyond the intrinsic support universe. The third former open problem has therefore been sharpened from a broad pathwise-and-quantitative gap into two concrete subproblems: strong realization and intrinsic coercivity.

Remark D.52. The present appendix now carries twenty-nine distinct downstream messages. The new twenty-ninth message is the strongest bridge presently justified between the abstract continuum Hunt process and an explicit stochastic-calculus / quantitative-mixing package: the intrinsic interaction field has coordinate martingale decompositions, and every genuine continuum coercivity estimate is now seen to be equivalent to a uniform family of finite-cylinder estimates. What remains open after this point is not the formal shape of the full-continuum stochastic calculus, but the stronger Brownian/SPDE realization and the intrinsic proof of the uniform coercivity bounds themselves.

D.17 Worked continuation XXX: monotone intrinsic retractions and beyond-intrinsic basin rigidity

The first remaining open problem after the intrinsic attainment theorem asks whether the preferred branch is an artifact of the manuscript's own doubled released parametrization, or whether it persists across genuinely broader interaction-type families. A full answer across arbitrary external ansatz universes would still be irresponsible. There is, however, one mathematically clean step that can be completed now. If a broader class admits a continuous reduction back to the intrinsic interaction-type basin, and if that reduction never increases the corresponding free energy, then all minimizers in the broader class are forced to have the same intrinsic shadow as the preferred intrinsic branch. If the reduction is strictly dissipative away from the intrinsic basin, then the broader class has exactly the same minimizer.

Definition D.53 (Intrinsic-retractable interaction-type enlargement)**.** Fix $\eta > 0$ and a compact intrinsic basin $\mathcal{B}_\eta \subset \mathfrak{S}_\eta^{(3)}$ satisfying the hypotheses of Theorem D.33, with unique intrinsic minimizer $x_\eta^\star$. A compact class $\mathfrak{E}_\eta$ containing $\mathcal{B}_\eta$ is called an *intrinsic-retractable interaction-type enlargement over* $\mathcal{B}_\eta$ if there exist:

(R1) a continuous retraction

$$\mathcal{R}_\eta : \mathfrak{E}_\eta \to \mathcal{B}_\eta, \qquad \mathcal{R}_\eta|_{\mathcal{B}_\eta} = \mathrm{id};$$

(R2) a continuous extended free energy

$$\mathfrak{G}_\eta^{\mathrm{ext}} : \mathfrak{E}_\eta \to \mathbb{R}$$

whose restriction to $\mathcal{B}_\eta$ is $\mathfrak{G}_R^{(\infty)}(\eta, \cdot)$;

(R3) the *monotone reduction property*

$$\mathfrak{G}_\eta^{\mathrm{ext}}(y) \geq \mathfrak{G}_R^{(\infty)}(\eta, \mathcal{R}_\eta y) \qquad \text{for every } y \in \mathfrak{E}_\eta; \tag{376}$$

(R4) the convention that every intrinsic cylinder observable $F = f \circ \Pi_N$ on $\mathcal{B}_\eta$ is extended to $\mathfrak{E}_\eta$ by pullback,

$$F^{\mathrm{ext}} := F \circ \mathcal{R}_\eta.$$

If, in addition,

$$\mathfrak{G}_\eta^{\mathrm{ext}}(y) > \mathfrak{G}_R^{(\infty)}(\eta, \mathcal{R}_\eta y) \qquad \text{for every } y \in \mathfrak{E}_\eta \setminus \mathcal{B}_\eta, \tag{377}$$

then the enlargement is called *strictly retractable.*

Theorem D.54 (Beyond-intrinsic basin rigidity under monotone retraction)**.** *Fix $\eta > 0$ and let $\mathcal{B}_\eta \subset \mathfrak{S}_\eta^{(3)}$ be a compact intrinsic basin satisfying the hypotheses of Theorem D.33. Let $\mathfrak{E}_\eta$ be an intrinsic-retractable interaction-type enlargement over $\mathcal{B}_\eta$ in the sense of Definition D.53. Then every global minimizer $y_\eta^\star$ of $\mathfrak{G}_\eta^{\mathrm{ext}}$ on $\mathfrak{E}_\eta$ satisfies*

$$\mathcal{R}_\eta y_\eta^\star = x_\eta^\star. \tag{378}$$

Consequently, every extended intrinsic cylinder observable $F^{\mathrm{ext}} = F \circ \mathcal{R}_\eta$ has the universal value

$$F^{\mathrm{ext}}(y_\eta^\star) = F(x_\eta^\star). \tag{379}$$

If the enlargement is strictly retractable in the sense of (377)*, then*

$$y_\eta^\star = x_\eta^\star, \tag{380}$$

and the broader class has exactly the same unique global minimizer as the intrinsic continuum basin.

Proof. Because $x_\eta^\star \in \mathcal{B}_\eta \subset \mathfrak{E}_\eta$ and the extended free energy agrees with the intrinsic one on $\mathcal{B}_\eta$, we have

$$\inf_{y \in \mathfrak{E}_\eta} \mathfrak{G}_\eta^{\mathrm{ext}}(y) \le \mathfrak{G}_\eta^{\mathrm{ext}}(x_\eta^\star) = \mathfrak{G}_R^{(\infty)}(\eta, x_\eta^\star).$$

Let $y_\eta^\star$ be any global minimizer of $\mathfrak{G}_\eta^{\mathrm{ext}}$ on $\mathfrak{E}_\eta$. The monotone reduction property (376) gives

$$\mathfrak{G}_\eta^{\mathrm{ext}}(y_\eta^\star) \ge \mathfrak{G}_R^{(\infty)}(\eta, \mathcal{R}_\eta y_\eta^\star) \ge \mathfrak{G}_R^{(\infty)}(\eta, x_\eta^\star),$$

because $x_\eta^\star$ is the unique minimizer of $\mathfrak{G}_R^{(\infty)}$ on $\mathcal{B}_\eta$. Combining the upper and lower bounds forces equality everywhere:

$$\mathfrak{G}_\eta^{\mathrm{ext}}(y_\eta^\star) = \mathfrak{G}_R^{(\infty)}(\eta, \mathcal{R}_\eta y_\eta^\star) = \mathfrak{G}_R^{(\infty)}(\eta, x_\eta^\star).$$

Hence $\mathcal{R}_\eta y_\eta^\star$ is an intrinsic minimizer of $\mathfrak{G}_R^{(\infty)}$ on $\mathcal{B}_\eta$. By uniqueness inside that basin, (378) follows. Formula (379) is immediate from the definition of the pulled-back observable.

If the enlargement is strictly retractable and $y_\eta^\star \notin \mathcal{B}_\eta$, then (377) would imply

$$\mathfrak{G}_\eta^{\mathrm{ext}}(y_\eta^\star) > \mathfrak{G}_R^{(\infty)}(\eta, \mathcal{R}_\eta y_\eta^\star) = \mathfrak{G}_R^{(\infty)}(\eta, x_\eta^\star) = \mathfrak{G}_\eta^{\mathrm{ext}}(x_\eta^\star),$$

contradicting the minimizing property of $y_\eta^\star$. Therefore $y_\eta^\star \in \mathcal{B}_\eta$, and then (378) together with $\mathcal{R}_\eta|_{\mathcal{B}_\eta} = \mathrm{id}$ gives $y_\eta^\star = x_\eta^\star$. Uniqueness follows because any second minimizer would satisfy the same identity. □

Remark D.55 (What the new theorem closes and what it leaves open)**.** Theorem D.54 does not claim ansatz independence across arbitrary external parametrizations. What it does show is sharper and more structural: every broader interaction-type class that admits a continuous energy-monotone reduction back to the intrinsic basin is already forced to share the same preferred intrinsic branch, and every strictly retractable class shares the same actual minimizer. Thus the remaining content of the beyond-intrinsic problem is no longer the full universe of external ansatz families, but the construction — or sharp obstruction — of monotone intrinsic reductions for genuinely broader classes.

Geometric reading of the retraction hypothesis. The monotone-retraction hypothesis is consistent with the manuscript's e-Arc-Helix geometry. The intrinsic basin records the localized interaction-type doubled core on the inertial plane, while a broader class may contain additional exterior or open-field degrees of freedom. The theorem says that whenever those extra degrees of freedom can be collapsed continuously back to the localized core without lowering the total free energy, the preferred branch is already determined by the intrinsic e-Arc-Helix sector. The theorem itself does not assume any particular external picture, but this is the correct geometric reading of its hypothesis.

Integrated downstream conclusion after the retraction theorem. This continuation adds a thirtieth downstream message. Positively, beyond-intrinsic ansatz independence is no longer completely open: every compact interaction-type enlargement carrying a continuous energy-monotone retraction onto the intrinsic basin shares the same preferred reduced branch, and every strictly retractable enlargement shares the same actual minimizer. Cautionarily, the theorem does not manufacture such retractions automatically for arbitrary external families. What remains open is therefore no longer ansatz independence in every possible sense, but the existence — or obstruction — of monotone intrinsic reductions for genuinely broader classes and broader tail universes.

Remark D.56. The present appendix now carries thirty distinct downstream messages. The new thirtieth message is the first genuine extension of intrinsic ansatz independence beyond the manuscript's own released ladder class: the preferred branch is universal across every broader interaction-type family that can be retracted monotonically onto the intrinsic continuum basin. What remains open after this point is not the retraction principle itself, but whether the concrete external families one might ultimately care about actually possess such a reduction.

D.18 Worked continuation XXXI: bounded renormalized tail classes and non-product continuum closure

The second remaining open problem after Theorems D.37 and D.54 is no longer whether the manuscript's own intrinsic product tail prior determines a continuum law; that point is already closed. The remaining question is whether any mathematically meaningful enlargement beyond the raw product prior can be absorbed without destroying projective closure, continuum support control, or uniqueness of the stationary law. The present continuation shows that there is a substantial class of such enlargements. Every bounded, uniformly cylinder-approximable tail reweighting — including genuinely non-product couplings among the released modes — still closes to the same compact continuum support and determines a unique renormalized continuum law there.

Status label. The renormalized tail class below has status **definition**. The closure and support-equivalence results have status **theorem-level downstream result**. Extension to singular reweightings, non-box support classes, or noncompact renormalized continuum limits remains an **open problem** and is not claimed here.

Definition D.57 (Tempered renormalized tail potential)**.** Fix $\eta > 0$ and the intrinsic product box $\mathcal{K}_\eta^{(\infty)}$ of (284). A bounded continuous function

$$\Psi : \mathcal{K}_\eta^{(\infty)} \to \mathbb{R}$$

is called a *tempered renormalized tail potential* if there exists a sequence of bounded continuous cylinder functions

$$\Psi_m(\mathbf{a}) = \psi_m(a_1, \dots, a_m), \qquad \psi_m \in C\left(\prod_{n=1}^{m} [-b_n(\eta), b_n(\eta)]\right),$$

such that

$$\sup_{\mathbf{a}\in\mathcal{K}_\eta^{(\infty)}} \big|\Psi(\mathbf{a}) - \Psi_m(P_m\mathbf{a})\big| \longrightarrow 0 \qquad (m\to\infty). \tag{381}$$

For such a potential, define the renormalized tail prior

$$d\nu_{\eta,\Psi}^{(\infty)}(\mathbf{a}) := \frac{1}{\mathcal{N}_{\eta,\Psi}} e^{-\Psi(\mathbf{a})}\, d\nu_\eta^{(\infty)}(\mathbf{a}), \qquad \mathcal{N}_{\eta,\Psi} := \int_{\mathcal{K}_\eta^{(\infty)}} e^{-\Psi(\mathbf{a})}\, d\nu_\eta^{(\infty)}(\mathbf{a}). \tag{382}$$

For each finite release level m, define the finite renormalized reduced measure

$$d\pi_{m,\Psi}(h,\mathbf{a}\mid\eta) := \frac{1}{\mathcal{Z}_{m,\Psi}(\eta)} \rho_*(h)\widetilde{\Delta}_R^{(m)}(\eta,h,P_m\mathbf{a})^{-1} e^{-\Psi_m(P_m\mathbf{a})} \mathbf{1}_{(h,\mathbf{a})\in\mathcal{U}_\eta}\, dh\, d\nu_\eta^{(\infty)}(\mathbf{a}), \tag{383}$$

and the infinite renormalized reduced measure

$$d\pi_{\infty,\Psi}(h,\mathbf{a}\mid\eta) := \frac{1}{\mathcal{Z}_{\infty,\Psi}(\eta)} \rho_*(h)\widetilde{\Delta}_R^{(\infty)}(\eta,h,\mathbf{a})^{-1} e^{-\Psi(\mathbf{a})} \mathbf{1}_{(h,\mathbf{a})\in\mathcal{U}_\eta}\, dh\, d\nu_\eta^{(\infty)}(\mathbf{a}). \tag{384}$$

Write

$$\widehat{\pi}_{m,\Psi} := (\Gamma_\eta)_\#\pi_{m,\Psi}, \qquad \widehat{\pi}_{\infty,\Psi} := (\Gamma_\eta)_\#\pi_{\infty,\Psi}$$

for the corresponding renormalized continuum laws on $\mathfrak{S}_\eta^{(3)}$.

Theorem D.58 (Renormalized non-product tail closure on the intrinsic continuum support). *Assume the hypotheses of Proposition D.6. Let Ψ be a tempered renormalized tail potential in the sense of Definition D.57. Then:*

(R1) the finite renormalized measures $\pi_{m,\Psi}(\cdot\mid\eta)$ converge weakly to $\pi_{\infty,\Psi}(\cdot\mid\eta)$ on $\mathcal{U}_\eta$;

(R2) the pushforward laws $\widehat{\pi}_{m,\Psi}$ converge weakly to $\widehat{\pi}_{\infty,\Psi}$ as Borel probability measures on the same compact intrinsic continuum support $\mathfrak{S}_\eta^{(3)}$;

(R3) $\widehat{\pi}_{\infty,\Psi}$ is the unique Borel probability measure on $\mathfrak{S}_\eta^{(3)}$ whose integrals against bounded continuous cylinder functionals agree with the renormalized projective cylinder limits generated by $\pi_{m,\Psi}$;

(R4) the renormalized continuum law is uniformly equivalent to the intrinsic continuum law:

$$e^{-\operatorname{osc}(\Psi)} \le \frac{d\widehat{\pi}_{\infty,\Psi}}{d\widehat{\pi}_\infty} \le e^{\operatorname{osc}(\Psi)} \qquad \widehat{\pi}_\infty\text{-a.e. on } \mathfrak{S}_\eta^{(3)}, \tag{385}$$

where

$$\operatorname{osc}(\Psi) := \sup_{\mathcal{K}_\eta^{(\infty)}} \Psi - \inf_{\mathcal{K}_\eta^{(\infty)}} \Psi.$$

In particular, $\widehat{\pi}_{\infty,\Psi}$ and $\widehat{\pi}_\infty$ have the same support, namely $\mathfrak{S}_\eta^{(3)}$.

Equivalently, the manuscript's projective-to-continuum control extends from the raw product prior to every bounded cylinder-approximable renormalized tail coupling on the same intrinsic tail-controlled support universe, including genuinely non-product tail priors.

Proof. The hypotheses of Proposition D.6 give uniform convergence

$$\widetilde{\Delta}_R^{(m)}(\eta,h,P_m\mathbf{a}) \longrightarrow \widetilde{\Delta}_R^{(\infty)}(\eta,h,\mathbf{a})$$

on $\mathcal{U}_\eta$. By (381), the reweighting factors also converge uniformly:

$$e^{-\Psi_m(P_m\mathbf{a})} \longrightarrow e^{-\Psi(\mathbf{a})} \qquad \text{uniformly on } \mathcal{K}_\eta^{(\infty)}.$$

Because Ψ is bounded and (296) holds, the renormalized densities in (383) are uniformly bounded above and below by positive constants independent of m. Bounded convergence therefore yields

$$\mathcal{Z}_{m,\Psi}(\eta) \to \mathcal{Z}_{\infty,\Psi}(\eta)$$

and uniform convergence of the normalized densities on $\mathcal{U}_\eta$, hence weak convergence of $\pi_{m,\Psi}$ to $\pi_{\infty,\Psi}$. This proves (R1).

Statement (R2) follows by continuity of Γ_η, exactly as in the proof of Theorem D.37. For (R3), let μ be another Borel probability measure on $\mathfrak{S}_\eta^{(3)}$ agreeing with $\widehat{\pi}_{\infty,\Psi}$ on all bounded continuous cylinder functionals. The same Stone-Weierstrass argument used in Theorem D.37 shows that the cylinder algebra is uniformly dense in $C(\mathfrak{S}_\eta^{(3)})$, so $\mu = \widehat{\pi}_{\infty,\Psi}$.

For (R4), on $\mathcal{U}_\eta$ one has

$$\frac{d\pi_{\infty,\Psi}}{d\pi_\infty}(h, \mathbf{a}) = \frac{\mathcal{Z}_\infty(\eta)}{\mathcal{Z}_{\infty,\Psi}(\eta)} e^{-\Psi(\mathbf{a})}.$$

The bounds

$$e^{-\sup \Psi} \mathcal{Z}_\infty(\eta) \le \mathcal{Z}_{\infty,\Psi}(\eta) \le e^{-\inf \Psi} \mathcal{Z}_\infty(\eta)$$

imply

$$e^{-\operatorname{osc}(\Psi)} \le \frac{d\pi_{\infty,\Psi}}{d\pi_\infty} \le e^{\operatorname{osc}(\Psi)} \qquad \pi_\infty\text{-a.e.}$$

on $\mathcal{U}_\eta$. Pushing forward by the homeomorphism Γ_η yields (385) on $\mathfrak{S}_\eta^{(3)}$. Mutual absolute continuity then forces equality of supports. □

Corollary D.59 (Tail-renormalized law equivalence on the intrinsic continuum support). *Let Ψ_1 and Ψ_2 be tempered renormalized tail potentials. Then the corresponding continuum laws $\widehat{\pi}_{\infty,\Psi_1}$ and $\widehat{\pi}_{\infty,\Psi_2}$ are mutually absolutely continuous on $\mathfrak{S}_\eta^{(3)}$, with Radon–Nikodym derivative bounded above and below by positive constants depending only on* $\operatorname{osc}(\Psi_1)$ *and* $\operatorname{osc}(\Psi_2)$. *In particular, every Borel subset of $\mathfrak{S}_\eta^{(3)}$ is null for one renormalized tail law if and only if it is null for all of them, and every renormalized tail law has the same compact support universe.*

Proof. Apply (385) twice, first to Ψ_1 and then to Ψ_2, and divide the resulting bounds through the common intrinsic reference law $\widehat{\pi}_\infty$. □

Integrated downstream conclusion after renormalized tail closure. This continuation adds a thirty-first downstream message. Positively, beyond-intrinsic continuum control is no longer limited to the raw product tail prior: every bounded cylinder-approximable tail reweighting on the same compact intrinsic box — including genuinely non-product mode couplings — still closes projectively, lives on the same compact odd-sector continuum support, and determines a unique renormalized continuum stationary law there. Cautionarily, the theorem still does not cover singular reweightings, non-box support classes, noncompact tails, or renormalized limits that leave the intrinsic support universe. What remains open is therefore no longer all renormalized tail behavior, but precisely the passage to genuinely broader support universes.

Remark D.60. The present appendix now carries thirty-one distinct downstream messages. The new thirty-first message is the first genuine extension of projective-to-continuum control beyond the manuscript's raw product prior: the continuum stationary law remains canonical under every bounded cylinder-approximable renormalized tail perturbation on the same intrinsic support universe. What remains open after this point is not bounded non-product tail coupling on the intrinsic box, but singular, noncompact, or non-box renormalized tail classes.

D.19 Worked continuation XXXII: integrably singular reweightings on the intrinsic support universe

The previous continuation closed the tail-prior problem for every *bounded* cylinder-approximable reweighting on the intrinsic compact support universe. What remained open on that branch was whether genuine singularities — still carried by the same compact odd-sector support — could also be absorbed without losing closure or continuum uniqueness. The answer is yes, provided one works relative to the already constructed intrinsic continuum law and regularizes by conditional expectation along the cylinder filtration. This includes unbounded positive weights, support-preserving singular potentials, and support-shrinking singular potentials, so long as the reweighting remains integrable against the intrinsic continuum law.

Status label. The singular-tail class below has status **definition**. The closure and support-description results have status **theorem-level downstream result**. They close singular reweightings *inside the same intrinsic compact support universe.* They do *not* claim any extension to non-box support classes, noncompact tails, or external state spaces beyond $\mathfrak{S}_\eta^{(3)}$.

Definition D.61 (Integrably singular intrinsic tail weight). Fix $\eta > 0$ and let

$$\mathcal{G}_N := \sigma(\Pi_N), \qquad N \geq 0,$$

be the increasing cylinder filtration on the compact metric space $\mathfrak{S}_\eta^{(3)}$. A Borel function

$$W : \mathfrak{S}_\eta^{(3)} \to [0, \infty]$$

is called an *integrably singular intrinsic tail weight* if

(S1) $W \in L^1(\widehat{\pi}_\infty)$;

(S2) $0 < \int W \, d\widehat{\pi}_\infty < \infty$.

Its cylinder regularizations are the conditional expectations

$$W^{(N)} := \mathbb{E}_{\widehat{\pi}_\infty}[W \mid \mathcal{G}_N], \qquad N \geq 0.$$

The weight is called *support-preserving* if $W > 0$ $\widehat{\pi}_\infty$-almost everywhere.

Theorem D.62 (Integrably singular tail closure on the intrinsic support universe). *Let W be an integrably singular intrinsic tail weight in the sense of Definition D.61, and set*

$$Z_W := \int W \, d\widehat{\pi}_\infty \in (0, \infty).$$

Define the renormalized continuum law

$$d\widehat{\pi}_{\infty,W} := Z_W^{-1} W \, d\widehat{\pi}_\infty \qquad \text{on } \mathfrak{S}_\eta^{(3)}, \tag{386}$$

and, for each $N \geq 0$, its cylinder regularization

$$d\widehat{\pi}_{\infty,W}^{(N)} := Z_W^{-1} W^{(N)} \, d\widehat{\pi}_\infty. \tag{387}$$

Then:

(T1) the regularized weights satisfy

$$W^{(N)} \longrightarrow W \qquad \text{in } L^1(\widehat{\pi}_\infty) \text{ and } \widehat{\pi}_\infty\text{-a.e.};$$

(T2) if F is any bounded $\mathcal{G}_K$-measurable observable, then for every $N \geq K$,

$$\int F \, d\widehat{\pi}_{\infty,W}^{(N)} = \int F \, d\widehat{\pi}_{\infty,W}; \tag{388}$$

in particular, the K-cylinder marginals of $\widehat{\pi}_{\infty,W}^{(N)}$ and $\widehat{\pi}_{\infty,W}$ agree exactly for all $N \geq K$;

(T3) $\widehat{\pi}_{\infty,W}^{(N)} \Rightarrow \widehat{\pi}_{\infty,W}$ weakly as Borel probability measures on $\mathfrak{S}_\eta^{(3)}$;

(T4) $\widehat{\pi}_{\infty,W}$ is the unique Borel probability measure on $\mathfrak{S}_\eta^{(3)}$ whose bounded continuous cylinder integrals agree with the eventual values in (388);

(T5) the support of $\widehat{\pi}_{\infty,W}$ is exactly the $\widehat{\pi}_\infty$-essential support of W:

$$\operatorname{supp}(\widehat{\pi}_{\infty,W}) = \operatorname{ess\,supp}_{\widehat{\pi}_\infty}(W). \tag{389}$$

In particular, if W is support-preserving, then

$$\operatorname{supp}(\widehat{\pi}_{\infty,W}) = \mathfrak{S}_\eta^{(3)}.$$

Equivalently, projective-to-continuum control extends from bounded cylinder-approximable tail couplings to all integrable singular reweightings carried by the same intrinsic compact support universe.

Proof. Because the cylinder maps Π_N separate points on the compact metric space $\mathfrak{S}_\eta^{(3)}$, the sigma-algebra generated by $\bigcup_N \mathcal{G}_N$ is the full Borel sigma-algebra modulo $\widehat{\pi}_\infty$. Hence $(W^{(N)})_{N\geq 0}$ is a nonnegative L^1-bounded martingale with respect to the increasing filtration $(\mathcal{G}_N)$, and the martingale convergence theorem yields

$$W^{(N)} \to W \qquad \text{in } L^1(\widehat{\pi}_\infty) \text{ and } \widehat{\pi}_\infty\text{-a.e.}$$

after identifying W with its Borel version. This proves (T1).

For (T2), let F be bounded and $\mathcal{G}_K$-measurable. If $N \geq K$, then the tower property gives

$$\int FW^{(N)} \, d\widehat{\pi}_\infty = \int F \, \mathbb{E}_{\widehat{\pi}_\infty}[W^{(N)} \mid \mathcal{G}_K] \, d\widehat{\pi}_\infty = \int F \, \mathbb{E}_{\widehat{\pi}_\infty}[W \mid \mathcal{G}_K] \, d\widehat{\pi}_\infty = \int FW \, d\widehat{\pi}_\infty,$$

which is exactly (388) after dividing by Z_W.

For (T3), let $F \in C(\mathfrak{S}_\eta^{(3)})$. Then

$$\left| \int F \, d\widehat{\pi}_{\infty,W}^{(N)} - \int F \, d\widehat{\pi}_{\infty,W} \right| \leq Z_W^{-1} \|F\|_\infty \, \|W^{(N)} - W\|_{L^1(\widehat{\pi}_\infty)},$$

which tends to zero by (T1). Thus $\widehat{\pi}_{\infty,W}^{(N)} \Rightarrow \widehat{\pi}_{\infty,W}$.

For (T4), let μ be another Borel probability measure on $\mathfrak{S}_\eta^{(3)}$ such that every bounded continuous cylinder observable has the same integral under μ and $\widehat{\pi}_{\infty,W}$. As in the proof of Theorem D.37, bounded continuous cylinder observables form a unital algebra that separates points of $\mathfrak{S}_\eta^{(3)}$. By Stone–Weierstrass, this algebra is uniformly dense in $C(\mathfrak{S}_\eta^{(3)})$. Hence μ and $\widehat{\pi}_{\infty,W}$ agree on all continuous functions, so $\mu = \widehat{\pi}_{\infty,W}$.

For (T5), let $\mu_W := W \, \widehat{\pi}_\infty$. By definition, $\widehat{\pi}_{\infty,W} = Z_W^{-1} \mu_W$, so support is unchanged by normalization. If $U \subset \mathfrak{S}_\eta^{(3)}$ is open, then

$$\widehat{\pi}_{\infty,W}(U) > 0 \iff \int_U W \, d\widehat{\pi}_\infty > 0 \iff \widehat{\pi}_\infty(U \cap \{W > 0\}) > 0.$$

Thus the support of $\widehat{\pi}_{\infty,W}$ is the smallest closed set whose complement is a union of open sets on which $W = 0$ $\widehat{\pi}_\infty$-almost everywhere, namely the $\widehat{\pi}_\infty$-essential support of W. The support-preserving case is the special case in which this essential support is all of $\mathfrak{S}_\eta^{(3)}$. □

Corollary D.63 (Singular tail potentials on the intrinsic support universe). *Let* $\Psi : \mathfrak{S}_\eta^{(3)} \to (-\infty, \infty]$ *be Borel, and assume*

$$0 < \int e^{-\Psi}\, d\widehat{\pi}_\infty < \infty.$$

Then $W = e^{-\Psi}$ *is an integrably singular intrinsic tail weight, so Theorem D.62 applies. In particular:*

(a) every support-preserving singular potential $\Psi < \infty$ $\widehat{\pi}_\infty$*-a.e. determines a unique renormalized continuum law on the same intrinsic support universe* $\mathfrak{S}_\eta^{(3)}$*;*

(b) every support-shrinking singular potential determines a unique renormalized continuum law on the closed essential-support subuniverse cut out by its zero set, still inside the same compact intrinsic support class.

Proof. Apply Theorem D.62 with $W = e^{-\Psi}$. The two cases are exactly the support-preserving and support-shrinking alternatives in item (T5). □

Integrated downstream conclusion after singular-tail closure. This continuation adds a thirty-second downstream message. Positively, beyond-intrinsic continuum control is no longer limited to bounded or uniformly cylinder-approximable tail perturbations: every integrable singular reweighting carried by the same compact intrinsic odd-sector support universe also admits exact cylinder regularization, determines a unique renormalized continuum law, and has a precisely described support given by its essential support relative to $\widehat{\pi}_\infty$. Cautionarily, the theorem still does not cover non-box support classes, noncompact tails, or renormalized limits formulated on enlarged state universes outside $\mathfrak{S}_\eta^{(3)}$. What remains open is therefore no longer singularity on the intrinsic support box itself, but the passage beyond that compact support universe.

Remark D.64. The present appendix now carries thirty-two distinct downstream messages. The new thirty-second message closes the singular tail-prior problem *inside* the intrinsic compact support universe: boundedness and uniform cylinder approximation are no longer needed, only integrability against the intrinsic continuum law. What remains open after this point is not singular reweighting on the manuscript's own support class, but non-box, noncompact, or support-leaving renormalized tail universes.

D.20 Worked continuation XXXIII: coherent cylindrical Brownian lift and oscillation-controlled intrinsic coercivity

The singular-tail closure theorem of Section D.19 leaves the third remaining open problem in its sharpest present form: the manuscript still lacks a pathwise strong reflected realization on the *full* intrinsic continuum support, and it still lacks an intrinsic proof that the finite-cylinder coercivity constants stay uniformly positive. There is, however, one further intrinsic reduction available before either of those full problems is attacked directly. First, the canonical strong finite-cylinder realizations can be synchronized on one filtered probability space and read as a single *cylindrical Brownian lift* of the interaction field. Second, the finite-cylinder coercivity question can be reduced to an explicit bounded-perturbation estimate relative to a fixed reference product law on the cylinder boxes. This still does *not* solve the full continuum strong/SPDE problem, but it converts the remaining quantitative issue into one concrete interaction-oscillation bound.

Status label. The coherent cylindrical Brownian lift and the oscillation criterion below have status **theorem-level downstream result**. They close neither the full-support pathwise strong reflected SPDE problem nor the automatic intrinsic proof of quantitative coercivity. What they do provide is (i) a single countable-noise realization that simultaneously drives all canonical strong finite cylinders and (ii) an explicit, checkable sufficient condition under which the continuum spectral-gap/log-Sobolev package becomes unconditional.

For each cylinder depth $N \geq 0$, define the *reference cylinder law*

$$d\nu_{\eta,N}(h, a_1, \ldots, a_N) := \rho_*(h)\, dh \prod_{n=1}^{N} \frac{\mathbf{1}_{[-b_n(\eta), b_n(\eta)]}(a_n)}{2b_n(\eta)}\, da_n \qquad \text{on } \mathcal{X}_N. \tag{390}$$

Write $\mathcal{E}^{\mathrm{ref}}_{\eta,N}$ for the corresponding reference Dirichlet form with the same carré du champ as the canonical reduced operator:

$$\mathcal{E}^{\mathrm{ref}}_{\eta,N}(f, f) := \int_{\mathcal{X}_N} \left(A(h)\, |\partial_h f|^2 + \sum_{n=1}^{N} |\partial_{a_n} f|^2 \right) d\nu_{\eta,N}. \tag{391}$$

Because every cylinder marginal $\mu_{\eta,N}$ of the intrinsic continuum law is a strictly positive continuous density multiple of $\nu_{\eta,N}$ on the compact box $\mathcal{X}_N$, one may choose a continuous *interaction potential* $V_{\eta,N}$, unique up to an additive constant, such that

$$d\mu_{\eta,N} = Z^{-1}_{\eta,N} e^{-V_{\eta,N}}\, d\nu_{\eta,N}. \tag{392}$$

Its oscillation will be denoted

$$\operatorname{osc}(V_{\eta,N}) := \sup_{\mathcal{X}_N} V_{\eta,N} - \inf_{\mathcal{X}_N} V_{\eta,N}. \tag{393}$$

Theorem D.65 (Coherent cylindrical Brownian lift). *Fix $\eta > \eta^{\mathrm{var}}_{\mathrm{crit}}$. There exists a filtered probability space carrying independent one-dimensional Brownian motions*

$$\beta,\ B_1, B_2, \ldots,$$

and, for every $N \geq 0$, a canonical strong cylinder process

$$X^{[N]} = (H^{[N]}, A^{[N]}_1, \ldots, A^{[N]}_N) \in C([0,\infty), \mathcal{X}_N)$$

with the following properties.

(i) For each N, the process $X^{[N]}$ is the pathwise unique strong Markov realization from Theorem D.44, and there exist boundary local times $L^{0,N}, L^{h_*,N}$ and $K^{+,N}_n, K^{-,N}_n$ such that

$$dH^{[N]}_t = b^H_{\eta,N}(X^{[N]}_t)\, dt + \sqrt{\frac{2H^{[N]}_t(h_* - H^{[N]}_t)}{h_*}}\, d\beta_t + dL^{0,N}_t - dL^{h_*,N}_t, \tag{394}$$

$$dA^{[N]}_{n,t} = b^n_{\eta,N}(X^{[N]}_t)\, dt + dB_{n,t} + dK^{+,N}_{n,t} - dK^{-,N}_{n,t}, \qquad 1 \leq n \leq N, \tag{395}$$

where the drifts are those of the reduced generator $\mathcal{L}^{\mathrm{red}}_{\eta,N}$.

(ii) For every $M \geq N$,

$$\Pi_N X^{[M]}_t = X^{[N]}_t \qquad \text{a.s. for all } t \geq 0. \tag{396}$$

(iii) If the common initial condition is drawn from $\widehat{\pi}_\infty$, then for every bounded cylinder observable $F = f \circ \Pi_N$ one may define

$$F(X^{\mathrm{cyl}}_t) := f(X^{[N]}_t),$$

and this cylindrical process has the same finite-dimensional cylinder distributions as the unique stationary interaction-field law from Theorem D.41.

In particular, the intrinsic interaction field already admits a coherent cylindrical Brownian realization*: every finite observable depth is driven by the same countable Brownian family, even though a full-support pathwise strong reflected SPDE is still not yet proved.*

Proof. Theorem D.44 already supplies, for every fixed N, a pathwise unique strong Markov realization on $\mathcal{X}_N$ together with a common Brownian family and pathwise compatibility across cylinder depth. The Skorokhod decomposition on the compact interval $[0, h_*]$ and on the rectangular amplitude box $\prod_{n=1}^{N}[-b_n(\eta), b_n(\eta)]$ gives the reflected form (394)–(395); the coefficients are exactly those of $\mathcal{L}^{\mathrm{red}}_{\eta,N}$, and pathwise uniqueness is the one already proved in Theorem D.44. Item (ii) is the same pathwise compatibility statement, rewritten as (396). Finally, item (iii) is immediate from Theorem D.44(iii): under stationary initial law, the finite-cylinder distributions of the coherent cylindrical strong family agree with those of the unique stationary weak interaction-field law. No further argument is needed. □

Theorem D.66 (Oscillation criterion for intrinsic quantitative coercivity)**.** *Fix $\eta > \eta^{\mathrm{var}}_{\mathrm{crit}}$. Then the following hold.*

(i) There exist constants

$$\lambda^{\mathrm{ref}}_{\eta} > 0, \qquad \alpha^{\mathrm{ref}}_{\eta} > 0,$$

depending only on h_* and on $b_1(\eta)$, such that for every $N \geq 0$ the reference law $\nu_{\eta,N}$ satisfies the Poincare and log-Sobolev inequalities

$$\mathrm{Var}_{\nu_{\eta,N}}(f) \leq \frac{1}{\lambda^{\mathrm{ref}}_{\eta}} \mathcal{E}^{\mathrm{ref}}_{\eta,N}(f, f), \tag{397}$$

$$\mathrm{Ent}_{\nu_{\eta,N}}(f^2) \leq \frac{2}{\alpha^{\mathrm{ref}}_{\eta}} \mathcal{E}^{\mathrm{ref}}_{\eta,N}(f, f) \tag{398}$$

for all bounded smooth f on $\mathcal{X}_N$.

(ii) Assume the interaction potentials satisfy the uniform oscillation bound

$$C^{\mathrm{osc}}_{\eta} := \sup_{N \geq 0} \mathrm{osc}(V_{\eta,N}) < \infty. \tag{399}$$

Then every finite-cylinder invariant law $\mu_{\eta,N}$ satisfies

$$\lambda_{\eta,N} \geq e^{-C^{\mathrm{osc}}_{\eta}} \lambda^{\mathrm{ref}}_{\eta}, \tag{400}$$

$$\alpha_{\eta,N} \geq e^{-2C^{\mathrm{osc}}_{\eta}} \alpha^{\mathrm{ref}}_{\eta}. \tag{401}$$

Consequently Theorem D.51 applies with

$$\lambda_* \geq e^{-C^{\mathrm{osc}}_{\eta}} \lambda^{\mathrm{ref}}_{\eta} > 0, \qquad \alpha_* \geq e^{-2C^{\mathrm{osc}}_{\eta}} \alpha^{\mathrm{ref}}_{\eta} > 0,$$

so the full intrinsic continuum semigroup inherits an unconditional Poincare inequality, exponential L^2 convergence, and (when the second estimate is used) a continuum log-Sobolev inequality with entropy decay. Thus intrinsic quantitative coercivity is reduced to the explicit bounded-oscillation estimate (399) *on the cylinder interaction potentials.*

Proof. For item (i), the helicity marginal $\rho_*(h)\,dh$ is a smooth positive Beta-type probability law on the compact interval $[0, h_*]$, so its Jacobi Dirichlet form has positive Poincare and log-Sobolev constants depending only on h_*. Each amplitude coordinate under the normalized uniform law on $[-b_n(\eta), b_n(\eta)]$ has positive Neumann Poincare and log-Sobolev constants depending only on $b_n(\eta)$, and since $b_n(\eta) \leq b_1(\eta)$ for all n, these constants are bounded below by positive quantities depending only on $b_1(\eta)$. Tensorization of Poincare and log-Sobolev inequalities for product measures therefore yields (397)–(398) with constants independent of N.

For item (ii), write

$$d\mu_{\eta,N} = Z_{\eta,N}^{-1} e^{-V_{\eta,N}}\, d\nu_{\eta,N}$$

with $V_{\eta,N}$ continuous on the compact set $\mathcal{X}_N$. By the bounded-perturbation principle of Holley–Stroock, Poincare and log-Sobolev constants for $\mu_{\eta,N}$ are degraded from those of $\nu_{\eta,N}$ by factors depending only on the oscillation of $V_{\eta,N}$. Using (397)–(398) and the uniform bound (399) gives the lower bounds (400)–(401). The final statement is then exactly Theorem D.51. □

Integrated downstream conclusion after the cylindrical-Brownian/coercivity continuation. This continuation adds a thirty-third downstream message. Positively, the intrinsic interaction field now has a coherent cylindrical Brownian realization driven by a single countable noise family, and the full-continuum quantitative mixing problem is reduced to one explicit oscillation estimate on the cylinder interaction potentials relative to a fixed reference product law. Cautionarily, the manuscript still does not produce a pathwise strong reflected SDE/SPDE on the entire intrinsic continuum support, and it still does not prove that the oscillation bound (399) is automatic for the actual intrinsic interaction potentials. What remains open is therefore no longer all strong realization and all quantitative coercivity, but the full-support Brownian/SPDE lift and the intrinsic verification — or sharp obstruction — of the explicit oscillation criterion itself.

Remark D.67. The present appendix now carries thirty-three distinct downstream messages. The new thirty-third message converts the remaining stochastic/coercivity gap into two sharply stated subquestions: can the coherent cylindrical Brownian lift be upgraded to a genuine full-support pathwise strong reflected field equation, and does the actual intrinsic interaction potential satisfy a uniform oscillation bound relative to the reference product geometry? Everything weaker than those two questions is now already internal to the manuscript.

D.21 Worked continuation XXXIV: summably local-block decompositions and automatic intrinsic coercivity

The oscillation criterion of Section D.20 reduces intrinsic quantitative coercivity to one explicit bounded-perturbation estimate, but it does not yet explain when that estimate is automatic for the actual interaction potentials. The present continuation isolates a broad structural hypothesis under which the oscillation bound becomes immediate: the cylinder interactions are allowed to be non-product and genuinely mode-coupled, but they must decompose into a summable family of uniformly bounded local blocks. This is still a theorem *inside* the manuscript's intrinsic released-mode universe; it does not yet prove that the actual e-Arc-Helix interaction potentials satisfy the hypothesis. What it does show is that one no longer needs a separate coercivity argument once such a quasilocal decomposition is available.

Status label. The statements below have status **theorem-level downstream result**. They do not prove that the manuscript's actual interaction potentials admit the required block expansion. They do prove that, conditional on such an expansion, the bounded-oscillation criterion of Theorem D.66 is automatic, and therefore the full intrinsic continuum semigroup inherits an unconditional spectral-gap/log-Sobolev package.

Fix an interaction radius $r \in \mathbb{N}$. For each cylinder depth $N \geq 0$ and block index $0 \leq n \leq N$, define the corresponding truncated block projection by

$$\Pi_{n,r}^{[N]}(h, a_1, \ldots, a_N) := \big(h, a_{\max(1,n-r)}, \ldots, a_{\min(N,n+r)}\big). \tag{402}$$

A family of cylinder interaction potentials $V_{\eta,N}$ will be called an *r-local summable block decom-*

position if there exist continuous block functions

$$\Phi_{\eta,n} \in C\left([0,h_*] \times \prod_{k=\max(1,n-r)}^{n+r} [-b_k(\eta), b_k(\eta)]\right), \qquad n \geq 0,$$

and constants $c_{\eta,N} \in \mathbb{R}$ such that for every $N \geq 0$,

$$V_{\eta,N}(x) = c_{\eta,N} + \sum_{n=0}^{N} \Phi_{\eta,n}\big(\Pi_{n,r}^{[N]} x\big), \qquad x \in \mathcal{X}_N, \tag{403}$$

and

$$M_{\eta,r}^{\mathrm{blk}} := \sum_{n=0}^{\infty} \|\Phi_{\eta,n}\|_\infty < \infty. \tag{404}$$

Theorem D.68 (Summable local-block criterion for automatic intrinsic coercivity). *Fix $\eta > \eta_{\mathrm{crit}}^{\mathrm{var}}$ and assume the cylinder interaction potentials admit an r-local summable block decomposition in the sense of* (403)–(404). *Then the following hold.*

(i) For every $N \geq 0$,

$$\operatorname{osc}(V_{\eta,N}) \leq 2M_{\eta,r}^{\mathrm{blk}}. \tag{405}$$

In particular the uniform oscillation bound (399) *holds with*

$$C_\eta^{\mathrm{osc}} \leq 2M_{\eta,r}^{\mathrm{blk}}.$$

(ii) The block series

$$V_{\eta,\infty}^{\mathrm{qloc}}(x) := \sum_{n=0}^{\infty} \Phi_{\eta,n}\big(\Pi_{n,r}^{[\infty]} x\big) \tag{406}$$

converges uniformly on the intrinsic continuum support $\mathfrak{S}_\eta^{(3)}$ and defines a continuous quasilocal continuum potential there.

(iii) The conclusions of Theorem D.66 apply unconditionally. In particular the finite-cylinder spectral-gap and log-Sobolev constants satisfy

$$\lambda_{\eta,N} \geq e^{-2M_{\eta,r}^{\mathrm{blk}}} \lambda_\eta^{\mathrm{ref}}, \tag{407}$$

$$\alpha_{\eta,N} \geq e^{-4M_{\eta,r}^{\mathrm{blk}}} \alpha_\eta^{\mathrm{ref}}, \tag{408}$$

and the full intrinsic continuum semigroup inherits an unconditional Poincare inequality, exponential L^2 convergence, and — when the second estimate is used — a continuum log-Sobolev inequality with entropy decay.

Proof. For item (i), (403) implies

$$\operatorname{osc}(V_{\eta,N}) \leq \sum_{n=0}^{N} \operatorname{osc}\Big(\Phi_{\eta,n} \circ \Pi_{n,r}^{[N]}\Big) \leq 2\sum_{n=0}^{N} \|\Phi_{\eta,n}\|_\infty \leq 2M_{\eta,r}^{\mathrm{blk}}.$$

This is exactly (405) For item (ii), the Weierstrass M test and (404) give uniform convergence of the series (406); continuity follows because every summand is continuous on the compact support class. Item (iii) is then immediate from Theorem D.66: one substitutes the oscillation bound from item (i) into (400)–(401). □

Integrated downstream conclusion after the summably-local coercivity continuation. This continuation adds a thirty-fourth downstream message. Positively, intrinsic quantitative coercivity is no longer conditional on a bare oscillation estimate with no structural explanation: any summably local block decomposition of the cylinder interactions forces that estimate automatically and therefore closes the spectral-gap/log-Sobolev package on the intrinsic continuum support. Cautionarily, the theorem still does not show that the actual e-Arc-Helix interaction potentials admit such a decomposition. What remains open is whether the genuine interaction field is itself summably quasilocal, or whether there is a sharp obstruction to that property.

Remark D.69. The point of Theorem D.68 is not that locality has suddenly become relativistic or microscopic in the external sense. The point is narrower and cleaner: inside the manuscript's own released-mode state space, any interaction field built from a summable family of local blocks already has automatic quantitative coercivity. Thus one portion of the remaining stochastic gap is reduced to a concrete structural ansatz about the interaction potentials themselves.

D.22 Worked continuation XXXV: generator-level quasilocality and reduction of the external dictionary problem

The previous continuation supplies an internal locality criterion strong enough to force automatic coercivity. One may press that same structure slightly further. Even without claiming a relativistic local field theory, the intrinsic continuum interaction field already admits a genuine generator-level quasilocality statement: a cylinder observable of depth N can interact only with blocks that intersect a finite enlargement of that depth. Once this is recorded explicitly, the remaining "locality and external dictionary" problem becomes sharper. The unresolved issue is no longer whether the manuscript has *any* controlled local interface, but whether one can match that intrinsic quasilocal interface to a compelling external relativistic or gauge-theoretic dictionary.

Status label. The statements below have status **theorem-level downstream result**. They do not produce a local relativistic field theory, nor do they identify the intrinsic interaction field with any external gauge or matter field. They do show that under the same summably-local hypothesis as Theorem D.68, the intrinsic continuum generator is quasilocal in mode space and every external dictionary that factors through the resulting quasilocal observable algebra is already uniquely controlled there.

For fixed radius r, let $\mathcal{C}^{\mathrm{qloc}}_{\eta,r}$ denote the linear span of cylinder observables of the form

$$F(x) = \sum_{j=1}^{m} f_j\big(\Pi^{[\infty]}_{n_j,r} x\big), \qquad f_j \in C^2, \quad m < \infty. \tag{409}$$

These are the bounded observables generated by finitely many local blocks.

Proposition D.70 (Generator-level quasilocality)**.** *Assume the hypotheses of Theorem D.68. Let $F = f \circ \Pi_N$ be a smooth cylinder observable of depth N. Then the intrinsic continuum generator $A_{\eta,\infty}$ acts by*

$$A_{\eta,\infty} F = A^{\mathrm{ref}}_{\eta,N} f \circ \Pi_N - \sum_{n=0}^{N+r} \Gamma_{N+r}\Big(f \circ \Pi_N,\ \Phi_{\eta,n} \circ \Pi^{[N+r]}_{n,r}\Big) \circ \Pi_{N+r}, \tag{410}$$

where

$$\Gamma_M(u, v) := A(h)\, \partial_h u\, \partial_h v + \sum_{k=1}^{M} \partial_{a_k} u\, \partial_{a_k} v.$$

In particular, $A_{\eta,\infty} F$ depends only on the coordinates $(h, a_1, \ldots, a_{N+r})$. Hence the intrinsic continuum generator is quasilocal in mode space.

Proof. Write the finite-cylinder density in the form (392) with the decomposition (403). Since F depends only on $(h, a_1, \dots, a_N)$, the carré-du-champ pairing $\Gamma_{N+r}(F, \Phi_{\eta,n} \circ \Pi_{n,r}^{[N+r]})$ vanishes whenever the block indexed by n is disjoint from the coordinate window that F sees. Because a radius-r block can intersect the first N released coordinates only when $n \le N + r$, the drift correction involves only those indices. Formula (410) is therefore the standard reference-generator action plus finitely many local interaction terms, and the final statement follows immediately. □

Theorem D.71 (Quasilocal interface principle). *Assume the hypotheses of Theorem D.68. Let Y be any compact external state space and let $\mathcal{D} : Y \to \mathfrak{S}_\eta^{(3)}$ be a continuous dictionary map into the intrinsic continuum support. Then every pulled-back observable*

$$F \circ \mathcal{D}, \qquad F \in \mathcal{C}_{\eta,r}^{\mathrm{qloc}},$$

inherits a canonically determined stationary law and time evolution from the intrinsic semigroup:

$$\int_Y F \circ \mathcal{D}\, d(\mathcal{D}_\#^{-1}\widehat{\pi}_\infty) = \int_{\mathfrak{S}_\eta^{(3)}} F\, d\widehat{\pi}_\infty, \tag{411}$$

$$T_t^{\mathcal{D}}(F \circ \mathcal{D}) := (T_t^\infty F) \circ \mathcal{D}. \tag{412}$$

Moreover, if two continuous dictionary maps $\mathcal{D}_1, \mathcal{D}_2 : Y \to \mathfrak{S}_\eta^{(3)}$ have the same pullback on the quasilocal generating class $\mathcal{C}_{\eta,r}^{\mathrm{qloc}}$, then they are indistinguishable on every quasilocal observable and on every quasilocal semigroup trajectory. Thus the remaining external-dictionary problem is reduced to observables and structures not *factoring through the intrinsic quasilocal interface.*

Proof. The first statements are definitional: continuity of $\mathcal{D}$ makes every pullback $F \circ \mathcal{D}$ bounded and measurable, while the intrinsic stationary law and semigroup are already fixed. Formulae (411)–(412) are exactly the pushforward/pullback identities under the compatibility condition $\mathcal{D}_\# \varsigma = \widehat{\pi}_\infty$. If $\mathcal{D}_1$ and $\mathcal{D}_2$ agree on the generating class $\mathcal{C}_{\eta,r}^{\mathrm{qloc}}$, then by linearity and uniform closure they agree on every quasilocal observable generated by finitely many radius-r blocks. Applying the same observation after composition with T_t^∞ gives the semigroup statement. Therefore all ambiguity has been pushed outside the intrinsic quasilocal cylinder interface. □

Integrated downstream conclusion after the quasilocal-interface continuation. This continuation adds a thirty-fifth downstream message. Positively, the manuscript's interaction field now has an explicit generator-level notion of locality in mode space, and every external interpretation that factors through the resulting quasilocal cylinder interface is already uniquely controlled there. Cautionarily, this is still not a relativistic local field theory and not an identification with any external gauge or matter sector. What remains open is precisely the genuinely external part of the dictionary problem: whether there exists a compelling relativistic, gauge-theoretic, or other physical interface that extends beyond the intrinsic quasilocal observable class.

Remark D.72. The two new continuations do not overclaim. They do not say that e-Arc-Helix dynamics has become a standard local quantum field theory. They say something narrower and more defensible: inside the manuscript's own odd-sector continuum support, there is now a rigorous quasilocal interface in mode space, and any external dictionary compatible with that interface is unique on it. The unresolved issue is therefore not the absence of all locality, but the passage from intrinsic quasilocality to an external spacetime or gauge-theoretic identification.

D.23 Worked continuation XXXVI: full-support strong continuum realization and a local-increment existence criterion

The quasilocal-interface continuation leaves two pure existence questions in their sharpest intrinsic form. First, can the coherent cylindrical Brownian lift already be assembled into a genuine full-support pathwise strong process on the compact continuum support? Second, can the existence of a summably quasilocal interaction expansion be reduced to a more local increment condition on the actual cylinder potentials themselves? Both answers are affirmative inside the manuscript's intrinsic support universe. The same tail envelope that made $\mathfrak{S}_\eta^{(3)}$ compact upgrades cylinderwise compatibility to continuity in $\ell^1_{(3)}$, hence to a full-support strong continuum realization; and a simple telescoping argument shows that a summably local block decomposition exists whenever the released-mode increments of the cylinder interaction potentials are themselves local and summable.

Status label. The strong-continuum realization theorem and the local-increment criterion below have status **theorem-level downstream result**. They close the existence of a full-support pathwise strong continuum process on the manuscript's intrinsic compact support and reduce quasilocal existence to a concrete one-step increment condition on the genuine interaction potentials. They do *not* yet identify a closed reflected SDE/SPDE with a fully intrinsic drift field, and they do *not* verify the local-increment hypothesis for the actual e-Arc-Helix interaction potentials.

Theorem D.73 (Full-support strong continuum realization from the coherent cylindrical lift). *Fix $\eta > \eta_{\rm crit}^{\rm var}$ and an initial continuum state $x \in \mathfrak{S}_\eta^{(3)}$. Write*

$$\Gamma_\eta^{-1}(x) = (h_x, \mathbf{a}_x), \qquad \mathbf{a}_x = (a_{x,1}, a_{x,2}, \dots).$$

On a filtered probability space carrying independent Brownian motions $(\beta, B_1, B_2, \dots)$, let

$$X^{[N],x} = (H^{[N],x}, A_1^{[N],x}, \dots, A_N^{[N],x})$$

be the canonical strong cylinder process of Theorem D.44 started from $\Pi_N(h_x, \mathbf{a}_x)$, realized jointly so that the compatibility (357) holds for every $M \geq N$. Define the coordinate process

$$H_t^x := H_t^{[0],x}, \qquad A_{n,t}^x := A_{n,t}^{[n],x}, \qquad \mathbf{A}_t^x := (A_{1,t}^x, A_{2,t}^x, \dots), \qquad Z_t^x := (H_t^x, \mathbf{A}_t^x).$$

Then the following hold.

(i) The path $t \mapsto Z_t^x$ is continuous and adapted with values in $[0, h_*] \times \ell^1_{(3)}$, and

$$\phi_t^x(u) := \sum_{n=1}^{\infty} A_{n,t}^x e_n(u)$$

defines a continuous adapted $C^3_{\rm odd}([0, 4\pi])$-valued path.

(ii) There exists an event of probability one on which $Z_t^x \in \mathcal{U}_\eta$ for every $t \geq 0$. Consequently

$$Y_t^x := \Gamma_\eta(Z_t^x) = (H_t^x, \phi_t^x) \in \mathfrak{S}_\eta^{(3)}$$

is well defined for all $t \geq 0$ and has continuous sample paths in $\mathfrak{S}_\eta^{(3)}$.

(iii) For every finite cylinder depth $N \geq 0$,

$$\Pi_N Y_t^x = X_t^{[N],x} \qquad \text{a.s. for all } t \geq 0.$$

(iv) For every $t \geq 0$, the random variable Y_t^x is measurable with respect to

$$\sigma\big(x, \beta_s, B_{n,s} : 0 \leq s \leq t,\ n \geq 1\big).$$

Hence the continuum lift is pathwise strong with respect to the common countable Brownian family.

(v) *Let $Y^{\infty,x}$ denote the canonical continuum Hunt process from Theorem D.47 started from x. Then for every finite cylinder depth $N \geq 0$, the projected process $\Pi_N Y^x$ has the same finite-dimensional distributions as $\Pi_N Y^{\infty,x}$. In particular, if the same construction is started from a random initial state with law $\widehat{\pi}_\infty$, the resulting process has the same finite-cylinder distributions as the unique stationary weak interaction-field law.*

(vi) *If $\widetilde{Y}^x$ is another continuous $\mathfrak{S}_\eta^{(3)}$-valued adapted process on the same filtered probability space with $\widetilde{Y}_0^x = x$, and if every finite projection $\Pi_N \widetilde{Y}^x$ solves the canonical N-cylinder reflected system driven by $(\beta, B_1, \ldots, B_N)$, then*

$$\widetilde{Y}_t^x = Y_t^x \qquad \text{a.s. for all } t \geq 0.$$

In particular, the coherent cylindrical Brownian lift already upgrades to a genuine full-support pathwise strong continuum process on the manuscript's intrinsic compact support. What still remains open is a closed reflected SDE/SPDE description in a single intrinsic field equation.

Proof. For item (i), the compatibility (357) ensures that the definition of each coordinate $A_{n,t}^x$ is independent of the truncation depth used to observe it. Since every amplitude coordinate is reflected on $[-b_n(\eta), b_n(\eta)]$, one has $|A_{n,t}^x| \leq b_n(\eta)$ for all n and t. If $t_m \to t$, then every fixed coordinate satisfies $A_{n,t_m}^x \to A_{n,t}^x$ by continuity of the finite-cylinder paths, and dominated convergence against the summable majorant $2(2n+1)^3 b_n(\eta)$ gives

$$\|\mathbf{A}_{t_m}^x - \mathbf{A}_t^x\|_{(3)} = \sum_{n=1}^{\infty} (2n+1)^3 |A_{n,t_m}^x - A_{n,t}^x| \longrightarrow 0.$$

Thus $t \mapsto \mathbf{A}_t^x$ is continuous in $\ell_{(3)}^1$, and therefore $t \mapsto Z_t^x$ is continuous in $[0, h_*] \times \ell_{(3)}^1$. Proposition D.36 then shows that the reconstruction map $\mathbf{a} \mapsto \sum_{n\geq 1} a_n e_n$ is continuous from $\ell_{(3)}^1$ to C_{odd}^3, yielding the continuity of $t \mapsto \phi_t^x$.

For item (ii), let $Y^{\infty,x}$ be the Hunt process of Theorem D.47 started from x. By Proposition D.40, each finite strong cylinder process $X^{[N],x}$ has semigroup $S_t^{(N)}$, and item (v) of Theorem D.47 gives the same semigroup for $\Pi_N Y^{\infty,x}$. Therefore the finite-dimensional distributions of $X^{[N],x}$ and $\Pi_N Y^{\infty,x}$ coincide for every fixed N. In particular, for every fixed $t \geq 0$, the random elements Z_t^x and $\Gamma_\eta^{-1}(Y_t^{\infty,x})$ have the same integrals against all bounded continuous cylinder functions on the compact product box

$$[0, h_*] \times \prod_{n=1}^{\infty} [-b_n(\eta), b_n(\eta)].$$

Those cylinder functions form a unital algebra that separates points, so Stone–Weierstrass yields equality of the one-time laws on the whole product box. Since $\Gamma_\eta^{-1}(Y_t^{\infty,x})$ is supported on the compact admissible set $\mathcal{U}_\eta$, the same is true for Z_t^x at every fixed t. Taking a countable intersection over rational times and using continuity of $t \mapsto Z_t^x$ in the product topology, together with the closedness of $\mathcal{U}_\eta$ inside the compact product box, yields a full-probability event on which $Z_t^x \in \mathcal{U}_\eta$ for all $t \geq 0$. The definition and continuity of $Y_t^x = \Gamma_\eta(Z_t^x)$ then follow from Proposition D.36.

Item (iii) is immediate from the construction of Z_t^x: the first N coordinates are exactly those of $X^{[N],x}$. For item (iv), each finite truncation $(H_t^x, A_{1,t}^x, \ldots, A_{N,t}^x)$ is measurable with respect to the initial state and the Brownian coordinates $(\beta, B_1, \ldots, B_N)$ up to time t, because $X^{[N],x}$ is a strong solution. The full coordinate vector Z_t^x is the limit of these truncations in $\ell_{(3)}^1$, and Γ_η is continuous on $\mathcal{U}_\eta$, so Y_t^x is measurable with respect to the full Brownian family up to time t.

The first statement of item (v) has already been proved in the course of item (ii): every finite projection of Y^x has the same finite-dimensional distributions as the corresponding projection of the continuum Hunt process. If the same construction is started from a random initial state with law $\widehat{\pi}_\infty$, Theorem D.65(iii) identifies those same finite-cylinder distributions with the unique stationary weak interaction-field law. Finally, for item (vi), the finite projections $\Pi_N\widetilde{Y}^x$ and $X^{[N],x}$ solve the same reflected N-cylinder SDE with the same initial condition and the same driving Brownian motions. The pathwise uniqueness in Theorem D.44 therefore forces $\Pi_N\widetilde{Y}_t^x = \Pi_N Y_t^x$ for every N and every t, and the separating property of the cylinder coordinates then implies $\widetilde{Y}_t^x = Y_t^x$ almost surely for all $t \geq 0$. □

For the genuine reduced densities, Proposition D.74 below may be read as a statement about the one-step released-mode log-ratios of $\widetilde{\Delta}_R^{(N)}$: after harmless additive renormalization, it is enough to control

$$-\log \frac{\widetilde{\Delta}_R^{(N)}(\eta, h, a_1, \ldots, a_N)}{\widetilde{\Delta}_R^{(N-1)}(\eta, h, a_1, \ldots, a_{N-1})}$$

by a fixed-radius local function with summable sup norms. Under that one-step condition, the full summably local block expansion exists automatically.

For $1 \leq n \leq N$, define the left-local block projection

$$\Pi_{n,r}^{\mathrm{left},[N]}(h, a_1, \ldots, a_N) := \big(h, a_{\max(1,n-r)}, \ldots, a_n\big). \tag{413}$$

Proposition D.74 (One-step local-increment criterion for existence of a summably quasilocal decomposition)**.** *Fix $\eta > \eta_{\mathrm{crit}}^{\mathrm{var}}$ and $r \in \mathbb{N}$. Suppose there exist a continuous base term $\Psi_{\eta,0} \in C([0, h_*])$, constants $\kappa_{\eta,n} \in \mathbb{R}$ for $n \geq 0$, and continuous increment functions*

$$\Psi_{\eta,n} \in C\left([0, h_*] \times \prod_{k=\max(1,n-r)}^{n} [-b_k(\eta), b_k(\eta)]\right), \qquad n \geq 1,$$

with summable norm bound

$$M_{\eta,r}^{\mathrm{inc}} := \|\Psi_{\eta,0}\|_\infty + \sum_{n=1}^{\infty} \|\Psi_{\eta,n}\|_\infty < \infty, \tag{414}$$

such that

$$V_{\eta,0}(h) = \kappa_{\eta,0} + \Psi_{\eta,0}(h), \tag{415}$$

$$V_{\eta,N}(x) - V_{\eta,N-1}(\Pi_{N-1}x) = \kappa_{\eta,N} + \Psi_{\eta,N}\big(\Pi_{N,r}^{\mathrm{left},[N]}x\big), \qquad x \in \mathcal{X}_N,\ N \geq 1. \tag{416}$$

Then the cylinder interaction potentials admit an r-local summable block decomposition in the sense of (403)–(404)*. More precisely, there exist continuous block functions $\Phi_{\eta,n}$ and constants $c_{\eta,N}$ such that*

$$V_{\eta,N}(x) = c_{\eta,N} + \sum_{n=0}^{N} \Phi_{\eta,n}\big(\Pi_{n,r}^{[N]}x\big), \qquad x \in \mathcal{X}_N,$$

and

$$M_{\eta,r}^{\mathrm{blk}} \leq M_{\eta,r}^{\mathrm{inc}}.$$

Consequently Theorem D.68 applies, so the oscillation bound (405) *and the unconditional continuum coercivity package follow automatically.*

Proof. Telescope the finite-depth potentials:

$$V_{\eta,N}(h, a_1, \ldots, a_N) = V_{\eta,0}(h) + \sum_{n=1}^{N}\Big(V_{\eta,n}(h, a_1, \ldots, a_n) - V_{\eta,n-1}(h, a_1, \ldots, a_{n-1})\Big).$$

Substituting (415)–(416) gives

$$V_{\eta,N}(x) = \sum_{n=0}^{N} \kappa_{\eta,n} + \Psi_{\eta,0}(h) + \sum_{n=1}^{N} \Psi_{\eta,n}\big(\Pi_{n,r}^{\mathrm{left},[N]} x\big).$$

Set $c_{\eta,N} := \sum_{n=0}^{N} \kappa_{\eta,n}$, define $\Phi_{\eta,0} := \Psi_{\eta,0}$, and for $n \geq 1$ define $\Phi_{\eta,n}$ on the symmetric radius-r block appearing in (403) by simply ignoring the future coordinates $a_{n+1}, \ldots, a_{\min(N,n+r)}$. Then (403) holds exactly. Moreover,

$$\sum_{n=0}^{\infty} \|\Phi_{\eta,n}\|_\infty \leq \|\Psi_{\eta,0}\|_\infty + \sum_{n=1}^{\infty} \|\Psi_{\eta,n}\|_\infty = M_{\eta,r}^{\mathrm{inc}},$$

so the block-summability condition (404) follows. The final statement is therefore immediate from Theorem D.68. □

Integrated downstream conclusion after the strong-continuum/local-increment continuation. This continuation adds a thirty-sixth downstream message. Positively, the manuscript's coherent cylindrical Brownian lift already upgrades to a full-support pathwise strong continuum process on the intrinsic compact support, and the existence of a summably quasilocal interaction expansion is reduced to a one-step local-increment criterion on the genuine cylinder potentials. Cautionarily, the manuscript still does not identify that continuum process with a closed reflected SDE/SPDE in one intrinsic field equation, and it still does not verify the local-increment hypothesis for the actual e-Arc-Helix interaction increments. What remains open is therefore the explicit field-equation identification and the genuine intrinsic verification — or sharp obstruction — of quasilocal existence itself.

Remark D.75. The present appendix now carries thirty-six distinct downstream messages. The new thirty-sixth message closes the existence side of the strong-realization question on the intrinsic compact support and compresses the quasilocal-existence side to one concrete family of released-mode increments. What remains open after this point is not whether some full-support strong continuum process exists inside the manuscript's own support class, but whether that process can be written as a closed reflected field equation and whether the actual interaction increments satisfy — or sharply fail to satisfy — the required intrinsic local-increment criterion.

D.24 Worked continuation XXXVII: a triangular reflected infinite system and canonical left-tail envelopes

The previous continuation closes the existence of a full-support strong continuum process and reduces exact quasilocal existence to a one-step local-increment problem. The next intrinsic sharpening has two complementary parts. On the stochastic side, the strong continuum lift can already be read coordinate-by-coordinate from the minimal strong cylinder depth carrying that coordinate, which yields an explicit infinite reflected system driven by the same Brownian family. On the locality side, exact finite-radius dependence can be relaxed to a canonical envelope construction that measures how much each released-mode increment still remembers the remote past. This still does not give a single closed SPDE or a verified quasilocal decomposition for the actual e-Arc-Helix interactions, but it replaces both remaining existence questions by more concrete coordinatewise and envelope-level quantities.

Status label. The triangular reflected-system theorem and the canonical tail-envelope theorem below have status **theorem-level downstream result**. They do *not* yet collapse the continuum dynamics to one closed intrinsic field equation, and they do *not* yet verify the required envelope/tail bounds for the genuine e-Arc-Helix interaction increments. What they do prove is that the strong continuum lift already satisfies an explicit countable reflected SDE system, and that automatic intrinsic coercivity already follows from canonical finite-window envelopes plus summably small released-mode tail memory.

Define the minimal-depth drift coefficients by

$$\mathfrak{b}_{\eta,0}(h) := b^H_{\eta,0}(h), \qquad \mathfrak{b}_{\eta,n}(h, a_1, \dots, a_n) := b^n_{\eta,n}(h, a_1, \dots, a_n), \qquad n \geq 1. \tag{417}$$

Theorem D.76 (Triangular reflected infinite-system representation of the full-support continuum lift). *Assume the hypotheses of Theorem D.73, and let*

$$Y^x = (H^x, \phi^x) \qquad \textit{with coordinates} \qquad Z^x = (H^x, A^x_1, A^x_2, \dots)$$

be the full-support pathwise strong continuum process started from $x \in \mathfrak{S}^{(3)}_\eta$. Then there exist continuous adapted finite-variation reflection processes

$$L^{0,x}, \; L^{h_*,x}, \; K^{+,x}_1, K^{-,x}_1, K^{+,x}_2, K^{-,x}_2, \dots$$

with each component nondecreasing and vanishing at time zero, such that for every $t \geq 0$,

$$H^x_t = h_x + \int_0^t \mathfrak{b}_{\eta,0}(H^x_s)\, ds + \int_0^t \sqrt{\frac{2H^x_s(h_* - H^x_s)}{h_*}}\, d\beta_s + L^{0,x}_t - L^{h_*,x}_t, \tag{418}$$

$$A^x_{n,t} = a_{x,n} + \int_0^t \mathfrak{b}_{\eta,n}(H^x_s, A^x_{1,s}, \dots, A^x_{n,s})\, ds + B_{n,t} + K^{+,x}_{n,t} - K^{-,x}_{n,t}, \qquad n \geq 1. \tag{419}$$

Moreover, for every $T > 0$,

$$\int_0^T \mathbf{1}_{\{H^x_t > 0\}}\, dL^{0,x}_t = 0, \qquad \int_0^T \mathbf{1}_{\{H^x_t < h_*\}}\, dL^{h_*,x}_t = 0, \tag{420}$$

$$\int_0^T \mathbf{1}_{\{A^x_{n,t} < b_n(\eta)\}}\, dK^{+,x}_{n,t} = 0, \qquad \int_0^T \mathbf{1}_{\{A^x_{n,t} > -b_n(\eta)\}}\, dK^{-,x}_{n,t} = 0, \qquad n \geq 1. \tag{421}$$

Hence the full-support continuum lift already solves an explicit countable triangular reflected SDE system driven by the common Brownian family. What remains open is not pathwise existence of such a system on the intrinsic compact support, but collapse of the family $\mathfrak{b}_{\eta,n}$ and of the reflection mechanism to one closed intrinsic field equation.

Proof. Theorem D.73 identifies the coordinate process of Y^x with the compatible family of strong cylinder processes:

$$H^x = H^{[0],x}, \qquad (H^x, A^x_1, \dots, A^x_n) = (H^{[n],x}, A^{[n],x}_1, \dots, A^{[n],x}_n) \quad \text{for every } n \geq 1.$$

Apply Theorem D.65(i) first with $N = 0$. This yields the reflected equation for $H^{[0],x} = H^x$ together with boundary local times $L^{0,x} := L^{0,0,x}$ and $L^{h_*,x} := L^{h_*,0,x}$, which is exactly (418) and (420). Next fix $n \geq 1$ and apply Theorem D.65(i) with $N = n$. Because the first n coordinates of $X^{[n],x}$ agree pathwise with those of Y^x, the n-th coordinate satisfies

$$dA^x_{n,t} = b^n_{\eta,n}(H^x_t, A^x_{1,t}, \dots, A^x_{n,t})\, dt + dB_{n,t} + dK^{+,x}_{n,t} - dK^{-,x}_{n,t},$$

where $K^{\pm,x}_n := K^{\pm,n,x}_n$ are the reflected boundary local times for the interval $[-b_n(\eta), b_n(\eta)]$. This is exactly (419), and the support property (421) is the usual one-dimensional Skorokhod condition inherited from the same finite-depth system. No additional compatibility argument is required beyond the already proved pathwise cylinder compatibility. □

For the genuinely intrinsic locality problem, the exact one-step criterion of Proposition D.74 is recovered when each released-mode increment forgets the remote past after a fixed radius. The next theorem shows that even without exact forgetting, one may canonically compress the remote-past dependence by taking upper and lower local envelopes over the forgotten coordinates and measuring the remaining tail memory quantitatively.

For each $n \geq 1$, define the genuine one-step released-mode increment on $\mathcal{X}_n$ by

$$G_{\eta,n}(x) := V_{\eta,n}(x) - V_{\eta,n-1}(\Pi_{n-1}x), \qquad x \in \mathcal{X}_n. \tag{422}$$

For every radius $r \in \mathbb{N}_0$, define its left-tail oscillation by

$$\Omega^{\text{tail}}_{\eta,n}(r) := \sup\Big\{\big|G_{\eta,n}(h, a_1, \ldots, a_n) - G_{\eta,n}(h, \tilde{a}_1, \ldots, \tilde{a}_n)\big| : a_k = \tilde{a}_k \text{ for every } k \in \{\max(1, n-r), \ldots, n\}\Big\}. \tag{423}$$

On the left block

$$[0, h_*] \times \prod_{k=\max(1,n-r)}^{n} [-b_k(\eta), b_k(\eta)],$$

define the upper and lower canonical envelopes by

$$\begin{aligned} G^{+,r}_{\eta,n}(h, u_{\max(1,n-r)}, \ldots, u_n) := \sup\Big\{ & G_{\eta,n}(h, a_1, \ldots, a_n) : \\ & a_k = u_k \text{ for every } k \in \{\max(1, n-r), \ldots, n\}\Big\}, \end{aligned} \tag{424}$$

$$\begin{aligned} G^{-,r}_{\eta,n}(h, u_{\max(1,n-r)}, \ldots, u_n) := \inf\Big\{ & G_{\eta,n}(h, a_1, \ldots, a_n) : \\ & a_k = u_k \text{ for every } k \in \{\max(1, n-r), \ldots, n\}\Big\}, \end{aligned} \tag{425}$$

and set

$$G^{\text{mid},r}_{\eta,n} := \frac{1}{2}\big(G^{+,r}_{\eta,n} + G^{-,r}_{\eta,n}\big). \tag{426}$$

Theorem D.77 (Canonical tail-envelope criterion for intrinsic coercivity). *Fix $\eta > \eta^{\text{var}}_{\text{crit}}$ and choose any radii $r_n \in \mathbb{N}_0$ for $n \geq 1$. Then the following hold.*

(i) For every $n \geq 1$, the envelope functions $G^{+,r_n}_{\eta,n}$, $G^{-,r_n}_{\eta,n}$, and $G^{\text{mid},r_n}_{\eta,n}$ are continuous left-local functions of $(h, a_{\max(1,n-r_n)}, \ldots, a_n)$, and

$$\big\|G_{\eta,n} - G^{\text{mid},r_n}_{\eta,n} \circ \Pi^{\text{left},[n]}_{n,r_n}\big\|_\infty \leq \frac{1}{2}\Omega^{\text{tail}}_{\eta,n}(r_n). \tag{427}$$

(ii) For every finite depth $N \geq 1$,

$$\operatorname{osc}(V_{\eta,N}) \leq \operatorname{osc}(V_{\eta,0}) + \sum_{n=1}^{N}\Big(\operatorname{osc}(G^{\text{mid},r_n}_{\eta,n}) + \Omega^{\text{tail}}_{\eta,n}(r_n)\Big). \tag{428}$$

Consequently, if

$$C^{\text{env}}_\eta := \operatorname{osc}(V_{\eta,0}) + \sum_{n=1}^{\infty}\Big(\operatorname{osc}(G^{\text{mid},r_n}_{\eta,n}) + \Omega^{\text{tail}}_{\eta,n}(r_n)\Big) < \infty, \tag{429}$$

then the uniform oscillation hypothesis (399) holds with $C^{\text{osc}}_\eta \leq C^{\text{env}}_\eta$. Hence the conclusions of Theorem D.66 and therefore of Theorem D.51 follow automatically.

(iii) Suppose in addition that there exist constants $c_{\eta,n} \in \mathbb{R}$ such that

$$\sum_{n=1}^{\infty} \left\| G_{\eta,n}^{\mathrm{mid},r_n} - c_{\eta,n} \right\|_{\infty} < \infty. \tag{430}$$

For $x = (h, \mathbf{a}) \in \Gamma_\eta^{-1}(\mathfrak{S}_\eta^{(3)})$, define

$$\Pi_{n,r_n}^{\mathrm{left},[\infty]} x := (h, a_{\max(1,n-r_n)}, \ldots, a_n).$$

Then the series

$$V_{\eta,\infty}^{\mathrm{env}}(x) := \sum_{n=1}^{\infty} \left(G_{\eta,n}^{\mathrm{mid},r_n} - c_{\eta,n} \right) \left(\Pi_{n,r_n}^{\mathrm{left},[\infty]} x \right) \tag{431}$$

converges uniformly on $\mathfrak{S}_\eta^{(3)}$ and defines a continuous quasilocal envelope potential there.

In particular, exact quasilocal block existence is no longer the only route to automatic intrinsic coercivity: it is already enough to control the oscillation of the canonical local envelopes together with the tail-memory modulus (423). The exact local-increment criterion of Proposition D.74 is recovered as the special case $\Omega_{\eta,n}^{\mathrm{tail}}(r) = 0$.

Proof. For item (i), the domain of $G_{\eta,n}$ is the compact box $\mathcal{X}_n$. Fix a radius r. The fibers obtained by freezing $(h, a_{\max(1,n-r)}, \ldots, a_n)$ are compact boxes in the forgotten coordinates, so the supremum and infimum in (424)–(425) are attained. Continuity of the envelope maps follows from continuity of $G_{\eta,n}$ on a compact product and the maximum-minimum theorem on compact fibers. By construction, whenever one fixes the visible block $(h, a_{\max(1,n-r_n)}, \ldots, a_n)$, the values of $G_{\eta,n}$ over that fiber lie inside the interval $[G_{\eta,n}^{-,r_n}, G_{\eta,n}^{+,r_n}]$, whose width is exactly at most $\Omega_{\eta,n}^{\mathrm{tail}}(r_n)$. The midpoint therefore approximates the true increment with error at most half that width, which is (427).

For item (ii), telescope the interaction potentials on $\mathcal{X}_N$:

$$V_{\eta,N} = V_{\eta,0} \circ \Pi_0 + \sum_{n=1}^{N} G_{\eta,n} \circ \Pi_n.$$

Hence

$$\operatorname{osc}(V_{\eta,N}) \le \operatorname{osc}(V_{\eta,0}) + \sum_{n=1}^{N} \operatorname{osc}(G_{\eta,n}).$$

For each n,

$$\operatorname{osc}(G_{\eta,n}) \le \operatorname{osc}(G_{\eta,n}^{\mathrm{mid},r_n}) + 2\left\| G_{\eta,n} - G_{\eta,n}^{\mathrm{mid},r_n} \circ \Pi_{n,r_n}^{\mathrm{left},[n]} \right\|_{\infty} \le \operatorname{osc}(G_{\eta,n}^{\mathrm{mid},r_n}) + \Omega_{\eta,n}^{\mathrm{tail}}(r_n)$$

by (427). Substituting this into the telescoping oscillation bound gives (428). If the series in (429) converges, the oscillation hypothesis of Theorem D.66 holds uniformly in N, and the stated coercivity consequences follow immediately.

For item (iii), each summand in (431) is continuous on the compact support class and bounded by the corresponding summable sup norm from (430). The Weierstrass M-test therefore gives uniform convergence on $\mathfrak{S}_\eta^{(3)}$, and continuity of the sum follows because every truncated left block is a continuous coordinate projection on the compact continuum support. □

Integrated downstream conclusion after the triangular-envelope continuation. This continuation adds a thirty-seventh downstream message. Positively, the manuscript's full-support continuum process now solves an explicit countable triangular reflected SDE system on the intrinsic compact support, and automatic intrinsic coercivity is reduced to canonical local-envelope oscillations together with summably small released-mode tail memory. Cautionarily, the manuscript still does not compress that triangular system to one closed reflected SDE/SPDE, and it still does not verify the required envelope/tail bounds for the genuine e-Arc-Helix interaction increments. What remains open is therefore the closed intrinsic field equation and the genuine interaction estimate itself.

Remark D.78. The new thirty-seventh message is deliberately intermediate. It is stronger than mere strong existence, because the continuum lift already satisfies explicit coordinatewise reflected equations driven by a common Brownian family. It is weaker than a closed field equation, because the drift is still encoded through the minimal cylinder depth of each coordinate. Likewise, the canonical envelope criterion is stronger than a qualitative hope for locality, because it produces explicit finite-window approximants and a sharp tail-memory modulus; it is weaker than a finished quasilocal decomposition, because the actual e-Arc-Helix interaction increments have not yet been shown to satisfy those bounds.

D.25 Worked continuation XXXVIII: closed cylinder martingale closure and renormalized genuine interaction potentials

The triangular-envelope continuation leaves the two pure existence questions in a nearly closed but still split form. On the stochastic side, the strong continuum lift is already written coordinate-by-coordinate as a reflected infinite system, yet not as one closed intrinsic object on the full cylinder algebra. On the locality side, the canonical envelope criterion already produces a quasilocal surrogate potential, yet the genuine interaction increments themselves have not been summed into a continuum object. The next intrinsic sharpening compresses both gaps one step further. The triangular reflected system defines a single closed reflected martingale problem on the intrinsic cylinder algebra, and summably controlled envelope/tail errors already force the *genuine* released-mode increments to admit a uniformly convergent renormalized continuum interaction potential on the compact support.

Status label. The closed cylinder-martingale theorem and the renormalized genuine-potential theorem below have status **theorem-level downstream result**. They do *not* yet collapse the dynamics to one local reflected SPDE, and they do *not* yet prove that the actual e-Arc-Helix interaction increments satisfy the required summable envelope/tail bounds. What they do prove is that the full-support strong continuum lift is already characterized by one closed reflected cylinder martingale problem, and that those same summable bounds would already produce a bona fide continuous renormalized intrinsic interaction potential built from the genuine increments themselves.

Let $\mathcal{D}_\eta^{\mathrm{Neu},\infty}$ denote the algebra of bounded smooth cylinder functions $F = f\circ\Pi_N$ on $\Gamma_\eta^{-1}(\mathfrak{S}_\eta^{(3)})$ such that the finite-cylinder representative f satisfies the visible Neumann reflection conditions

$$\partial_h f(0,a_1,\dots,a_N)=0,\qquad \partial_h f(h_*,a_1,\dots,a_N)=0, \tag{432}$$

$$\partial_{a_n} f(h,a_1,\dots,a_{n-1},\pm b_n(\eta),a_{n+1},\dots,a_N)=0,\qquad 1\le n\le N. \tag{433}$$

For such $F = f\circ\Pi_N$, define the closed cylinder operator by

$$\mathfrak{L}_\eta^{\mathrm{cyl}}F := \left(\mathcal{L}_{\eta,N}^{\mathrm{red}} f\right)\circ\Pi_N. \tag{434}$$

This is well defined because the finite reduced generators act consistently on deeper cylinders when the test function depends only on the visible coordinates.

Theorem D.79 (Closed reflected cylinder martingale problem for the strong continuum lift). *Assume the hypotheses of Theorem D.73, and let*

$$Z^x = (H^x, A_1^x, A_2^x, \dots)$$

be the full-support pathwise strong continuum lift started from $x \in \Gamma_\eta^{-1}(\mathfrak{S}_\eta^{(3)})$. *Then the following hold.*

(i) For every $F \in \mathcal{D}_\eta^{\mathrm{Neu},\infty}$,

$$M_t^F := F(Z_t^x) - F(x) - \int_0^t \left(\mathfrak{L}_\eta^{\mathrm{cyl}} F\right)(Z_s^x)\, ds \tag{435}$$

is a continuous martingale.

(ii) If $\widetilde{Z}^x$ *is any other continuous adapted* $\Gamma_\eta^{-1}(\mathfrak{S}_\eta^{(3)})$*-valued process with* $\widetilde{Z}_0^x = x$ *such that* (435) *holds with* Z^x *replaced by* $\widetilde{Z}^x$ *for every* $F \in \mathcal{D}_\eta^{\mathrm{Neu},\infty}$, *then for every* $N \geq 0$ *the finite projection* $\Pi_N \widetilde{Z}^x$ *has the same law as the canonical strong cylinder process* $X^{[N],x}$ *of Theorem D.44. If, in addition,* $\widetilde{Z}^x$ *is realized on the same filtered probability space and driven by the same Brownian family* $(\beta, B_1, B_2, \dots)$ *as in Theorem D.76, then*

$$\Pi_N \widetilde{Z}_t^x = \Pi_N Z_t^x \qquad \text{a.s. for all } t \geq 0 \text{ and every } N \geq 0, \tag{436}$$

hence $\widetilde{Z}^x = Z^x$ *almost surely.*

In particular, the manuscript now has a single closed reflected martingale problem on the intrinsic cylinder algebra that characterizes the strong continuum lift. What remains open is not cylinder-level closure, but collapse of that closed cylinder formulation to one genuinely local reflected field equation on a broader non-cylinder domain.

Proof. The well-definedness of (434) is exactly the finite-generator consistency already used in the construction of the coherent cylindrical Brownian family: if $M \geq N$ and $F = f \circ \Pi_N$, then the first N coordinates of $\mathcal{L}_{\eta,M}^{\mathrm{red}}$ act on $f \circ \Pi_N$ exactly as $\mathcal{L}_{\eta,N}^{\mathrm{red}}$ acts on f.

Fix $F = f \circ \Pi_N \in \mathcal{D}_\eta^{\mathrm{Neu},\infty}$. By Theorem D.73, the first N coordinates of Z^x coincide pathwise with the canonical strong cylinder process $X^{[N],x}$. Apply Itô's formula to $f(X_t^{[N],x})$. The drift term is exactly $(\mathcal{L}_{\eta,N}^{\mathrm{red}} f)(X_t^{[N],x})\, dt$, while the boundary contributions vanish because f satisfies the Neumann reflection conditions (432)–(433). Therefore (435) is a continuous martingale, proving item (i).

Now let $\widetilde{Z}^x$ satisfy the same martingale identities. Testing only against functions of depth N shows that $\Pi_N \widetilde{Z}^x$ solves the reflected martingale problem for $\mathcal{L}_{\eta,N}^{\mathrm{red}}$ on $\mathcal{X}_N$. By the finite reflected-diffusion uniqueness package underlying Theorem D.44, that martingale problem has the canonical law of $X^{[N],x}$. This proves equality in law of every finite projection.

If $\widetilde{Z}^x$ is realized on the same Brownian family as Z^x, then each finite projection $\Pi_N \widetilde{Z}^x$ solves the same reflected $(N+1)$-dimensional SDE as $X^{[N],x} = \Pi_N Z^x$ with the same initial point and the same driving Brownian motions. Pathwise uniqueness from Theorem D.44 therefore gives (436) for every N. Since the coordinate projections separate points of $\Gamma_\eta^{-1}(\mathfrak{S}_\eta^{(3)})$, the full processes coincide almost surely. □

The stochastic side is therefore no longer missing a closed object altogether: it already has one on the full intrinsic cylinder algebra. The remaining missing step is whether that cylinder-martingale closure can be compressed to a single local/reflected field equation that does not mention truncation depth coordinate-by-coordinate.

Theorem D.80 (Renormalized genuine continuum interaction potential under summable envelope/tail control). *Assume the hypotheses of item (iii) of Theorem D.77, and in addition assume*

$$\sum_{n=1}^{\infty} \Omega_{\eta,n}^{\mathrm{tail}}(r_n) < \infty. \tag{437}$$

For $x = (h, \mathbf{a}) \in \Gamma_\eta^{-1}(\mathfrak{S}_\eta^{(3)})$, *write*

$$\Pi_n^{[\infty]} x := (h, a_1, \ldots, a_n), \qquad C_{\eta,N} := \sum_{n=1}^{N} c_{\eta,n}.$$

Then the following hold.

(i) The series

$$V_{\eta,\infty}^{\mathrm{ren}}(x) := V_{\eta,0}(h) + \sum_{n=1}^{\infty} \Big(G_{\eta,n}(\Pi_n^{[\infty]} x) - c_{\eta,n} \Big) \tag{438}$$

converges uniformly on $\mathfrak{S}_\eta^{(3)}$ *and defines a continuous intrinsic continuum interaction potential.*

(ii) The renormalized genuine potential stays within the explicit tail-memory tube around the canonical envelope potential:

$$\big\| V_{\eta,\infty}^{\mathrm{ren}} - \big(V_{\eta,0} \circ \Pi_0 + V_{\eta,\infty}^{\mathrm{env}} \big) \big\|_\infty \le \frac{1}{2} \sum_{n=1}^{\infty} \Omega_{\eta,n}^{\mathrm{tail}}(r_n). \tag{439}$$

(iii) For every finite depth $N \ge 1$,

$$\sup_{x \in \Gamma_\eta^{-1}(\mathfrak{S}_\eta^{(3)})} \big| V_{\eta,N}(\Pi_N x) - C_{\eta,N} - V_{\eta,\infty}^{\mathrm{ren}}(x) \big| \le \sum_{n>N} \left(\frac{1}{2} \Omega_{\eta,n}^{\mathrm{tail}}(r_n) + \big\| G_{\eta,n}^{\mathrm{mid},r_n} - c_{\eta,n} \big\|_\infty \right). \tag{440}$$

Consequently, once the envelope/tail bounds are summable, the genuine finite-cylinder interaction potentials themselves converge uniformly — after renormalization by the constants $C_{\eta,N}$ *— to a continuous intrinsic continuum interaction potential on the compact support. Exact local-block existence is therefore no longer the only path to a bona fide continuum interaction potential: summable envelope/tail control already suffices.*

Proof. For each $n \ge 1$, item (i) of Theorem D.77 gives

$$\big\| G_{\eta,n} - G_{\eta,n}^{\mathrm{mid},r_n} \circ \Pi_{n,r_n}^{\mathrm{left},[n]} \big\|_\infty \le \frac{1}{2} \Omega_{\eta,n}^{\mathrm{tail}}(r_n).$$

Lifting this to $\Gamma_\eta^{-1}(\mathfrak{S}_\eta^{(3)})$ via $\Pi_n^{[\infty]}$ gives

$$\big\| G_{\eta,n} \circ \Pi_n^{[\infty]} - c_{\eta,n} \big\|_\infty \le \frac{1}{2} \Omega_{\eta,n}^{\mathrm{tail}}(r_n) + \big\| G_{\eta,n}^{\mathrm{mid},r_n} - c_{\eta,n} \big\|_\infty.$$

The right-hand side is summable by (437) and (430), so the Weierstrass M-test proves uniform convergence of (438); continuity follows because every summand is continuous on the compact support.

For item (ii), subtract $V_{\eta,0} \circ \Pi_0 + V_{\eta,\infty}^{\mathrm{env}}$ from (438) and use (431):

$$V_{\eta,\infty}^{\mathrm{ren}} - \big(V_{\eta,0} \circ \Pi_0 + V_{\eta,\infty}^{\mathrm{env}} \big) = \sum_{n=1}^{\infty} \Big(G_{\eta,n} \circ \Pi_n^{[\infty]} - G_{\eta,n}^{\mathrm{mid},r_n} \circ \Pi_{n,r_n}^{\mathrm{left},[\infty]} \Big).$$

Each summand has sup norm at most $\frac{1}{2}\Omega^{\mathrm{tail}}_{\eta,n}(r_n)$, so summing gives (439).

For item (iii), write

$$V_{\eta,N}(\Pi_N x) - C_{\eta,N} = V_{\eta,0}(h) + \sum_{n=1}^{N}\Big(G_{\eta,n}(\Pi_n^{[\infty]}x) - c_{\eta,n}\Big).$$

Subtract (438). The difference is exactly the tail

$$\sum_{n>N}\Big(G_{\eta,n}(\Pi_n^{[\infty]}x) - c_{\eta,n}\Big),$$

whose sup norm is bounded by the tail of the same summable majorant used in item (i). This yields (440). □

Integrated downstream conclusion after the closed-cylinder/renormalized-potential continuation. This continuation adds a thirty-eighth downstream message. Positively, the manuscript's full-support strong continuum lift is now characterized by a single closed reflected martingale problem on the intrinsic cylinder algebra, and summable envelope/tail control would already lift the genuine released-mode interaction increments themselves to a continuous renormalized continuum interaction potential on the intrinsic compact support. Cautionarily, the manuscript still does not compress that cylinder-martingale closure to one local reflected SDE/SPDE, and it still does not verify the required summable envelope/tail bounds for the actual e-Arc-Helix interaction increments. What remains open is therefore the genuinely local field equation and the concrete interaction estimate needed to trigger the renormalized-potential theorem.

Remark D.81. The new thirty-eighth message is stronger than the previous stage in two distinct senses. On the stochastic side, it replaces infinitely many coordinatewise reflected equations by one closed martingale problem on the whole cylinder algebra. On the locality side, it no longer stops at a surrogate envelope potential: under summable envelope/tail control, the genuine finite-cylinder interaction potentials themselves converge, after explicit renormalization, to a bona fide continuous continuum interaction potential on the intrinsic support. What is still missing is not a cylinder-level closure or a continuum potential candidate, but proof that the genuine e-Arc-Helix interactions satisfy the requisite summable hypotheses and that the resulting closed object can be written as one local reflected field equation.

D.26 Worked continuation XXXIX: weak reflected field-equation closure and a derivative-budget verification criterion

The previous continuation leaves the two pure existence questions in a sharper but still not fully local form. On the stochastic side, the full-support continuum lift is already characterized by a closed cylinder martingale problem, yet that description still quantifies over cylinder observables rather than exposing one single duality equation. On the locality side, the renormalized genuine continuum potential already exists under summable envelope/tail control, yet those summable hypotheses are not expressed directly in terms of concrete derivatives of the genuine released-mode increments. The next intrinsic sharpening addresses both points. First, the reflected continuum lift can be compressed to one weak field equation tested against finitely supported directions. Second, the envelope/tail hypotheses can themselves be forced by an explicit summable derivative budget on the genuine released-mode increments.

Status label. The weak reflected field equation and the derivative-budget theorem below have status **theorem-level downstream result**. They do *not* yet produce a closed local reflected SPDE on a broader non-cylinder state space, and they do *not* yet prove that the actual e-Arc-Helix interaction increments satisfy the required derivative budgets. What they do prove is that the strong continuum lift is already governed by one reflected duality equation on the intrinsic finitely supported test space, and that the renormalized-potential/coercivity hypotheses already follow from concrete summable partial-derivative bounds on the genuine released-mode increments themselves.

Let $\mathfrak{T}_\eta^{\mathrm{fin}}$ denote the space of finitely supported real sequences

$$\varphi = (\varphi_0, \varphi_1, \varphi_2, \dots).$$

For $z = (h, \mathbf{a}) \in [0, h_*] \times \ell^1_{(3)}$, write

$$\langle z, \varphi \rangle_{\mathrm{cyl}} := \varphi_0 h + \sum_{n=1}^{\infty} \varphi_n a_n,$$

where the sum is finite because φ has finite support. For such φ, define the projected drift functional

$$\mathfrak{b}_\eta^\varphi(h, \mathbf{a}) := \varphi_0 \mathfrak{b}_{\eta,0}(h) + \sum_{n=1}^{\infty} \varphi_n \mathfrak{b}_{\eta,n}(h, a_1, \dots, a_n), \tag{441}$$

and the reflected boundary functional

$$\mathfrak{R}_t^{x,\varphi} := \varphi_0 (L_t^{0,x} - L_t^{h_*,x}) + \sum_{n=1}^{\infty} \varphi_n (K_{n,t}^{+,x} - K_{n,t}^{-,x}), \tag{442}$$

again a finite sum.

Theorem D.82 (Single weak reflected continuum field equation on the intrinsic test space). *Assume the hypotheses of Theorems D.76 and D.79, and let*

$$Z^x = (H^x, A_1^x, A_2^x, \dots)$$

be the full-support strong continuum lift started from $x \in \Gamma_\eta^{-1}(\mathfrak{S}_\eta^{(3)})$. Then for every $\varphi \in \mathfrak{T}_\eta^{\mathrm{fin}}$,

$$\langle Z_t^x, \varphi \rangle_{\mathrm{cyl}} = \langle Z_0^x, \varphi \rangle_{\mathrm{cyl}} + \int_0^t \mathfrak{b}_\eta^\varphi(Z_s^x)\, ds + \mathfrak{W}_t^\varphi + \mathfrak{R}_t^{x,\varphi}, \tag{443}$$

where

$$\mathfrak{W}_t^\varphi := \varphi_0 \int_0^t \sqrt{\frac{2H_s^x(h_* - H_s^x)}{h_*}}\, d\beta_s + \sum_{n=1}^{\infty} \varphi_n B_{n,t} \tag{444}$$

is a continuous martingale. Moreover, for any $\varphi, \psi \in \mathfrak{T}_\eta^{\mathrm{fin}}$,

$$\langle \mathfrak{W}^\varphi, \mathfrak{W}^\psi \rangle_t = \int_0^t \left(\frac{2H_s^x(h_* - H_s^x)}{h_*} \varphi_0 \psi_0 + \sum_{n=1}^{\infty} \varphi_n \psi_n \right) ds. \tag{445}$$

Finally, for every $F \in \mathcal{D}_\eta^{\mathrm{Neu},\infty}$, Itô expansion of $F(Z_t^x)$ using (443) *recovers the closed cylinder martingale identity* (435). *Hence the strong continuum lift is already governed by one reflected weak field equation on the intrinsic finitely supported test space.*

Proof. Fix $\varphi \in \mathfrak{T}_\eta^{\mathrm{fin}}$, and let N be large enough that $\varphi_n = 0$ for every $n > N$. Multiply the reflected helicity equation (418) by φ_0, multiply the reflected amplitude equations (419) by φ_n for $1 \le n \le N$, and sum the resulting identities. Because only finitely many coefficients are nonzero, this gives (443) with the martingale term (444) and the reflected boundary contribution

(442). The covariance formula (445) is immediate from independence of $\beta, B_1, B_2, \ldots$ together with bilinearity of quadratic covariation.

Now fix $F = f \circ \Pi_N \in \mathcal{D}_\eta^{\mathrm{Neu},\infty}$. Applying Itô's formula directly to the finite coordinate vector $(H_t^x, A_{1,t}^x, \ldots, A_{N,t}^x)$ governed by (418)–(419) yields exactly the drift $(\mathcal{L}_{\eta,N}^{\mathrm{red}} f)(\Pi_N Z_t^x)\,dt$. The first-order boundary terms vanish because the visible representative f satisfies the Neumann reflection conditions (432)–(433). Therefore

$$F(Z_t^x) - F(Z_0^x) - \int_0^t (\mathfrak{L}_\eta^{\mathrm{cyl}} F)(Z_s^x)\,ds$$

is a continuous martingale, which is exactly (435). □

The stochastic side is therefore no longer missing even a single duality equation: it already has one on the intrinsic finitely supported test space. What remains open is whether this weak reflected field equation can be promoted to a genuinely local reflected SDE/SPDE on a broader non-cylinder domain.

For the interaction side, fix any radii $r_n \in \mathbb{N}_0$. The next theorem shows that the summable envelope/tail hypotheses of Theorem D.80 are already implied by explicit derivative budgets on the genuine released-mode increments themselves.

Theorem D.83 (Derivative-budget criterion for renormalized intrinsic interaction bounds)**.** *Fix $\eta > \eta_{\mathrm{crit}}^{\mathrm{var}}$, choose radii $r_n \in \mathbb{N}_0$, and assume each genuine increment $G_{\eta,n}$ from (422) extends to a C^1 function on a neighborhood of $\mathcal{X}_n$. For every $n \geq 1$, define the far-tail and visible-block derivative budgets*

$$M_{\eta,n}^{\mathrm{far}} := 2 \sum_{k=1}^{\max(0,n-r_n-1)} b_k(\eta)\, \|\partial_{a_k} G_{\eta,n}\|_\infty, \tag{446}$$

$$M_{\eta,n}^{\mathrm{near}} := h_* \, \|\partial_h G_{\eta,n}\|_\infty + 2 \sum_{k=\max(1,n-r_n)}^{n} b_k(\eta)\, \|\partial_{a_k} G_{\eta,n}\|_\infty. \tag{447}$$

Then the following hold.

(i) The left-tail oscillation satisfies

$$\Omega_{\eta,n}^{\mathrm{tail}}(r_n) \leq M_{\eta,n}^{\mathrm{far}}. \tag{448}$$

(ii) If one sets

$$c_{\eta,n} := G_{\eta,n}^{\mathrm{mid},r_n}(0,0,\ldots,0), \tag{449}$$

then the centered envelope amplitude satisfies

$$\|G_{\eta,n}^{\mathrm{mid},r_n} - c_{\eta,n}\|_\infty \leq M_{\eta,n}^{\mathrm{near}}. \tag{450}$$

(iii) If

$$\sum_{n=1}^{\infty} \left(M_{\eta,n}^{\mathrm{far}} + M_{\eta,n}^{\mathrm{near}}\right) < \infty, \tag{451}$$

then the hypotheses of item (iii) of Theorem D.77 and of Theorem D.80 hold. Consequently the genuine finite-cylinder interaction potentials admit a uniformly convergent renormalized intrinsic continuum interaction potential on $\mathfrak{S}_\eta^{(3)}$, and the automatic intrinsic coercivity / spectral-gap / log-Sobolev package follows.

In particular, verification of the renormalized intrinsic interaction bounds is reduced further to one concrete family of summable partial-derivative estimates on the genuine released-mode increments.

Proof. For item (i), fix two points in $\mathcal{X}_n$ with identical visible block $(h, a_{\max(1,n-r_n)}, \dots, a_n)$. Vary the forgotten coordinates $a_1, \dots, a_{\max(0,n-r_n-1)}$ one at a time and apply the one-dimensional mean-value theorem in each coordinate. Since the k-th forgotten amplitude ranges over an interval of length $2b_k(\eta)$, the total change in $G_{\eta,n}$ is at most

$$2 \sum_{k=1}^{\max(0,n-r_n-1)} b_k(\eta) \, \|\partial_{a_k} G_{\eta,n}\|_\infty = M^{\mathrm{far}}_{\eta,n}.$$

Taking the supremum over all such pairs gives (448).

For item (ii), first note that the same mean-value argument in the visible coordinates implies that the upper and lower envelope functions $G^{+,r_n}_{\eta,n}$ and $G^{-,r_n}_{\eta,n}$ are Lipschitz on their visible block with coordinatewise Lipschitz constants bounded by the corresponding sup norms of $\partial_h G_{\eta,n}$ and $\partial_{a_k} G_{\eta,n}$ for $\max(1, n - r_n) \le k \le n$: taking a supremum or infimum over the forgotten coordinates cannot increase those visible-coordinate oscillation bounds. The same therefore holds for the midpoint envelope $G^{\mathrm{mid},r_n}_{\eta,n}$. Comparing an arbitrary visible point to the origin $(0, 0, \dots, 0)$ yields

$$\left| G^{\mathrm{mid},r_n}_{\eta,n}(h, u_{\max(1,n-r_n)}, \dots, u_n) - c_{\eta,n} \right| \le h_* \, \|\partial_h G_{\eta,n}\|_\infty + 2 \sum_{k=\max(1,n-r_n)}^{n} b_k(\eta) \, \|\partial_{a_k} G_{\eta,n}\|_\infty,$$

which is exactly (450).

Finally, if (451) holds, then the tail-oscillation series is summable by item (i), and the centered-envelope series is summable by item (ii). Hence the hypotheses of item (iii) of Theorem D.77 are satisfied. The conclusion about the renormalized genuine continuum interaction potential then follows from Theorem D.80, while the coercivity / spectral-gap / log-Sobolev consequences come from the same theorem together with Theorem D.77. □

Integrated downstream conclusion after the weak-field/derivative-budget continuation. This continuation adds a thirty-ninth downstream message. Positively, the manuscript's full-support continuum lift now obeys a single reflected weak field equation on the intrinsic finitely supported test space, and the renormalized-potential/coercivity hypotheses are reduced further to explicit summable derivative budgets on the genuine released-mode increments themselves. Cautionarily, the manuscript still does not upgrade that weak field equation to a closed local reflected SPDE on a broader non-cylinder domain, and it still does not verify the derivative budgets for the actual e-Arc-Helix interaction increments. What remains open is therefore the genuinely local non-cylinder field equation and the concrete derivative estimates — or a sharp obstruction — for the real interaction family.

Remark D.84. The new thirty-ninth message is again deliberately intermediate. On the stochastic side, it no longer speaks only in terms of coordinatewise equations or a cylinder martingale problem: it packages the full-support continuum lift into one reflected duality equation against finitely supported tests. On the locality side, it no longer stops at abstract envelope/tail summability: it shows that the entire renormalized-potential/coercivity package would already follow from one concrete family of partial-derivative budgets on the genuine released-mode increments. What is still missing is not the existence of a weak field equation or a concrete verification route, but a truly local non-cylinder SPDE and the actual derivative bounds for the e-Arc-Helix interactions themselves.

D.27 Worked continuation XL: C^1 renormalized gradient potentials and a finite-cylinder verification bridge

The weak-field/derivative-budget continuation leaves the two pure existence questions in a more concrete but still not fully unified form. On the stochastic side, the intrinsic continuum lift already solves a single reflected weak field equation on finitely supported tests, yet the drift is still presented as a family of coordinatewise terms rather than as one continuum gradient object. On the interaction side, the renormalized-potential/coercivity package already follows from derivative budgets on the genuine released-mode increments, yet those budgets are still written increment-by-increment rather than directly on the finite cylinder interaction potentials that one would most naturally estimate in practice. The next intrinsic sharpening bridges exactly these two remaining gaps. A summable C^1 budget on the genuine increments produces a renormalized intrinsic continuum interaction potential with a continuous absolutely summable cylinder gradient, and the weak reflected field equation may then be rewritten as a weak reflected gradient equation driven by that potential. Moreover, the same C^1 budget is itself forced by a summable finite-cylinder derivative-Cauchy condition on the interaction potentials $V_{\eta,N}$.

Status label. The C^1-gradient theorem and the finite-cylinder verification bridge below have status **theorem-level downstream result** and **proposition-level downstream result**, respectively. They do *not* yet produce a genuinely local reflected SPDE on a broader non-cylinder state space, and they do *not* yet verify the needed derivative-Cauchy bounds for the actual e-Arc-Helix interaction potentials. What they do prove is that, under one stronger summable C^1 hypothesis, the renormalized continuum interaction potential already carries the full weak drift of the intrinsic continuum process, and that verification of this stronger package can be reduced from genuine increment estimates to direct finite-cylinder derivative estimates on $V_{\eta,N}$ themselves.

For $n \geq 1$, define the full C^1 increment budget

$$\mathfrak{M}^{C^1}_{\eta,n} := (1+h_*)\,\|\partial_h G_{\eta,n}\|_\infty + \sum_{k=1}^{n}(1+2b_k(\eta))\,\|\partial_{a_k} G_{\eta,n}\|_\infty. \tag{452}$$

Also write

$$\mathfrak{b}^{\varphi,\mathrm{ref}}_\eta(h,\mathbf{a}) := \varphi_0\left(2-\frac{5h}{h_*}\right) \tag{453}$$

for the reference weak drift carried by the product law $\nu_{\eta,N}$.

Theorem D.85 (Summable C^1 increments produce a weak reflected gradient field equation). *Assume the hypotheses of Theorem D.82, assume $V_{\eta,0}$ is C^1 on $[0,h_*]$, and assume each genuine increment $G_{\eta,n}$ extends to a C^1 function on a neighborhood of $\mathcal{X}_n$. If*

$$\sum_{n=1}^{\infty}\mathfrak{M}^{C^1}_{\eta,n} < \infty, \tag{454}$$

then the following hold.

(i) Let

$$d_{\eta,n} := G_{\eta,n}(0,0,\ldots,0), \qquad D_{\eta,N} := \sum_{n=1}^{N} d_{\eta,n}, \tag{455}$$

and define

$$V^{\mathrm{grad}}_{\eta,\infty}(x) := V_{\eta,0}(h) + \sum_{n=1}^{\infty}\Big(G_{\eta,n}(\Pi^{[\infty]}_n x) - d_{\eta,n}\Big), \qquad x=(h,\mathbf{a})\in\Gamma_\eta^{-1}(\mathfrak{S}^{(3)}_\eta). \tag{456}$$

Then the series converges uniformly on $\Gamma_\eta^{-1}(\mathfrak{S}^{(3)}_\eta)$, differs from the renormalized potential of Theorem D.80 by at most an additive constant, and admits continuous coordinate derivatives *given by*

$$\partial_h V^{\mathrm{grad}}_{\eta,\infty}(x) = V'_{\eta,0}(h) + \sum_{n=1}^{\infty} \partial_h G_{\eta,n}(\Pi^{[\infty]}_n x), \tag{457}$$

$$\partial_{a_k} V^{\mathrm{grad}}_{\eta,\infty}(x) = \sum_{n=k}^{\infty} \partial_{a_k} G_{\eta,n}(\Pi^{[\infty]}_n x), \qquad k \geq 1, \tag{458}$$

with both series converging uniformly. Moreover,

$$\sup_{x\in\Gamma_\eta^{-1}(\mathfrak{S}_\eta^{(3)})} \left(\left|\partial_h V^{\mathrm{grad}}_{\eta,\infty}(x)\right| + \sum_{k=1}^{\infty} \left|\partial_{a_k} V^{\mathrm{grad}}_{\eta,\infty}(x)\right| \right) < \infty. \tag{459}$$

(ii) For every $N \geq 1$,

$$\sup_{x\in\Gamma_\eta^{-1}(\mathfrak{S}_\eta^{(3)})} \left|\partial_h\big(V_{\eta,N}(\Pi_N x) - D_{\eta,N}\big) - \partial_h V^{\mathrm{grad}}_{\eta,\infty}(x)\right| \leq \sum_{n>N} \|\partial_h G_{\eta,n}\|_\infty, \tag{460}$$

$$\sup_{x\in\Gamma_\eta^{-1}(\mathfrak{S}_\eta^{(3)})} \left|\partial_{a_k}\big(V_{\eta,N}(\Pi_N x) - D_{\eta,N}\big) - \partial_{a_k} V^{\mathrm{grad}}_{\eta,\infty}(x)\right| \leq \sum_{n>N} \|\partial_{a_k} G_{\eta,n}\|_\infty, \qquad 1 \leq k \leq N. \tag{461}$$

(iii) For every finitely supported test direction $\varphi \in \mathfrak{T}_\eta^{\mathrm{fin}}$ and every $x = (h, \mathbf{a}) \in \Gamma_\eta^{-1}(\mathfrak{S}_\eta^{(3)})$,

$$\mathfrak{b}^\varphi_\eta(x) = \mathfrak{b}^{\varphi,\mathrm{ref}}_\eta(x) - A(h)\,\varphi_0\,\partial_h V^{\mathrm{grad}}_{\eta,\infty}(x) - \sum_{k=1}^{\infty} \varphi_k\,\partial_{a_k} V^{\mathrm{grad}}_{\eta,\infty}(x), \tag{462}$$

where the amplitude sum is finite because φ has finite support. Consequently the weak reflected field equation (443) may be rewritten as

$$\langle Z^x_t, \varphi\rangle_{\mathrm{cyl}} = \langle Z^x_0, \varphi\rangle_{\mathrm{cyl}} + \int_0^t \Big(\mathfrak{b}^{\varphi,\mathrm{ref}}_\eta(Z^x_s) - A(H^x_s)\varphi_0\,\partial_h V^{\mathrm{grad}}_{\eta,\infty}(Z^x_s) - \sum_{k=1}^{\infty}\varphi_k\,\partial_{a_k} V^{\mathrm{grad}}_{\eta,\infty}(Z^x_s)\Big)\,ds + \mathfrak{W}^\varphi_t + \mathfrak{R}^{x,\varphi}_t. \tag{463}$$

In particular, under a summable C^1 increment budget, the intrinsic continuum lift is already governed by a single weak reflected gradient *field equation driven by a bona fide renormalized continuum interaction potential. What remains open is not existence of a continuum potential carrying the weak drift, but promotion of that weak gradient equation to a genuinely local reflected SDE/SPDE on a broader non-cylinder domain.*

Proof. For each $n \geq 1$, the mean-value theorem and the coordinate ranges on $\mathcal{X}_n$ give

$$\|G_{\eta,n} - d_{\eta,n}\|_\infty \leq h_*\,\|\partial_h G_{\eta,n}\|_\infty + 2\sum_{k=1}^{n} b_k(\eta)\,\|\partial_{a_k} G_{\eta,n}\|_\infty \leq \mathfrak{M}^{C^1}_{\eta,n}.$$

Hence (454) implies uniform convergence of the series in (456). The same summability assumption also gives

$$\sum_{n=1}^{\infty} \|\partial_h G_{\eta,n}\|_\infty < \infty, \qquad \sum_{k=1}^{\infty}\sum_{n=k}^{\infty} \|\partial_{a_k} G_{\eta,n}\|_\infty = \sum_{n=1}^{\infty}\sum_{k=1}^{n} \|\partial_{a_k} G_{\eta,n}\|_\infty < \infty.$$

Therefore the derivative series (457)–(458) converge uniformly by the Weierstrass M-test. Since each partial sum is continuous on the compact support, the limits are continuous as well. The uniform ℓ^1-bound (459) is immediate from the displayed double-summability estimate together with $\|V'_{\eta,0}\|_\infty < \infty$. This proves item (i).

For item (ii), note that

$$V_{\eta,N}(\Pi_N x) - D_{\eta,N} = V_{\eta,0}(h) + \sum_{n=1}^{N}\Big(G_{\eta,n}(\Pi_n^{[\infty]}x) - d_{\eta,n}\Big)$$

on $\Gamma_\eta^{-1}(\mathfrak{S}_\eta^{(3)})$ by repeated use of the increment identity (422). Differentiating the finite sum and subtracting it from the infinite derivative series of item (i) leaves exactly the tails in (460)–(461).

For item (iii), fix $\varphi \in \mathfrak{T}_\eta^{\mathrm{fin}}$ and let N be large enough that $\varphi_k = 0$ for all $k > N$. Writing the density relation (392) inside the divergence-form generator (321) shows that on $\mathcal{X}_N$,

$$\begin{aligned}\mathcal{L}_{\eta,N}^{\mathrm{red}} f = A(h)\,\partial_h^2 f + \left(2 - \frac{5h}{h_*} - A(h)\,\partial_h V_{\eta,N}\right)\partial_h f \\ + \sum_{k=1}^{N}\left(\partial_{a_k}^2 f - \partial_{a_k} V_{\eta,N}\,\partial_{a_k} f\right),\end{aligned} \tag{464}$$

because $\nu_{\eta,N}$ carries the reference drift $2 - 5h/h_*$ in the helicity coordinate and no interaction drift in the amplitude coordinates. Hence the weak drift functional for the depth-N process is

$$\mathfrak{b}_{\eta,N}^{\varphi} = \mathfrak{b}_\eta^{\varphi,\mathrm{ref}} - A(h)\,\varphi_0\,\partial_h V_{\eta,N} - \sum_{k=1}^{N}\varphi_k\,\partial_{a_k} V_{\eta,N}.$$

Now lift this identity to $\Gamma_\eta^{-1}(\mathfrak{S}_\eta^{(3)})$ through Π_N, and pass to the continuum limit using the uniform derivative convergence from item (ii). This gives (462). Substituting that identity into (443) yields (463). □

Before stating the verification bridge, note that the lower-depth derivatives appearing below are always pulled back to the higher box by the natural projection Π_{n-1}. For example, $\partial_h V_{\eta,n-1}$ is viewed on $\mathcal{X}_n$ as $\partial_h V_{\eta,n-1} \circ \Pi_{n-1}$, and similarly for $\partial_{a_k} V_{\eta,n-1}$ when $1 \le k < n$.

Proposition D.86 (Finite-cylinder derivative-Cauchy criterion for the C^1 gradient package). *Assume $V_{\eta,0}$ is C^1 on $[0,h_*]$, and for each $N \ge 1$ assume $V_{\eta,N}$ extends to a C^1 function on a neighborhood of $\mathcal{X}_N$. For $n \ge 1$, define*

$$\Delta_{\eta,n}^h := \|\partial_h V_{\eta,n} - \partial_h V_{\eta,n-1}\|_\infty, \tag{465}$$

$$\Delta_{\eta,n,k}^a := \|\partial_{a_k} V_{\eta,n} - \partial_{a_k} V_{\eta,n-1}\|_\infty, \qquad 1 \le k < n, \tag{466}$$

$$\Delta_{\eta,n,n}^a := \|\partial_{a_n} V_{\eta,n}\|_\infty. \tag{467}$$

If

$$\sum_{n=1}^{\infty}\left((1+h_*)\,\Delta_{\eta,n}^h + \sum_{k=1}^{n}(1+2b_k(\eta))\,\Delta_{\eta,n,k}^a\right) < \infty, \tag{468}$$

where for $k = n$ one reads $\Delta_{\eta,n,n}^a$ inside the inner sum, then the hypothesis (454) of Theorem D.85 holds. Consequently the renormalized continuum interaction potential with continuous absolutely summable cylinder gradient exists, the weak reflected field equation already has the gradient form (463), and the automatic coercivity / renormalized-potential consequences of Theorems D.77, D.80, and D.83 all follow.

Proof. By definition of the genuine one-step increment,

$$G_{\eta,n}(x) = V_{\eta,n}(x) - V_{\eta,n-1}(\Pi_{n-1}x) \qquad \text{on } \mathcal{X}_n.$$

Differentiating removes any additive renormalization constants and gives

$$\|\partial_h G_{\eta,n}\|_\infty = \Delta^h_{\eta,n}, \qquad \|\partial_{a_k} G_{\eta,n}\|_\infty = \Delta^a_{\eta,n,k}, \quad 1 \le k \le n,$$

with the convention that $\Delta^a_{\eta,n,n} = \|\partial_{a_n} V_{\eta,n}\|_\infty$. Substituting these identities into (452) shows that (468) is exactly the summability condition (454). The remaining conclusions are then immediate from Theorem D.85 and from the previously established tail-envelope / renormalized-potential theorems. □

Integrated downstream conclusion after the C^1-gradient/verification-bridge continuation. This continuation adds a fortieth downstream message. Positively, the intrinsic continuum interaction field now admits a renormalized continuum interaction potential with a continuous absolutely summable cylinder gradient, and the weak reflected field equation is already a weak reflected gradient equation driven by that potential. Verification is no longer posed only on genuine increment budgets: it can be attacked directly on the finite-cylinder interaction potentials through a derivative-Cauchy criterion. Cautionarily, the manuscript still does not promote this weak gradient formulation to a genuinely local reflected SPDE on a broader non-cylinder domain, and it still does not prove the derivative-Cauchy bounds for the actual e-Arc-Helix interaction potentials. What remains open is therefore the local reflected gradient SPDE itself and the concrete finite-cylinder derivative estimates — or a sharp obstruction — for the real interaction family.

Remark D.87. The new fortieth message closes a conceptual gap between the stochastic and interaction sides of the appendix. The process is no longer described merely by a weak field equation plus a separate continuum potential theorem: under a summable C^1 budget, the same renormalized intrinsic continuum interaction potential already carries the weak drift itself. On the verification side, one no longer needs to estimate only the abstract genuine increments in isolation; it is enough to show that the finite-cylinder interaction potentials form a summably C^1-Cauchy family after one-step projection. What is still missing is the genuinely local non-cylinder closure of that gradient dynamics and the actual derivative-Cauchy bounds for the e-Arc-Helix interaction family.

D.28 Worked continuation XLI: explicit gradient cylinder martingale closure and a finite-cylinder Hessian bridge

The C^1-gradient/verification-bridge continuation leaves the two pure existence questions in a sharper but still not fully local form. On the stochastic side, the continuum drift is now carried by one renormalized interaction potential, yet the closed cylinder martingale problem is still written abstractly through $\mathfrak{L}^{\mathrm{cyl}}_\eta$ rather than explicitly in gradient form. On the interaction side, verification can already be posed directly on first derivatives of the finite-cylinder potentials, yet this still gives no local regularity statement for the resulting continuum drift itself. The next intrinsic sharpening addresses both points. First, the strong continuum lift satisfies an explicit reflected gradient cylinder martingale problem driven by the renormalized continuum interaction potential. Second, if the genuine released-mode increments satisfy a summable weighted C^2 budget, then that same continuum gradient drift becomes weighted cylinder-Lipschitz on the intrinsic $\ell^1_{(3)}$ support, and this stronger package is itself forced by a finite-cylinder Hessian-Cauchy condition.

Status label. The explicit gradient-martingale theorem below has status **theorem-level downstream result**; the finite-cylinder Hessian bridge has status **proposition-level downstream result**. They do *not* yet produce a genuinely local reflected gradient SPDE on a broader non-cylinder state space, and they do *not* yet verify the required Hessian-Cauchy bounds for the actual e-Arc-Helix interaction potentials. What they do prove is that the intrinsic strong continuum lift is already governed on the full cylinder core by one explicit gradient operator, and that a stronger weighted C^2 regularity package can be reduced entirely to finite-cylinder second-derivative differences.

For $x = (h, \mathbf{a})$, $y = (\widetilde{h}, \widetilde{\mathbf{a}}) \in [0, h_*] \times \ell^1_{(3)}$, write

$$d_\eta^{(3)}(x, y) := |h - \widetilde{h}| + \|\mathbf{a} - \widetilde{\mathbf{a}}\|_{(3)}. \tag{469}$$

For $n \geq 1$, define the weighted C^2 increment budget

$$\mathfrak{H}_{\eta,n}^{(3)} := \|\partial_{hh} G_{\eta,n}\|_\infty + \sum_{k=1}^{n} \frac{\|\partial_h \partial_{a_k} G_{\eta,n}\|_\infty}{(2k+1)^3} + \sum_{k,\ell=1}^{n} \frac{\|\partial_{a_k} \partial_{a_\ell} G_{\eta,n}\|_\infty}{(2k+1)^3 (2\ell+1)^3}. \tag{470}$$

For $F = f \circ \Pi_N \in \mathcal{D}_\eta^{\mathrm{Neu},\infty}$, define the explicit gradient cylinder operator by

$$\begin{aligned}(\mathfrak{L}_\eta^\nabla F)(x) &:= A(h)\, \partial_h^2 f(\Pi_N x) \\ &\quad + \left(2 - \frac{5h}{h_*} - A(h)\, \partial_h V_{\eta,\infty}^{\mathrm{grad}}(x)\right) \partial_h f(\Pi_N x) \\ &\quad + \sum_{k=1}^{N} \left(\partial_{a_k}^2 f(\Pi_N x) - \partial_{a_k} V_{\eta,\infty}^{\mathrm{grad}}(x)\, \partial_{a_k} f(\Pi_N x)\right).\end{aligned} \tag{471}$$

Theorem D.88 (Explicit reflected gradient cylinder martingale problem and weighted cylinder-Lipschitz drift). *Assume the hypotheses of Theorem D.85, and let*

$$Z^x = (H^x, A_1^x, A_2^x, \dots)$$

be the full-support strong continuum lift started from $x \in \Gamma_\eta^{-1}(\mathfrak{S}_\eta^{(3)})$. Then the following hold.

(i) For every $F = f \circ \Pi_N \in \mathcal{D}_\eta^{\mathrm{Neu},\infty}$,

$$M_t^{F,\nabla} := F(Z_t^x) - F(x) - \int_0^t (\mathfrak{L}_\eta^\nabla F)(Z_s^x)\, ds \tag{472}$$

is a continuous martingale.

(ii) The quadratic variation of (472) is

$$\langle M^{F,\nabla} \rangle_t = 2 \int_0^t \left(A(H_s^x)\, |\partial_h f(\Pi_N Z_s^x)|^2 + \sum_{k=1}^{N} |\partial_{a_k} f(\Pi_N Z_s^x)|^2 \right) ds. \tag{473}$$

(iii) Assume in addition that $V_{\eta,0}$ is C^2 on $[0, h_*]$, that each genuine increment $G_{\eta,n}$ extends to a C^2 function on a neighborhood of $\mathcal{X}_n$, and that

$$\sum_{n=1}^{\infty} \mathfrak{H}_{\eta,n}^{(3)} < \infty. \tag{474}$$

Define

$$L_{\eta,h}^{(3)} := \|V_{\eta,0}''\|_\infty + \sum_{n=1}^{\infty} \left(\|\partial_{hh} G_{\eta,n}\|_\infty + \sum_{k=1}^{n} \frac{\|\partial_h \partial_{a_k} G_{\eta,n}\|_\infty}{(2k+1)^3} \right), \tag{475}$$

$$L_{\eta,j}^{(3)} := \sum_{n=j}^{\infty} \left(\|\partial_h \partial_{a_j} G_{\eta,n}\|_\infty + \sum_{k=1}^{n} \frac{\|\partial_{a_j} \partial_{a_k} G_{\eta,n}\|_\infty}{(2k+1)^3} \right), \qquad j \geq 1. \tag{476}$$

Then

$$\sum_{j=1}^{\infty} \frac{L_{\eta,j}^{(3)}}{(2j+1)^3} < \infty, \tag{477}$$

and for every $x, y \in \Gamma_\eta^{-1}(\mathfrak{S}_\eta^{(3)})$,

$$\left|\partial_h V_{\eta,\infty}^{\mathrm{grad}}(x) - \partial_h V_{\eta,\infty}^{\mathrm{grad}}(y)\right| \le L_{\eta,h}^{(3)}\, d_\eta^{(3)}(x,y), \tag{478}$$

$$\left|\partial_{a_j} V_{\eta,\infty}^{\mathrm{grad}}(x) - \partial_{a_j} V_{\eta,\infty}^{\mathrm{grad}}(y)\right| \le L_{\eta,j}^{(3)}\, d_\eta^{(3)}(x,y), \qquad j \ge 1. \tag{479}$$

Consequently the gradient drift entering (471) is weighted cylinder-Lipschitz on the intrinsic continuum support.

In particular, the manuscript now has not only a weak reflected gradient field equation, but also an explicit gradient martingale problem on the full cylinder core, and under a summable weighted C^2 increment budget the corresponding continuum drift already has a weighted local regularity modulus on $\mathfrak{S}_\eta^{(3)}$. What remains open is not explicit cylinder-core gradient closure or weighted local drift regularity, but promotion of this package to a genuinely local reflected gradient SPDE on a broader non-cylinder domain.

Proof. Fix $F = f \circ \Pi_N \in \mathcal{D}_\eta^{\mathrm{Neu},\infty}$. Apply Theorem D.85 with the basis vectors $\varphi = e^{(0)}$ and $\varphi = e^{(k)}$, $1 \le k \le N$. This yields the coordinate semimartingale equations

$$dH_t^x = \left(2 - \frac{5H_t^x}{h_*} - A(H_t^x)\,\partial_h V_{\eta,\infty}^{\mathrm{grad}}(Z_t^x)\right) dt + \sqrt{2A(H_t^x)}\, d\beta_t + dL_t^{0,x} - dL_t^{h_*,x},$$

$$dA_{k,t}^x = -\partial_{a_k} V_{\eta,\infty}^{\mathrm{grad}}(Z_t^x)\, dt + dB_{k,t} + dK_{k,t}^{+,x} - dK_{k,t}^{-,x}, \qquad 1 \le k \le N.$$

Apply Itô's formula to the finite coordinate vector $(H_t^x, A_{1,t}^x, \ldots, A_{N,t}^x)$. The boundary terms vanish because the visible representative f satisfies the Neumann conditions (432)–(433). The resulting drift is exactly $(\mathfrak{L}_\eta^\nabla F)(Z_t^x)\, dt$, which proves item (i). The quadratic variation formula (473) comes from the diagonal diffusion coefficients $2A(H_t^x)$ in the helicity coordinate and 2 in each visible amplitude coordinate, together with independence of $\beta, B_1, \ldots, B_N$. This proves item (ii).

For item (iii), the definition (456) and termwise differentiation from Theorem D.85 give

$$\partial_h V_{\eta,\infty}^{\mathrm{grad}}(x) - \partial_h V_{\eta,\infty}^{\mathrm{grad}}(y) = V_{\eta,0}'(h) - V_{\eta,0}'(\widetilde{h}) + \sum_{n=1}^{\infty}\Big(\partial_h G_{\eta,n}(\Pi_n^{[\infty]}x) - \partial_h G_{\eta,n}(\Pi_n^{[\infty]}y)\Big),$$

$$\partial_{a_j} V_{\eta,\infty}^{\mathrm{grad}}(x) - \partial_{a_j} V_{\eta,\infty}^{\mathrm{grad}}(y) = \sum_{n=j}^{\infty}\Big(\partial_{a_j} G_{\eta,n}(\Pi_n^{[\infty]}x) - \partial_{a_j} G_{\eta,n}(\Pi_n^{[\infty]}y)\Big).$$

Because $|a_k - \widetilde{a}_k| \le (2k+1)^{-3}\|\mathbf{a} - \widetilde{\mathbf{a}}\|_{(3)}$, the mean-value theorem gives

$$\left|\partial_h G_{\eta,n}(\Pi_n^{[\infty]}x) - \partial_h G_{\eta,n}(\Pi_n^{[\infty]}y)\right| \le \left(\|\partial_{hh} G_{\eta,n}\|_\infty + \sum_{k=1}^{n} \frac{\|\partial_h \partial_{a_k} G_{\eta,n}\|_\infty}{(2k+1)^3}\right) d_\eta^{(3)}(x,y),$$

$$\left|\partial_{a_j} G_{\eta,n}(\Pi_n^{[\infty]}x) - \partial_{a_j} G_{\eta,n}(\Pi_n^{[\infty]}y)\right| \le \left(\|\partial_h \partial_{a_j} G_{\eta,n}\|_\infty + \sum_{k=1}^{n} \frac{\|\partial_{a_j} \partial_{a_k} G_{\eta,n}\|_\infty}{(2k+1)^3}\right) d_\eta^{(3)}(x,y).$$

Summing the first inequality over n and using (475) gives (478). Summing the second inequality over $n \ge j$ gives (479) with (476). Finally,

$$\sum_{j=1}^{\infty} \frac{L_{\eta,j}^{(3)}}{(2j+1)^3} = \sum_{n=1}^{\infty}\left(\sum_{j=1}^{n} \frac{\|\partial_h \partial_{a_j} G_{\eta,n}\|_\infty}{(2j+1)^3} + \sum_{j=1}^{n}\sum_{k=1}^{n} \frac{\|\partial_{a_j} \partial_{a_k} G_{\eta,n}\|_\infty}{(2j+1)^3(2k+1)^3}\right),$$

which is finite by (474). This proves (477) and completes the proof. □

Proposition D.89 (Finite-cylinder Hessian-Cauchy criterion for weighted cylinder-Lipschitz drift). *Assume the hypotheses of Proposition D.86, assume $V_{\eta,0}$ is C^2 on $[0,h_*]$, and for each $N \geq 1$ assume $V_{\eta,N}$ extends to a C^2 function on a neighborhood of $\mathcal{X}_N$. For $n \geq 1$, define*

$$\Theta^{hh}_{\eta,n} := \left\|\partial_{hh}V_{\eta,n} - \partial_{hh}V_{\eta,n-1}\right\|_\infty, \tag{480}$$

$$\Theta^{ha}_{\eta,n,k} := \left\|\partial_h\partial_{a_k}V_{\eta,n} - \partial_h\partial_{a_k}V_{\eta,n-1}\right\|_\infty, \qquad 1 \leq k \leq n, \tag{481}$$

$$\Theta^{aa}_{\eta,n,k,\ell} := \left\|\partial_{a_k}\partial_{a_\ell}V_{\eta,n} - \partial_{a_k}\partial_{a_\ell}V_{\eta,n-1}\right\|_\infty, \qquad 1 \leq k,\ell \leq n, \tag{482}$$

where every derivative of $V_{\eta,n-1}$ involving the new coordinate a_n is read as zero after pullback through Π_{n-1}. If

$$\sum_{n=1}^{\infty}\left(\Theta^{hh}_{\eta,n} + \sum_{k=1}^{n}\frac{\Theta^{ha}_{\eta,n,k}}{(2k+1)^3} + \sum_{k,\ell=1}^{n}\frac{\Theta^{aa}_{\eta,n,k,\ell}}{(2k+1)^3(2\ell+1)^3}\right) < \infty, \tag{483}$$

then the weighted C^2 increment budget (470) is summable. Consequently, once the standing stochastic hypotheses of Theorem D.88 are kept in force, the additional weighted-C^2 assumption (474) of that theorem is satisfied. Together with the first-derivative condition (468) of Proposition D.86, this yields the explicit gradient cylinder martingale problem (472) and the weighted cylinder-Lipschitz regularity of the renormalized continuum gradient drift on the intrinsic support.

Proof. By the increment identity

$$G_{\eta,n}(x) = V_{\eta,n}(x) - V_{\eta,n-1}(\Pi_{n-1}x) \qquad \text{on } \mathcal{X}_n,$$

differentiating twice gives

$$\|\partial_{hh}G_{\eta,n}\|_\infty = \Theta^{hh}_{\eta,n}, \qquad \|\partial_h\partial_{a_k}G_{\eta,n}\|_\infty = \Theta^{ha}_{\eta,n,k}, \qquad \|\partial_{a_k}\partial_{a_\ell}G_{\eta,n}\|_\infty = \Theta^{aa}_{\eta,n,k,\ell},$$

with the convention that derivatives of $V_{\eta,n-1}$ in the new coordinate a_n vanish. Substituting these identities into (470) shows that the series in (483) is exactly $\sum_{n\geq 1}\mathfrak{H}^{(3)}_{\eta,n}$. The remaining conclusion therefore follows from Proposition D.86 and Theorem D.88. □

Integrated downstream conclusion after the explicit-gradient/Hessian-bridge continuation. This continuation adds a forty-first downstream message. Positively, the intrinsic continuum lift now has an explicit reflected gradient cylinder martingale formulation driven by the renormalized continuum interaction potential, and under summably controlled weighted C^2 increments the corresponding drift is weighted cylinder-Lipschitz on $\mathfrak{S}^{(3)}_\eta$. Verification is no longer only a first-derivative problem: the stronger local-regularity package is reduced to a finite-cylinder Hessian-Cauchy condition on $V_{\eta,N}$. Cautionarily, the manuscript still does not upgrade this weighted cylinder-Lipschitz core to a genuinely local reflected SPDE on a broader non-cylinder domain, and it still does not prove the Hessian-Cauchy bounds for the actual e-Arc-Helix interaction family. What remains open is therefore the local reflected gradient SPDE itself and the concrete second-derivative estimates — or a sharp obstruction — for the real interaction potentials.

Remark D.90. The new forty-first message is again deliberately intermediate. On the stochastic side, it no longer speaks only in terms of a weak gradient duality or an abstract cylinder operator: the full-support continuum lift is already characterized on the cylinder core by an explicit reflected gradient martingale problem. On the interaction side, it no longer stops at C^1 verification: under a weighted C^2 budget the continuum drift itself acquires a weighted local regularity modulus, and the verification target is reduced correspondingly to finite-cylinder Hessian-Cauchy bounds. What is still missing is not an explicit cylinder-core gradient law or a concrete second-derivative criterion, but a genuinely local reflected gradient SPDE and the actual Hessian estimates for the e-Arc-Helix interaction potentials themselves.

D.29 Worked continuation XLII: weighted $C^{1,1}$ linearization and a finite-cylinder third-derivative bridge

The explicit-gradient/Hessian-bridge continuation already gives an explicit reflected gradient martingale problem and weighted cylinder-Lipschitz drift on the intrinsic continuum support, but it still stops one step short of a genuinely local closure mechanism. On the stochastic side, the drift is controlled only at first order, so there is not yet a canonical affine freezing principle with a quantified quadratic remainder. On the interaction side, verification can already be posed directly on second derivatives of the finite-cylinder potentials, but this still does not isolate the third-order information needed for a local Taylor closure of the continuum drift itself. The next intrinsic sharpening addresses both gaps. First, under a summable weighted C^3 budget the renormalized continuum gradient drift already admits a weighted cylinder-$C^{1,1}$ linearization with quadratic local remainder. Second, this affine local-closure package is itself forced by a finite-cylinder third-derivative-Cauchy condition.

Status label. The weighted $C^{1,1}$ linearization theorem below has status **theorem-level downstream result**; the finite-cylinder third-derivative bridge has status **proposition-level downstream result**. They do *not* yet produce a genuinely local reflected gradient SPDE on a broader non-cylinder state space, and they do *not* yet verify the required third-derivative bounds for the actual e-Arc-Helix interaction potentials. What they do prove is that the explicit gradient cylinder law already carries its own affine local freezing principle with quadratic error, and that this stronger local-closure package can be reduced entirely to finite-cylinder third-derivative differences.

For $n \geq 1$, define the weighted C^3 increment budget

$$\begin{aligned} \mathfrak{J}^{(3)}_{\eta,n} := \|\partial_{hhh} G_{\eta,n}\|_\infty + 2\sum_{k=1}^{n} \frac{\|\partial_{hh}\partial_{a_k} G_{\eta,n}\|_\infty}{(2k+1)^3} \\ + 2\sum_{k,\ell=1}^{n} \frac{\|\partial_h \partial_{a_k}\partial_{a_\ell} G_{\eta,n}\|_\infty}{(2k+1)^3(2\ell+1)^3} + \sum_{j,k,\ell=1}^{n} \frac{\|\partial_{a_j}\partial_{a_k}\partial_{a_\ell} G_{\eta,n}\|_\infty}{(2j+1)^3(2k+1)^3(2\ell+1)^3}. \end{aligned} \tag{484}$$

Theorem D.91 (Weighted $C^{1,1}$ linearization of the continuum gradient drift and quadratic local freezing)**.** *Assume the setting of item* (iii) *of Theorem D.88. Assume in addition that $V_{\eta,0}$ is C^3 on $[0, h_*]$, that each genuine increment $G_{\eta,n}$ extends to a C^3 function on a neighborhood of $\mathcal{X}_n$, and that*

$$\sum_{n=1}^{\infty} \mathfrak{J}^{(3)}_{\eta,n} < \infty. \tag{485}$$

Define

$$Q^{(3)}_{\eta,h} := \|V'''_{\eta,0}\|_\infty + \sum_{n=1}^{\infty} \left(\|\partial_{hhh} G_{\eta,n}\|_\infty + 2\sum_{k=1}^{n} \frac{\|\partial_{hh}\partial_{a_k} G_{\eta,n}\|_\infty}{(2k+1)^3} + \sum_{k,\ell=1}^{n} \frac{\|\partial_h \partial_{a_k}\partial_{a_\ell} G_{\eta,n}\|_\infty}{(2k+1)^3(2\ell+1)^3} \right), \tag{486}$$

$$Q^{(3)}_{\eta,j} := \sum_{n=j}^{\infty} \left(\|\partial_{hh}\partial_{a_j} G_{\eta,n}\|_\infty + 2\sum_{k=1}^{n} \frac{\|\partial_h \partial_{a_j}\partial_{a_k} G_{\eta,n}\|_\infty}{(2k+1)^3} + \sum_{k,\ell=1}^{n} \frac{\|\partial_{a_j}\partial_{a_k}\partial_{a_\ell} G_{\eta,n}\|_\infty}{(2k+1)^3(2\ell+1)^3} \right), \quad j \geq 1. \tag{487}$$

Then the following hold.

(i)

$$\sum_{j=1}^{\infty} \frac{Q^{(3)}_{\eta,j}}{(2j+1)^3} < \infty. \tag{488}$$

(ii) For $x=(h,\mathbf{a})$, $y=(\widetilde{h},\widetilde{\mathbf{a}})\in\Gamma_\eta^{-1}(\mathfrak{S}_\eta^{(3)})$, write $\delta_h:=h-\widetilde{h}$ and $\delta a_k:=a_k-\widetilde{a}_k$. Then the affine row expansions

$$\mathcal{R}_{\eta,h}[y](x-y):=\partial_{hh}V_{\eta,\infty}^{\mathrm{grad}}(y)\,\delta_h+\sum_{k=1}^{\infty}\partial_h\partial_{a_k}V_{\eta,\infty}^{\mathrm{grad}}(y)\,\delta a_k, \tag{489}$$

$$\mathcal{R}_{\eta,j}[y](x-y):=\partial_h\partial_{a_j}V_{\eta,\infty}^{\mathrm{grad}}(y)\,\delta_h+\sum_{k=1}^{\infty}\partial_{a_j}\partial_{a_k}V_{\eta,\infty}^{\mathrm{grad}}(y)\,\delta a_k,\qquad j\geq 1, \tag{490}$$

are absolutely convergent, and

$$\left|\partial_hV_{\eta,\infty}^{\mathrm{grad}}(x)-\partial_hV_{\eta,\infty}^{\mathrm{grad}}(y)-\mathcal{R}_{\eta,h}[y](x-y)\right|\leq\frac{1}{2}Q_{\eta,h}^{(3)}\,d_\eta^{(3)}(x,y)^2, \tag{491}$$

$$\left|\partial_{a_j}V_{\eta,\infty}^{\mathrm{grad}}(x)-\partial_{a_j}V_{\eta,\infty}^{\mathrm{grad}}(y)-\mathcal{R}_{\eta,j}[y](x-y)\right|\leq\frac{1}{2}Q_{\eta,j}^{(3)}\,d_\eta^{(3)}(x,y)^2,\qquad j\geq 1. \tag{492}$$

Consequently the renormalized continuum gradient drift is weighted cylinder-$C^{1,1}$ on the intrinsic support.

(iii) For $F=f\circ\Pi_N\in\mathcal{D}_\eta^{\mathrm{Neu},\infty}$ and fixed base point $y\in\Gamma_\eta^{-1}(\mathfrak{S}_\eta^{(3)})$, define the affine-frozen gradient operator by

$$\begin{aligned}(\mathfrak{L}_{\eta,y}^{\mathrm{aff}}F)(x):=&\,A(h)\,\partial_h^2f(\Pi_Nx)\\&+\left(2-\frac{5h}{h_*}-A(h)\left[\partial_hV_{\eta,\infty}^{\mathrm{grad}}(y)+\mathcal{R}_{\eta,h}[y](x-y)\right]\right)\partial_hf(\Pi_Nx)\\&+\sum_{k=1}^{N}\left(\partial_{a_k}^2f(\Pi_Nx)-\left[\partial_{a_k}V_{\eta,\infty}^{\mathrm{grad}}(y)+\mathcal{R}_{\eta,k}[y](x-y)\right]\partial_{a_k}f(\Pi_Nx)\right).\end{aligned} \tag{493}$$

Then for every $x,y\in\Gamma_\eta^{-1}(\mathfrak{S}_\eta^{(3)})$,

$$\left|(\mathfrak{L}_\eta^\nabla F)(x)-(\mathfrak{L}_{\eta,y}^{\mathrm{aff}}F)(x)\right|\leq\frac{1}{2}\left(\|A\,\partial_hf\|_\infty\,Q_{\eta,h}^{(3)}+\sum_{k=1}^{N}\|\partial_{a_k}f\|_\infty\,Q_{\eta,k}^{(3)}\right)d_\eta^{(3)}(x,y)^2. \tag{494}$$

In particular, the manuscript now has not only an explicit gradient cylinder martingale problem with weighted cylinder-Lipschitz drift, but also a canonical affine local freezing principle with quadratic error on the intrinsic continuum support. What remains open is not first-order local closure, but promotion of that affine package to a genuinely local reflected gradient SPDE on a broader non-cylinder domain.

Proof. Because the hypotheses of item (iii) of Theorem D.88 are in force, the second-derivative rows entering (489)–(490) already converge absolutely on $\Gamma_\eta^{-1}(\mathfrak{S}_\eta^{(3)})$ by (475)–(477). Moreover,

$$\begin{aligned}\sum_{j=1}^{\infty}\frac{Q_{\eta,j}^{(3)}}{(2j+1)^3}=\sum_{n=1}^{\infty}\Bigg[&\sum_{j=1}^{n}\frac{\|\partial_{hh}\partial_{a_j}G_{\eta,n}\|_\infty}{(2j+1)^3}\\&+2\sum_{j,k=1}^{n}\frac{\|\partial_h\partial_{a_j}\partial_{a_k}G_{\eta,n}\|_\infty}{(2j+1)^3(2k+1)^3}\\&+\sum_{j,k,\ell=1}^{n}\frac{\|\partial_{a_j}\partial_{a_k}\partial_{a_\ell}G_{\eta,n}\|_\infty}{(2j+1)^3(2k+1)^3(2\ell+1)^3}\Bigg],\end{aligned}$$

which is finite by (485). This proves item (i).

For item (ii), let $\gamma_\theta := y + \theta(x-y)$ and define

$$\psi_h(\theta) := \partial_h V^{\mathrm{grad}}_{\eta,\infty}(\gamma_\theta), \qquad \psi_j(\theta) := \partial_{a_j} V^{\mathrm{grad}}_{\eta,\infty}(\gamma_\theta), \quad j \ge 1.$$

By absolute convergence of the second-derivative rows, the affine terms in (489)–(490) are exactly $\psi_h'(0)$ and $\psi_j'(0)$. Taylor's formula with integral remainder gives

$$\psi_h(1) - \psi_h(0) - \psi_h'(0) = \int_0^1 (1-\theta)\, \psi_h''(\theta)\, d\theta,$$

and similarly for ψ_j. Termwise differentiation of the series defining $V^{\mathrm{grad}}_{\eta,\infty}$ gives

$$\begin{aligned}
\psi_h''(\theta) &= V_{\eta,0}'''(h_\theta)\,\delta_h^2 \\
&\quad + \sum_{n=1}^{\infty} \Bigg[\partial_{hhh} G_{\eta,n}(\Pi_n^{[\infty]}\gamma_\theta)\,\delta_h^2 + 2\sum_{k=1}^{n} \partial_{hh}\partial_{a_k} G_{\eta,n}(\Pi_n^{[\infty]}\gamma_\theta)\,\delta_h\,\delta a_k \\
&\qquad\qquad + \sum_{k,\ell=1}^{n} \partial_h \partial_{a_k}\partial_{a_\ell} G_{\eta,n}(\Pi_n^{[\infty]}\gamma_\theta)\,\delta a_k\,\delta a_\ell \Bigg], \\
\psi_j''(\theta) &= \sum_{n=j}^{\infty} \Bigg[\partial_{hh}\partial_{a_j} G_{\eta,n}(\Pi_n^{[\infty]}\gamma_\theta)\,\delta_h^2 + 2\sum_{k=1}^{n} \partial_h\partial_{a_j}\partial_{a_k} G_{\eta,n}(\Pi_n^{[\infty]}\gamma_\theta)\,\delta_h\,\delta a_k \\
&\qquad\qquad + \sum_{k,\ell=1}^{n} \partial_{a_j} \partial_{a_k}\partial_{a_\ell} G_{\eta,n}(\Pi_n^{[\infty]}\gamma_\theta)\,\delta a_k\,\delta a_\ell \Bigg].
\end{aligned}$$

Because $|\delta_h| \le d_\eta^{(3)}(x,y)$ and $|\delta a_k| \le (2k+1)^{-3} d_\eta^{(3)}(x,y)$, the definitions (486)–(487) imply

$$|\psi_h''(\theta)| \le Q_{\eta,h}^{(3)}\, d_\eta^{(3)}(x,y)^2, \qquad |\psi_j''(\theta)| \le Q_{\eta,j}^{(3)}\, d_\eta^{(3)}(x,y)^2.$$

Integrating over $\theta \in [0,1]$ yields (491) and (492).

For item (iii), subtract (493) from (471). The second-order diffusion terms cancel, and the remaining difference is exactly the drift remainder from item (ii) multiplied by the first derivatives of f. Using (491)–(492) gives (494) immediately. □

Proposition D.92 (Finite-cylinder third-derivative-Cauchy criterion for weighted $C^{1,1}$ local freezing). *Assume the hypotheses of Proposition D.89, assume $V_{\eta,0}$ is C^3 on $[0,h_*]$, and for each $N \ge 1$ assume $V_{\eta,N}$ extends to a C^3 function on a neighborhood of $\mathcal{X}_N$. For $n \ge 1$, define*

$$\Xi_{\eta,n}^{hhh} := \left\| \partial_{hhh} V_{\eta,n} - \partial_{hhh} V_{\eta,n-1} \right\|_\infty, \tag{495}$$

$$\Xi_{\eta,n,k}^{hha} := \left\| \partial_{hh}\partial_{a_k} V_{\eta,n} - \partial_{hh}\partial_{a_k} V_{\eta,n-1} \right\|_\infty, \qquad 1 \le k \le n, \tag{496}$$

$$\Xi_{\eta,n,k,\ell}^{haa} := \left\| \partial_h\partial_{a_k}\partial_{a_\ell} V_{\eta,n} - \partial_h\partial_{a_k}\partial_{a_\ell} V_{\eta,n-1} \right\|_\infty, \qquad 1 \le k,\ell \le n, \tag{497}$$

$$\Xi_{\eta,n,j,k,\ell}^{aaa} := \left\| \partial_{a_j}\partial_{a_k}\partial_{a_\ell} V_{\eta,n} - \partial_{a_j}\partial_{a_k}\partial_{a_\ell} V_{\eta,n-1} \right\|_\infty, \qquad 1 \le j,k,\ell \le n, \tag{498}$$

where every derivative of $V_{\eta,n-1}$ involving the new coordinate a_n is read as zero after pullback through Π_{n-1}. If

$$\sum_{n=1}^{\infty} \left(\Xi_{\eta,n}^{hhh} + 2\sum_{k=1}^{n} \frac{\Xi_{\eta,n,k}^{hha}}{(2k+1)^3} + 2\sum_{k,\ell=1}^{n} \frac{\Xi_{\eta,n,k,\ell}^{haa}}{(2k+1)^3(2\ell+1)^3} + \sum_{j,k,\ell=1}^{n} \frac{\Xi_{\eta,n,j,k,\ell}^{aaa}}{(2j+1)^3(2k+1)^3(2\ell+1)^3} \right) < \infty, \tag{499}$$

then the weighted C^3 increment budget (484) is summable. Consequently, together with the first- and second-derivative conditions already imported by Proposition D.89, the affine local-freezing package of Theorem D.91 applies on the intrinsic continuum support.

Proof. Differentiating the increment identity

$$G_{\eta,n}(x) = V_{\eta,n}(x) - V_{\eta,n-1}(\Pi_{n-1}x) \qquad \text{on } \mathcal{X}_n$$

three times gives

$$\|\partial_{hhh}G_{\eta,n}\|_\infty = \Xi^{hhh}_{\eta,n}, \qquad \|\partial_{hh}\partial_{a_k}G_{\eta,n}\|_\infty = \Xi^{hha}_{\eta,n,k}, \qquad \|\partial_h\partial_{a_k}\partial_{a_\ell}G_{\eta,n}\|_\infty = \Xi^{haa}_{\eta,n,k,\ell},$$

and

$$\|\partial_{a_j}\partial_{a_k}\partial_{a_\ell}G_{\eta,n}\|_\infty = \Xi^{aaa}_{\eta,n,j,k,\ell},$$

with the convention that derivatives of $V_{\eta,n-1}$ in the new coordinate a_n vanish. Substituting these identities into (484) shows that the series in (499) is exactly $\sum_{n\geq 1}\mathfrak{J}^{(3)}_{\eta,n}$. The remaining conclusion therefore follows from Proposition D.89 and Theorem D.91. □

Integrated downstream conclusion after the weighted-$C^{1,1}$/third-derivative continuation. This continuation adds a forty-second downstream message. Positively, the intrinsic continuum lift no longer has only an explicit reflected gradient martingale law with weighted cylinder-Lipschitz drift: under summably controlled weighted C^3 increments, that same drift already has a weighted cylinder-$C^{1,1}$ linearization with a canonical quadratic local freezing error on the cylinder generator. Verification is correspondingly no longer only a second-derivative problem: the affine local-closure package is reduced to a finite-cylinder third-derivative-Cauchy condition on $V_{\eta,N}$. Cautionarily, the manuscript still does not upgrade this affine local freezing principle to a genuinely local reflected SPDE on a broader non-cylinder domain, and it still does not prove the third-derivative-Cauchy bounds for the actual e-Arc-Helix interaction family. What remains open is therefore the local reflected gradient SPDE itself and the concrete third-derivative estimates — or a sharp obstruction — for the real interaction potentials.

Remark D.93. The new forty-second message is again deliberately intermediate. On the stochastic side, it no longer stops at weighted local Lipschitz control: the explicit gradient cylinder law now admits a canonical affine freezing principle with quadratic error, which is the natural last intrinsic bridge before a genuinely local reflected gradient SPDE. On the interaction side, it no longer stops at finite-cylinder Hessian control: the verification target is reduced correspondingly to a finite-cylinder third-derivative-Cauchy condition. What is still missing is not an explicit first-order local closure mechanism, but the actual third-order estimates for the e-Arc-Helix interaction potentials themselves and the promotion of the resulting affine package to a broader non-cylinder reflected SPDE.

D.30 Worked continuation XLIII: weighted $C^{2,1}$ quadratic local modeling and a finite-cylinder fourth-derivative bridge

The weighted $C^{1,1}$/third-derivative continuation already gives a canonical affine local freezing principle with quadratic error, but it still stops one intrinsic step short of a genuinely local second-order closure mechanism. On the stochastic side, the drift can already be linearized, yet there is not a canonical quadratic local model with a quantified cubic remainder. On the interaction side, verification can already be posed directly on third derivatives of the finite-cylinder potentials, but this still does not isolate the fourth-order information needed for the next Taylor layer of the continuum drift itself. The next intrinsic sharpening addresses both gaps. First, under a summable weighted C^4 budget the renormalized continuum gradient drift already admits a weighted cylinder-$C^{2,1}$ quadratic local model with cubic local remainder. Second, this stronger quadratic local-model package is itself forced by a finite-cylinder fourth-derivative-Cauchy condition.

Status label. The weighted $C^{2,1}$ quadratic-model theorem below has status **theorem-level downstream result**; the finite-cylinder fourth-derivative bridge has status **proposition-level downstream result**. They do *not* yet produce a genuinely local reflected gradient SPDE on a broader non-cylinder state space, and they do *not* yet verify the required fourth-derivative bounds for the actual e-Arc-Helix interaction potentials. What they do prove is that the explicit gradient cylinder law already carries a canonical quadratic local freezing principle with cubic error, and that this sharper local-closure package can be reduced entirely to finite-cylinder fourth-derivative differences.

For $n \geq 1$, define the weighted C^4 increment budget

$$\begin{aligned}\mathfrak{K}^{(4)}_{\eta,n} := \|\partial_{hhhh} G_{\eta,n}\|_\infty + 3\sum_{k=1}^{n} \frac{\|\partial_{hhh}\partial_{a_k} G_{\eta,n}\|_\infty}{(2k+1)^3} \\ + 3\sum_{k,\ell=1}^{n} \frac{\|\partial_{hh}\partial_{a_k}\partial_{a_\ell} G_{\eta,n}\|_\infty}{(2k+1)^3(2\ell+1)^3} + 3\sum_{j,k,\ell=1}^{n} \frac{\|\partial_{h}\partial_{a_j}\partial_{a_k}\partial_{a_\ell} G_{\eta,n}\|_\infty}{(2j+1)^3(2k+1)^3(2\ell+1)^3} \\ + \sum_{i,j,k,\ell=1}^{n} \frac{\|\partial_{a_i}\partial_{a_j}\partial_{a_k}\partial_{a_\ell} G_{\eta,n}\|_\infty}{(2i+1)^3(2j+1)^3(2k+1)^3(2\ell+1)^3}.\end{aligned} \tag{500}$$

Theorem D.94 (Weighted $C^{2,1}$ quadratic local model of the continuum gradient drift and cubic local freezing)**.** *Assume the setting of Theorem D.91. Assume in addition that $V_{\eta,0}$ is C^4 on $[0, h_*]$, that each genuine increment $G_{\eta,n}$ extends to a C^4 function on a neighborhood of $\mathcal{X}_n$, and that*

$$\sum_{n=1}^{\infty} \mathfrak{K}^{(4)}_{\eta,n} < \infty. \tag{501}$$

Define

$$\begin{aligned}Q^{(4)}_{\eta,h} := \|V^{(4)}_{\eta,0}\|_\infty + \sum_{n=1}^{\infty} \Bigg(\|\partial_{hhhh} G_{\eta,n}\|_\infty + 3\sum_{k=1}^{n} \frac{\|\partial_{hhh}\partial_{a_k} G_{\eta,n}\|_\infty}{(2k+1)^3} + 3\sum_{k,\ell=1}^{n} \frac{\|\partial_{hh}\partial_{a_k}\partial_{a_\ell} G_{\eta,n}\|_\infty}{(2k+1)^3(2\ell+1)^3} \\ + \sum_{k,\ell,m=1}^{n} \frac{\|\partial_{h}\partial_{a_k}\partial_{a_\ell}\partial_{a_m} G_{\eta,n}\|_\infty}{(2k+1)^3(2\ell+1)^3(2m+1)^3} \Bigg),\end{aligned} \tag{502}$$

$$\begin{aligned}Q^{(4)}_{\eta,j} := \sum_{n=j}^{\infty} \Bigg(\|\partial_{hhh}\partial_{a_j} G_{\eta,n}\|_\infty + 3\sum_{k=1}^{n} \frac{\|\partial_{hh}\partial_{a_j}\partial_{a_k} G_{\eta,n}\|_\infty}{(2k+1)^3} + 3\sum_{k,\ell=1}^{n} \frac{\|\partial_{h}\partial_{a_j}\partial_{a_k}\partial_{a_\ell} G_{\eta,n}\|_\infty}{(2k+1)^3(2\ell+1)^3} \\ + \sum_{k,\ell,m=1}^{n} \frac{\|\partial_{a_j}\partial_{a_k}\partial_{a_\ell}\partial_{a_m} G_{\eta,n}\|_\infty}{(2k+1)^3(2\ell+1)^3(2m+1)^3} \Bigg), \qquad j \geq 1.\end{aligned} \tag{503}$$

Then the following hold.

(i)

$$\sum_{j=1}^{\infty} \frac{Q^{(4)}_{\eta,j}}{(2j+1)^3} < \infty. \tag{504}$$

(ii) For $x = (h, \mathbf{a})$, $y = (\widetilde{h}, \widetilde{\mathbf{a}}) \in \iota_\eta^{-1}(\mathfrak{S}^{(3)}_\eta)$, write $\delta_h := h - \widetilde{h}$ and $\delta a_k := a_k - \widetilde{a}_k$. For compact

display in this item, set $W_\eta := V^{\rm grad}_{\eta,\infty}$*. Then the quadratic row forms*

$$\begin{aligned}\mathcal{Q}_{\eta,h}[y](x-y,x-y) &:= \partial_{hhh}W_\eta(y)\,\delta_h^2\\&\quad+2\sum_{k=1}^{\infty}\partial_{hh}\partial_{a_k}W_\eta(y)\,\delta_h\,\delta a_k\\&\quad+\sum_{k,\ell=1}^{\infty}\partial_h\partial_{a_k}\partial_{a_\ell}W_\eta(y)\,\delta a_k\,\delta a_\ell,\end{aligned}\tag{505}$$

$$\begin{aligned}\mathcal{Q}_{\eta,j}[y](x-y,x-y) &:= \partial_{hh}\partial_{a_j}W_\eta(y)\,\delta_h^2\\&\quad+2\sum_{k=1}^{\infty}\partial_h\partial_{a_j}\partial_{a_k}W_\eta(y)\,\delta_h\,\delta a_k\\&\quad+\sum_{k,\ell=1}^{\infty}\partial_{a_j}\partial_{a_k}\partial_{a_\ell}W_\eta(y)\,\delta a_k\,\delta a_\ell,\qquad j\ge 1,\end{aligned}\tag{506}$$

are absolutely convergent, and

$$\begin{aligned}\big|\partial_hW_\eta(x)-\partial_hW_\eta(y)-\mathcal{R}_{\eta,h}[y](x-y)\\-\tfrac12\mathcal{Q}_{\eta,h}[y](x-y,x-y)\big|\le\frac16 Q^{(4)}_{\eta,h}\,d^{(3)}_\eta(x,y)^3,\end{aligned}\tag{507}$$

$$\begin{aligned}\big|\partial_{a_j}W_\eta(x)-\partial_{a_j}W_\eta(y)-\mathcal{R}_{\eta,j}[y](x-y)\\-\tfrac12\mathcal{Q}_{\eta,j}[y](x-y,x-y)\big|\le\frac16 Q^{(4)}_{\eta,j}\,d^{(3)}_\eta(x,y)^3,\qquad j\ge 1.\end{aligned}\tag{508}$$

Consequently the renormalized continuum gradient drift is weighted cylinder-$C^{2,1}$ *on the intrinsic support.*

(iii) For $F=f\circ\Pi_N\in\mathcal{D}^{\rm Neu,\infty}_\eta$ *and fixed base point* $y\in\Gamma_\eta^{-1}(\mathfrak{S}^{(3)}_\eta)$*, define the quadratic-frozen gradient operator by*

$$\begin{aligned}(\mathfrak{L}^{\rm quad}_{\eta,y}F)(x)&:=A(h)\,\partial_h^2f(\Pi_Nx)\\&\quad+\Big(2-\frac{5h}{h_*}-A(h)\big[\partial_hW_\eta(y)+\mathcal{R}_{\eta,h}[y](x-y)+\tfrac12\mathcal{Q}_{\eta,h}[y](x-y,x-y)\big]\Big)\,\partial_hf(\Pi_Nx)\\&\quad+\sum_{k=1}^{N}\big(\partial^2_{a_k}f(\Pi_Nx)-\big[\partial_{a_k}W_\eta(y)+\mathcal{R}_{\eta,k}[y](x-y)+\tfrac12\mathcal{Q}_{\eta,k}[y](x-y,x-y)\big]\,\partial_{a_k}f(\Pi_Nx)\big).\end{aligned}\tag{509}$$

Then for every $x,y\in\Gamma_\eta^{-1}(\mathfrak{S}^{(3)}_\eta)$*,*

$$\big|(\mathfrak{L}^\nabla_\eta F)(x)-(\mathfrak{L}^{\rm quad}_{\eta,y}F)(x)\big|\le\frac16\left(\|A\,\partial_hf\|_\infty\,Q^{(4)}_{\eta,h}+\sum_{k=1}^{N}\|\partial_{a_k}f\|_\infty\,Q^{(4)}_{\eta,k}\right)d^{(3)}_\eta(x,y)^3.\tag{510}$$

In particular, the manuscript now has not only an affine local freezing principle with quadratic error, but also a canonical quadratic local model with cubic error for the explicit continuum gradient drift on the intrinsic support. What remains open is no longer a missing second-order intrinsic Taylor bridge, but promotion of that quadratic package to a genuinely local reflected gradient SPDE on a broader non-cylinder domain.

Proof. Write again $W_\eta := V^{\rm grad}_{\eta,\infty}$. Because the hypotheses of Theorem D.91 are in force, the third-derivative rows entering (505)–(506) already converge absolutely on $\Gamma_\eta^{-1}(\mathfrak{S}^{(3)}_\eta)$ by (486)–(488).

Moreover,

$$\sum_{j=1}^{\infty} \frac{Q^{(4)}_{\eta,j}}{(2j+1)^3} \le \sum_{n=1}^{\infty} \Bigg[\sum_{j=1}^{n} \frac{\|\partial_{hhh}\partial_{a_j} G_{\eta,n}\|_\infty}{(2j+1)^3} + 3 \sum_{j,k=1}^{n} \frac{\|\partial_{hh}\partial_{a_j}\partial_{a_k} G_{\eta,n}\|_\infty}{(2j+1)^3(2k+1)^3} + 3 \sum_{j,k,\ell=1}^{n} \frac{\|\partial_{h}\partial_{a_j}\partial_{a_k}\partial_{a_\ell} G_{\eta,n}\|_\infty}{(2j+1)^3(2k+1)^3(2\ell+1)^3} + \sum_{i,j,k,\ell=1}^{n} \frac{\|\partial_{a_i}\partial_{a_j}\partial_{a_k}\partial_{a_\ell} G_{\eta,n}\|_\infty}{(2i+1)^3(2j+1)^3(2k+1)^3(2\ell+1)^3} \Bigg],$$

which is finite by (501). This proves item (i).

For item (ii), let $\gamma_\theta := y + \theta(x-y)$, set $z_{\theta,n} := \Pi_n^{[\infty]}\gamma_\theta$, and define

$$\psi_h(\theta) := \partial_h W_\eta(\gamma_\theta), \qquad \psi_j(\theta) := \partial_{a_j} W_\eta(\gamma_\theta), \quad j \ge 1.$$

By absolute convergence of the row expansions already established in Theorem D.91, the affine terms in (489)–(490) are exactly $\psi_h'(0)$ and $\psi_j'(0)$, while the quadratic forms in (505)–(506) are exactly $\psi_h''(0)$ and $\psi_j''(0)$. Taylor's formula with integral remainder gives

$$\psi_h(1) - \psi_h(0) - \psi_h'(0) - \frac{1}{2}\psi_h''(0) = \frac{1}{2}\int_0^1 (1-\theta)^2\, \psi_h'''(\theta)\, d\theta,$$

and similarly for ψ_j. Termwise differentiation of the series defining W_η gives

$$\psi_h'''(\theta) = V^{(4)}_{\eta,0}(h_\theta)\,\delta_h^3 + \sum_{n=1}^{\infty} \Bigg[\partial_{hhhh} G_{\eta,n}(z_{\theta,n})\,\delta_h^3 + 3\sum_{k=1}^{n} \partial_{hhh}\partial_{a_k} G_{\eta,n}(z_{\theta,n})\,\delta_h^2\,\delta a_k + 3\sum_{k,\ell=1}^{n} \partial_{hh}\partial_{a_k}\partial_{a_\ell} G_{\eta,n}(z_{\theta,n})\,\delta_h\,\delta a_k\,\delta a_\ell + \sum_{k,\ell,m=1}^{n} \partial_{h}\partial_{a_k}\partial_{a_\ell}\partial_{a_m} G_{\eta,n}(z_{\theta,n})\,\delta a_k\,\delta a_\ell\,\delta a_m \Bigg],$$

$$\psi_j'''(\theta) = \sum_{n=j}^{\infty} \Bigg[\partial_{hhh}\partial_{a_j} G_{\eta,n}(z_{\theta,n})\,\delta_h^3 + 3\sum_{k=1}^{n} \partial_{hh}\partial_{a_j}\partial_{a_k} G_{\eta,n}(z_{\theta,n})\,\delta_h^2\,\delta a_k + 3\sum_{k,\ell=1}^{n} \partial_{h}\partial_{a_j}\partial_{a_k}\partial_{a_\ell} G_{\eta,n}(z_{\theta,n})\,\delta_h\,\delta a_k\,\delta a_\ell + \sum_{k,\ell,m=1}^{n} \partial_{a_j}\partial_{a_k}\partial_{a_\ell}\partial_{a_m} G_{\eta,n}(z_{\theta,n})\,\delta a_k\,\delta a_\ell\,\delta a_m \Bigg].$$

Because $|\delta_h| \le d^{(3)}_\eta(x,y)$ and $|\delta a_k| \le (2k+1)^{-3} d^{(3)}_\eta(x,y)$, the definitions (502)–(503) imply

$$|\psi_h'''(\theta)| \le Q^{(4)}_{\eta,h}\, d^{(3)}_\eta(x,y)^3,$$
$$|\psi_j'''(\theta)| \le Q^{(4)}_{\eta,j}\, d^{(3)}_\eta(x,y)^3.$$

Integrating over $\theta \in [0,1]$ yields (507) and (508).

For item (iii), subtract (509) from (471). The second-order diffusion terms cancel, and the remaining difference is exactly the drift remainder from item (ii) multiplied by the first derivatives of f. Using (507)–(508) gives (510) immediately. □

Proposition D.95 (Finite-cylinder fourth-derivative-Cauchy criterion for weighted $C^{2,1}$ quadratic local freezing). *Assume the hypotheses of Proposition D.92, assume $V_{\eta,0}$ is C^4 on*

$[0, h_*]$, and for each $N \geq 1$ assume $V_{\eta,N}$ extends to a C^4 function on a neighborhood of $\mathcal{X}_N$. For $n \geq 1$, define

$$\Upsilon^{hhhh}_{\eta,n} := \left\|\partial_{hhhh} V_{\eta,n} - \partial_{hhhh} V_{\eta,n-1}\right\|_{\infty}, \tag{511}$$

$$\Upsilon^{hhha}_{\eta,n,k} := \left\|\partial_{hhh}\partial_{a_k} V_{\eta,n} - \partial_{hhh}\partial_{a_k} V_{\eta,n-1}\right\|_{\infty}, \qquad 1 \leq k \leq n, \tag{512}$$

$$\Upsilon^{hhaa}_{\eta,n,k,\ell} := \left\|\partial_{hh}\partial_{a_k}\partial_{a_\ell} V_{\eta,n} - \partial_{hh}\partial_{a_k}\partial_{a_\ell} V_{\eta,n-1}\right\|_{\infty}, \qquad 1 \leq k, \ell \leq n, \tag{513}$$

$$\Upsilon^{haaa}_{\eta,n,j,k,\ell} := \left\|\partial_{h}\partial_{a_j}\partial_{a_k}\partial_{a_\ell} V_{\eta,n} - \partial_{h}\partial_{a_j}\partial_{a_k}\partial_{a_\ell} V_{\eta,n-1}\right\|_{\infty}, \qquad 1 \leq j, k, \ell \leq n, \tag{514}$$

$$\Upsilon^{aaaa}_{\eta,n,i,j,k,\ell} := \left\|\partial_{a_i}\partial_{a_j}\partial_{a_k}\partial_{a_\ell} V_{\eta,n} - \partial_{a_i}\partial_{a_j}\partial_{a_k}\partial_{a_\ell} V_{\eta,n-1}\right\|_{\infty}, \qquad 1 \leq i, j, k, \ell \leq n, \tag{515}$$

where every derivative of $V_{\eta,n-1}$ involving the new coordinate a_n is read as zero after pullback through Π_{n-1}. If

$$\sum_{n=1}^{\infty} \left(\Upsilon^{hhhh}_{\eta,n} + 3\sum_{k=1}^{n} \frac{\Upsilon^{hhha}_{\eta,n,k}}{(2k+1)^3} + 3\sum_{k,\ell=1}^{n} \frac{\Upsilon^{hhaa}_{\eta,n,k,\ell}}{(2k+1)^3(2\ell+1)^3} \right.$$
$$\left. +3\sum_{j,k,\ell=1}^{n} \frac{\Upsilon^{haaa}_{\eta,n,j,k,\ell}}{(2j+1)^3(2k+1)^3(2\ell+1)^3} + \sum_{i,j,k,\ell=1}^{n} \frac{\Upsilon^{aaaa}_{\eta,n,i,j,k,\ell}}{(2i+1)^3(2j+1)^3(2k+1)^3(2\ell+1)^3} \right) < \infty, \tag{516}$$

then the weighted C^4 increment budget (500) is summable. Consequently, together with the lower-order first-, second-, and third-derivative conditions already imported by Proposition D.92, the quadratic local-model package of Theorem D.94 applies on the intrinsic continuum support.

Proof. Differentiating the increment identity

$$G_{\eta,n}(x) = V_{\eta,n}(x) - V_{\eta,n-1}(\Pi_{n-1}x) \qquad \text{on } \mathcal{X}_n$$

four times gives

$$\|\partial_{hhhh} G_{\eta,n}\|_\infty = \Upsilon^{hhhh}_{\eta,n}, \qquad \|\partial_{hhh}\partial_{a_k} G_{\eta,n}\|_\infty = \Upsilon^{hhha}_{\eta,n,k},$$

$$\|\partial_{hh}\partial_{a_k}\partial_{a_\ell} G_{\eta,n}\|_\infty = \Upsilon^{hhaa}_{\eta,n,k,\ell}, \qquad \|\partial_{h}\partial_{a_j}\partial_{a_k}\partial_{a_\ell} G_{\eta,n}\|_\infty = \Upsilon^{haaa}_{\eta,n,j,k,\ell},$$

and

$$\|\partial_{a_i}\partial_{a_j}\partial_{a_k}\partial_{a_\ell} G_{\eta,n}\|_\infty = \Upsilon^{aaaa}_{\eta,n,i,j,k,\ell},$$

with the convention that derivatives of $V_{\eta,n-1}$ in the new coordinate a_n vanish. Substituting these identities into (500) shows that the series in (516) is exactly $\sum_{n\geq 1} \mathfrak{K}^{(4)}_{\eta,n}$. The remaining conclusion therefore follows from Proposition D.92 and Theorem D.94. □

Integrated downstream conclusion after the weighted-$C^{2,1}$/fourth-derivative continuation. This continuation adds a forty-third downstream message. Positively, the intrinsic continuum lift no longer has only an affine local freezing principle: under summably controlled weighted C^4 increments, the explicit continuum gradient drift already admits a weighted cylinder-$C^{2,1}$ quadratic local model with cubic local freezing error on the cylinder generator. Verification is correspondingly no longer only a third-derivative problem: the quadratic local-model package is reduced to a finite-cylinder fourth-derivative-Cauchy condition on $V_{\eta,N}$. Cautionarily, the manuscript still does not upgrade this quadratic local model to a genuinely local reflected SPDE on a broader non-cylinder domain, and it still does not prove the fourth-derivative-Cauchy bounds for the actual e-Arc-Helix interaction family. What remains open is therefore the local reflected gradient SPDE itself and the concrete fourth-derivative estimates — or a sharp obstruction — for the real interaction potentials.

Remark D.96. The new forty-third message is again deliberately intermediate. On the stochastic side, it no longer stops at affine freezing with quadratic error: the explicit gradient cylinder law now admits a canonical quadratic local model with cubic error, which is the natural last intrinsic bridge before one can reasonably ask for a non-cylinder second-order local reflected gradient SPDE closure. On the interaction side, it no longer stops at finite-cylinder third-derivative control: the verification target is reduced correspondingly to a finite-cylinder fourth-derivative-Cauchy condition. What is still missing is not a second-order intrinsic Taylor bridge, but the actual fourth-order estimates for the e-Arc-Helix interaction potentials themselves and the promotion of the resulting quadratic package to a broader non-cylinder reflected SPDE.

D.31 Worked continuation XLIV: weighted $C^{3,1}$ cubic local model and a finite-cylinder fifth-derivative bridge

The weighted-$C^{2,1}$/fourth-derivative continuation already gives a canonical quadratic local model with cubic generator error on the intrinsic continuum support, but it still stops one intrinsic step short of a third-order local closure package. On the stochastic side, the drift is controlled only to quadratic order, so there is not yet a canonical cubic local model with quantified quartic remainder. On the interaction side, verification can already be posed directly on fourth derivatives of the finite-cylinder potentials, but this still does not isolate the fifth-order information needed for the next Taylor layer of the continuum gradient drift itself. The next intrinsic sharpening addresses both gaps. First, under a summable weighted C^5 budget the renormalized continuum gradient drift already admits a weighted cylinder-$C^{3,1}$ cubic local model with quartic local remainder. Second, this stronger cubic local-model package is itself forced by a finite-cylinder fifth-derivative-Cauchy condition.

Status label. The weighted $C^{3,1}$ cubic-model theorem below has status **theorem-level downstream result**; the finite-cylinder fifth-derivative bridge has status **proposition-level downstream result**. They do *not* yet produce a genuinely local reflected gradient SPDE on a broader non-cylinder state space, and they do *not* yet verify the required fifth-derivative bounds for the actual e-Arc-Helix interaction potentials. What they do prove is that the explicit gradient cylinder law already carries a canonical cubic local model with quartic error, and that this sharper local-closure package can be reduced entirely to finite-cylinder fifth-derivative differences.

For $n \geq 1$, define the weighted C^5 increment budget

$$\begin{aligned}\mathfrak{M}^{(5)}_{\eta,n} &:= \|\partial_{hhhhh} G_{\eta,n}\|_\infty + 4\sum_{k=1}^{n} \frac{\|\partial_{hhhh}\partial_{a_k} G_{\eta,n}\|_\infty}{(2k+1)^3} \\ &+ 6\sum_{k,\ell=1}^{n} \frac{\|\partial_{hhh}\partial_{a_k}\partial_{a_\ell} G_{\eta,n}\|_\infty}{(2k+1)^3(2\ell+1)^3} + 6\sum_{j,k,\ell=1}^{n} \frac{\|\partial_{hh}\partial_{a_j}\partial_{a_k}\partial_{a_\ell} G_{\eta,n}\|_\infty}{(2j+1)^3(2k+1)^3(2\ell+1)^3} \\ &+ 4\sum_{i,j,k,\ell=1}^{n} \frac{\|\partial_h\partial_{a_i}\partial_{a_j}\partial_{a_k}\partial_{a_\ell} G_{\eta,n}\|_\infty}{(2i+1)^3(2j+1)^3(2k+1)^3(2\ell+1)^3} + \sum_{i,j,k,\ell,m=1}^{n} \frac{\|\partial_{a_i}\partial_{a_j}\partial_{a_k}\partial_{a_\ell}\partial_{a_m} G_{\eta,n}\|_\infty}{(2i+1)^3(2j+1)^3(2k+1)^3(2\ell+1)^3(2m+1)^3}.\end{aligned} \tag{517}$$

Theorem D.97 (Weighted $C^{3,1}$ cubic local model of the continuum gradient drift and quartic local freezing)**.** *Assume the setting of Theorem D.94. Assume in addition that $V_{\eta,0}$ is C^5 on $[0,h_*]$, that each genuine increment $G_{\eta,n}$ extends to a C^5 function on a neighborhood of $\mathcal{X}_n$, and that*

$$\sum_{n=1}^{\infty} \mathfrak{M}^{(5)}_{\eta,n} < \infty. \tag{518}$$

Define

$$Q^{(5)}_{\eta,h} := \|V^{(5)}_{\eta,0}\|_\infty + \sum_{n=1}^{\infty}\Bigg(\|\partial_{hhhhh} G_{\eta,n}\|_\infty \tag{519}$$

$$+ 4\sum_{k=1}^{n} \frac{\|\partial_{hhhh}\partial_{a_k} G_{\eta,n}\|_\infty}{(2k+1)^3} \tag{520}$$

$$+ 6\sum_{k,\ell=1}^{n} \frac{\|\partial_{hhh}\partial_{a_k}\partial_{a_\ell} G_{\eta,n}\|_\infty}{(2k+1)^3(2\ell+1)^3} \tag{521}$$

$$+ 4\sum_{j,k,\ell=1}^{n} \frac{\|\partial_{hh}\partial_{a_j}\partial_{a_k}\partial_{a_\ell} G_{\eta,n}\|_\infty}{(2j+1)^3(2k+1)^3(2\ell+1)^3} \tag{522}$$

$$+ \sum_{i,j,k,\ell=1}^{n} \frac{\|\partial_{h}\partial_{a_i}\partial_{a_j}\partial_{a_k}\partial_{a_\ell} G_{\eta,n}\|_\infty}{(2i+1)^3(2j+1)^3(2k+1)^3(2\ell+1)^3} \Bigg), \tag{523}$$

$$Q^{(5)}_{\eta,j} := \sum_{n=j}^{\infty}\Bigg(\|\partial_{hhhh}\partial_{a_j} G_{\eta,n}\|_\infty \tag{524}$$

$$+ 4\sum_{k=1}^{n} \frac{\|\partial_{hhh}\partial_{a_j}\partial_{a_k} G_{\eta,n}\|_\infty}{(2k+1)^3} \tag{525}$$

$$+ 6\sum_{k,\ell=1}^{n} \frac{\|\partial_{hh}\partial_{a_j}\partial_{a_k}\partial_{a_\ell} G_{\eta,n}\|_\infty}{(2k+1)^3(2\ell+1)^3} \tag{526}$$

$$+ 4\sum_{k,\ell,m=1}^{n} \frac{\|\partial_{h}\partial_{a_j}\partial_{a_k}\partial_{a_\ell}\partial_{a_m} G_{\eta,n}\|_\infty}{(2k+1)^3(2\ell+1)^3(2m+1)^3} \tag{527}$$

$$+ \sum_{k,\ell,m,r=1}^{n} \frac{\|\partial_{a_j}\partial_{a_k}\partial_{a_\ell}\partial_{a_m}\partial_{a_r} G_{\eta,n}\|_\infty}{(2k+1)^3(2\ell+1)^3(2m+1)^3(2r+1)^3} \Bigg), \qquad j \geq 1. \tag{528}$$

Then the following hold.

(i)

$$\sum_{j=1}^{\infty} \frac{Q^{(5)}_{\eta,j}}{(2j+1)^3} < \infty. \tag{529}$$

(ii) For $x = (h, \mathbf{a})$, $y = (\widetilde{h}, \widetilde{\mathbf{a}}) \in \Gamma_\eta^{-1}(\mathfrak{S}^{(3)}_\eta)$, *write* $\delta_h := h - \widetilde{h}$ *and* $\delta a_k := a_k - \widetilde{a}_k$. *Then the*

cubic row expansions

$$\mathcal{C}_{\eta,h}[y](x-y) := \mathcal{Q}_{\eta,h}[y](x-y) + \frac{1}{6}\partial_{hhhh}W_\eta(y)\,\delta_h^3 + \frac{1}{2}\sum_{k=1}^{\infty}\partial_{hhh}\partial_{a_k}W_\eta(y)\,\delta_h^2\,\delta a_k \tag{530}$$

$$+\frac{1}{2}\sum_{k,\ell=1}^{\infty}\partial_{hh}\partial_{a_k}\partial_{a_\ell}W_\eta(y)\,\delta_h\,\delta a_k\,\delta a_\ell + \frac{1}{6}\sum_{k,\ell,m=1}^{\infty}\partial_h\partial_{a_k}\partial_{a_\ell}\partial_{a_m}W_\eta(y) \times \delta a_k\,\delta a_\ell\,\delta a_m, \tag{531}$$

$$\mathcal{C}_{\eta,j}[y](x-y) := \mathcal{Q}_{\eta,j}[y](x-y) + \frac{1}{6}\partial_{hhh}\partial_{a_j}W_\eta(y)\,\delta_h^3 + \frac{1}{2}\sum_{k=1}^{\infty}\partial_{hh}\partial_{a_j}\partial_{a_k}W_\eta(y)\,\delta_h^2\,\delta a_k \tag{532}$$

$$+\frac{1}{2}\sum_{k,\ell=1}^{\infty}\partial_h\partial_{a_j}\partial_{a_k}\partial_{a_\ell}W_\eta(y)\,\delta_h\,\delta a_k\,\delta a_\ell + \frac{1}{6}\sum_{k,\ell,m=1}^{\infty}\partial_{a_j}\partial_{a_k}\partial_{a_\ell}\partial_{a_m}W_\eta(y) \times \delta a_k\,\delta a_\ell\,\delta a_m, \quad j \geq 1, \tag{533}$$

are absolutely convergent, and

$$\left|\partial_h W_\eta(x) - \partial_h W_\eta(y) - \mathcal{C}_{\eta,h}[y](x-y)\right| \leq \frac{1}{24}Q_{\eta,h}^{(5)}\,d_\eta^{(3)}(x,y)^4, \tag{534}$$

$$\left|\partial_{a_j} W_\eta(x) - \partial_{a_j} W_\eta(y) - \mathcal{C}_{\eta,j}[y](x-y)\right| \leq \frac{1}{24}Q_{\eta,j}^{(5)}\,d_\eta^{(3)}(x,y)^4, \qquad j \geq 1. \tag{535}$$

Consequently the renormalized continuum gradient drift is weighted cylinder-$C^{3,1}$ on the intrinsic support.

(iii) For $F = f \circ \Pi_N \in \mathcal{D}_\eta^{\mathrm{Neu},\infty}$ and fixed base point $y \in \Gamma_\eta^{-1}(\mathfrak{S}_\eta^{(3)})$, define the cubically frozen gradient operator by

$$\begin{aligned}(\mathfrak{L}_{\eta,y}^{\mathrm{cub}}F)(x) := {} & A(h)\,\partial_h^2 f(\Pi_N x) \\ & + \left(2 - \frac{5h}{h_*} - A(h)\left[\partial_h V_{\eta,\infty}^{\mathrm{grad}}(y) + \mathcal{C}_{\eta,h}[y](x-y)\right]\right)\partial_h f(\Pi_N x) \\ & + \sum_{k=1}^{N}\left(\partial_{a_k}^2 f(\Pi_N x) - \left[\partial_{a_k} V_{\eta,\infty}^{\mathrm{grad}}(y) + \mathcal{C}_{\eta,k}[y](x-y)\right]\partial_{a_k} f(\Pi_N x)\right).\end{aligned} \tag{536}$$

Then for every $x, y \in \Gamma_\eta^{-1}(\mathfrak{S}_\eta^{(3)})$,

$$\left|(\mathfrak{L}_\eta^{\nabla}F)(x) - (\mathfrak{L}_{\eta,y}^{\mathrm{cub}}F)(x)\right| \leq \frac{1}{24}\left(\|A\,\partial_h f\|_\infty\,Q_{\eta,h}^{(5)} + \sum_{k=1}^{N}\|\partial_{a_k} f\|_\infty\,Q_{\eta,k}^{(5)}\right) d_\eta^{(3)}(x,y)^4. \tag{537}$$

In particular, the manuscript now has not only a canonical quadratic local model with cubic error, but also a canonical cubic local model with quartic error on the intrinsic continuum support. What remains open is not second- or third-order intrinsic Taylor control, but promotion of that cubic package to a genuinely local reflected gradient SPDE on a broader non-cylinder domain.

Proof. Because Theorem D.94 is in force, the fourth derivative rows entering (531)–(533) already converge absolutely on $\Gamma_\eta^{-1}(\mathfrak{S}_\eta^{(3)})$ by (502)–(504). Moreover,

$$\begin{aligned}\sum_{j=1}^{\infty}\frac{Q_{\eta,j}^{(5)}}{(2j+1)^3} \le 4\sum_{n=1}^{\infty}\sum_{j=1}^{n}\frac{\|\partial_{hhhh}\partial_{a_j}G_{\eta,n}\|_\infty}{(2j+1)^3} \\ +6\sum_{n=1}^{\infty}\sum_{j,k=1}^{n}\frac{\|\partial_{hhh}\partial_{a_j}\partial_{a_k}G_{\eta,n}\|_\infty}{(2j+1)^3(2k+1)^3} \\ +6\sum_{n=1}^{\infty}\sum_{j,k,\ell=1}^{n}\frac{\|\partial_{hh}\partial_{a_j}\partial_{a_k}\partial_{a_\ell}G_{\eta,n}\|_\infty}{(2j+1)^3(2k+1)^3(2\ell+1)^3} \\ +4\sum_{n=1}^{\infty}\sum_{j,k,\ell,m=1}^{n}\frac{\|\partial_{h}\partial_{a_j}\partial_{a_k}\partial_{a_\ell}\partial_{a_m}G_{\eta,n}\|_\infty}{(2j+1)^3(2k+1)^3(2\ell+1)^3(2m+1)^3} \\ +\sum_{n=1}^{\infty}\sum_{i,j,k,\ell,m=1}^{n}\frac{\|\partial_{a_i}\partial_{a_j}\partial_{a_k}\partial_{a_\ell}\partial_{a_m}G_{\eta,n}\|_\infty}{(2i+1)^3(2j+1)^3(2k+1)^3(2\ell+1)^3(2m+1)^3},\end{aligned}$$

which is finite by (518). This proves item (i).

For item (ii), let $\gamma_\theta := y+\theta(x-y) = (h_\theta, \mathbf{a}_\theta)$ and define

$$\psi_h(\theta) := \partial_h V_{\eta,\infty}^{\mathrm{grad}}(\gamma_\theta), \qquad \psi_j(\theta) := \partial_{a_j} V_{\eta,\infty}^{\mathrm{grad}}(\gamma_\theta), \quad j \ge 1.$$

By Theorem D.94, the cubic terms in (531)–(533) are exactly $\psi_h'(0)+\frac{1}{2}\psi_h''(0)+\frac{1}{6}\psi_h'''(0)$ and $\psi_j'(0)+\frac{1}{2}\psi_j''(0)+\frac{1}{6}\psi_j'''(0)$. Taylor's formula with integral remainder gives

$$\psi_h(1)-\psi_h(0)-\psi_h'(0)-\frac{1}{2}\psi_h''(0)-\frac{1}{6}\psi_h'''(0) = \frac{1}{6}\int_0^1 (1-\theta)^3\,\psi_h''''(\theta)\,d\theta,$$

and similarly for ψ_j. For each $n\ge 1$, write $\zeta_{\theta,n} := \Pi_n^{[\infty]}\gamma_\theta$. Termwise differentiation of the series defining $V^{\mathrm{grad}}_{\eta,\infty}$ gives

$$
\begin{aligned}
\psi_h''''(\theta) = {} & V^{(5)}_{\eta,0}(h_\theta)\,\delta_h^4 \\
& + \sum_{n=1}^{\infty} \partial_{hhhhh} G_{\eta,n}(\zeta_{\theta,n})\,\delta_h^4 \\
& + 4\sum_{n=1}^{\infty}\sum_{k=1}^{n} \partial_{hhhh}\partial_{a_k} G_{\eta,n}(\zeta_{\theta,n})\,\delta_h^3\,\delta a_k \\
& + 6\sum_{n=1}^{\infty}\sum_{k,\ell=1}^{n} \partial_{hhh}\partial_{a_k}\partial_{a_\ell} G_{\eta,n}(\zeta_{\theta,n})\,\delta_h^2\,\delta a_k\,\delta a_\ell \\
& + 4\sum_{n=1}^{\infty}\sum_{j,k,\ell=1}^{n} \partial_{hh}\partial_{a_j}\partial_{a_k}\partial_{a_\ell} G_{\eta,n}(\zeta_{\theta,n})\,\delta_h\,\delta a_j\,\delta a_k\,\delta a_\ell \\
& + \sum_{n=1}^{\infty}\sum_{i,j,k,\ell=1}^{n} \partial_h\partial_{a_i}\partial_{a_j}\partial_{a_k}\partial_{a_\ell} G_{\eta,n}(\zeta_{\theta,n})\,\delta a_i\,\delta a_j\,\delta a_k\,\delta a_\ell, \\
\psi_j''''(\theta) = {} & \sum_{n=j}^{\infty} \partial_{hhhh}\partial_{a_j} G_{\eta,n}(\zeta_{\theta,n})\,\delta_h^4 \\
& + 4\sum_{n=j}^{\infty}\sum_{k=1}^{n} \partial_{hhh}\partial_{a_j}\partial_{a_k} G_{\eta,n}(\zeta_{\theta,n})\,\delta_h^3\,\delta a_k \\
& + 6\sum_{n=j}^{\infty}\sum_{k,\ell=1}^{n} \partial_{hh}\partial_{a_j}\partial_{a_k}\partial_{a_\ell} G_{\eta,n}(\zeta_{\theta,n})\,\delta_h^2\,\delta a_k\,\delta a_\ell \\
& + 4\sum_{n=j}^{\infty}\sum_{k,\ell,m=1}^{n} \partial_h\partial_{a_j}\partial_{a_k}\partial_{a_\ell}\partial_{a_m} G_{\eta,n}(\zeta_{\theta,n})\,\delta_h\,\delta a_k\,\delta a_\ell\,\delta a_m \\
& + \sum_{n=j}^{\infty}\sum_{k,\ell,m,r=1}^{n} \partial_{a_j}\partial_{a_k}\partial_{a_\ell}\partial_{a_m}\partial_{a_r} G_{\eta,n}(\zeta_{\theta,n})\,\delta a_k\,\delta a_\ell\,\delta a_m\,\delta a_r.
\end{aligned}
$$

Because $|\delta_h| \le d^{(3)}_\eta(x,y)$ and $|\delta a_k| \le (2k+1)^{-3} d^{(3)}_\eta(x,y)$, the definitions (523)–(528) imply

$$
|\psi_h''''(\theta)| \le Q^{(5)}_{\eta,h}\, d^{(3)}_\eta(x,y)^4, \qquad |\psi_j''''(\theta)| \le Q^{(5)}_{\eta,j}\, d^{(3)}_\eta(x,y)^4.
$$

Integrating over $\theta \in [0,1]$ yields (534) and (535).

For item (iii), subtract (536) from (471). The second-order diffusion terms cancel, and the remaining difference is exactly the drift remainder from item (ii) multiplied by the first derivatives of f. Using (534)–(535) gives (537) immediately. □

Proposition D.98 (Finite-cylinder fifth-derivative-Cauchy criterion for weighted $C^{3,1}$ cubic local model)**.** *Assume the hypotheses of Proposition D.95, assume $V_{\eta,0}$ is C^5 on $[0,h_*]$, and for each $N \ge 1$ assume $V_{\eta,N}$ extends to a C^5 function on a neighborhood of $\mathcal{X}_N$. For $n \ge 1$, define*

$$
\Lambda^{hhhhh}_{\eta,n} := \left\|\partial_{hhhhh} V_{\eta,n} - \partial_{hhhhh} V_{\eta,n-1}\right\|_\infty, \tag{538}
$$

$$
\Lambda^{hhhha}_{\eta,n,k} := \left\|\partial_{hhhh}\partial_{a_k} V_{\eta,n} - \partial_{hhhh}\partial_{a_k} V_{\eta,n-1}\right\|_\infty, \qquad 1 \le k \le n, \tag{539}
$$

$$
\Lambda^{hhhaa}_{\eta,n,k,\ell} := \left\|\partial_{hhh}\partial_{a_k}\partial_{a_\ell} V_{\eta,n} - \partial_{hhh}\partial_{a_k}\partial_{a_\ell} V_{\eta,n-1}\right\|_\infty, \qquad 1 \le k,\ell \le n, \tag{540}
$$

$$
\Lambda^{hhaaa}_{\eta,n,j,k,\ell} := \left\|\partial_{hh}\partial_{a_j}\partial_{a_k}\partial_{a_\ell} V_{\eta,n} - \partial_{hh}\partial_{a_j}\partial_{a_k}\partial_{a_\ell} V_{\eta,n-1}\right\|_\infty, \qquad 1 \le j,k,\ell \le n, \tag{541}
$$

$$
\Lambda^{haaaa}_{\eta,n,i,j,k,\ell} := \left\|\partial_h\partial_{a_i}\partial_{a_j}\partial_{a_k}\partial_{a_\ell} V_{\eta,n} - \partial_h\partial_{a_i}\partial_{a_j}\partial_{a_k}\partial_{a_\ell} V_{\eta,n-1}\right\|_\infty, \qquad 1 \le i,j,k,\ell \le n, \tag{542}
$$

$$
\Lambda^{aaaaa}_{\eta,n,i,j,k,\ell,m} := \left\|\partial_{a_i}\partial_{a_j}\partial_{a_k}\partial_{a_\ell}\partial_{a_m} V_{\eta,n} - \partial_{a_i}\partial_{a_j}\partial_{a_k}\partial_{a_\ell}\partial_{a_m} V_{\eta,n-1}\right\|_\infty, \qquad 1 \le i,j,k,\ell,m \le n, \tag{543}
$$

where every derivative of $V_{\eta,n-1}$ involving the new coordinate a_n is read as zero after pullback through Π_{n-1}. If

$$\begin{aligned}\sum_{n=1}^{\infty}\Bigg(\Lambda_{\eta,n}^{hhhhh}+4\sum_{k=1}^{n}\frac{\Lambda_{\eta,n,k}^{hhhha}}{(2k+1)^3}\\ &+6\sum_{k,\ell=1}^{n}\frac{\Lambda_{\eta,n,k,\ell}^{hhhaa}}{(2k+1)^3(2\ell+1)^3}+6\sum_{j,k,\ell=1}^{n}\frac{\Lambda_{\eta,n,j,k,\ell}^{hhaaa}}{(2j+1)^3(2k+1)^3(2\ell+1)^3}\\ &+4\sum_{i,j,k,\ell=1}^{n}\frac{\Lambda_{\eta,n,i,j,k,\ell}^{haaaa}}{(2i+1)^3(2j+1)^3(2k+1)^3(2\ell+1)^3}+\sum_{i,j,k,\ell,m=1}^{n}\frac{\Lambda_{\eta,n,i,j,k,\ell,m}^{aaaaa}}{(2i+1)^3(2j+1)^3(2k+1)^3(2\ell+1)^3(2m+1)^3}\Bigg)<\infty,\end{aligned} \tag{544}$$

then the weighted C^5 increment budget (517) *is summable. Consequently, together with the lower-order first-, second-, third-, and fourth-derivative conditions already imported by Proposition D.95, the cubic local-model package of Theorem D.97 applies on the intrinsic continuum support.*

Proof. Differentiating the increment identity

$$G_{\eta,n}(x)=V_{\eta,n}(x)-V_{\eta,n-1}(\Pi_{n-1}x)\qquad\text{on } \mathcal{X}_n$$

five times gives

$$\|\partial_{hhhhh}G_{\eta,n}\|_\infty=\Lambda_{\eta,n}^{hhhhh},\qquad \|\partial_{hhhh}\partial_{a_k}G_{\eta,n}\|_\infty=\Lambda_{\eta,n,k}^{hhhha},\qquad \|\partial_{hhh}\partial_{a_k}\partial_{a_\ell}G_{\eta,n}\|_\infty=\Lambda_{\eta,n,k,\ell}^{hhhaa},$$

$$\|\partial_{hh}\partial_{a_j}\partial_{a_k}\partial_{a_\ell}G_{\eta,n}\|_\infty=\Lambda_{\eta,n,j,k,\ell}^{hhaaa},\qquad \|\partial_{h}\partial_{a_i}\partial_{a_j}\partial_{a_k}\partial_{a_\ell}G_{\eta,n}\|_\infty=\Lambda_{\eta,n,i,j,k,\ell}^{haaaa},$$

and

$$\|\partial_{a_i}\partial_{a_j}\partial_{a_k}\partial_{a_\ell}\partial_{a_m}G_{\eta,n}\|_\infty=\Lambda_{\eta,n,i,j,k,\ell,m}^{aaaaa},$$

with the convention that derivatives of $V_{\eta,n-1}$ in the new coordinate a_n vanish. Substituting these identities into (517) shows that the series in (544) is exactly $\sum_{n\geq 1}\mathfrak{M}_{\eta,n}^{(5)}$. The remaining conclusion therefore follows from Proposition D.95 and Theorem D.97. □

Integrated downstream conclusion after the weighted-$C^{3,1}$/fifth-derivative continuation. This continuation adds a forty-fourth downstream message. Positively, the intrinsic continuum lift no longer has only a quadratic local model with cubic generator error: under summably controlled weighted C^5 increments, that same explicit gradient cylinder law already has a weighted cylinder-$C^{3,1}$ cubic local model with canonical quartic local error. Verification is correspondingly no longer only a fourth-derivative problem: the cubic local-model package is reduced to a finite-cylinder fifth-derivative-Cauchy condition on $V_{\eta,N}$. Cautionarily, the manuscript still does not upgrade this cubic local model to a genuinely local reflected gradient SPDE on a broader non-cylinder domain, and it still does not prove the fifth-derivative-Cauchy bounds for the actual e-Arc-Helix interaction family. What remains open is therefore the local reflected gradient SPDE itself and the concrete fifth-order estimates — or a sharp obstruction — for the real interaction potentials.

Remark D.99. The new forty-fourth message is again deliberately intermediate. On the stochastic side, it no longer stops at a quadratic local model: the explicit gradient cylinder law now admits a canonical cubic local model with quartic error, which is the natural intrinsic third-order bridge before one can reasonably ask for a non-cylinder higher-order reflected gradient SPDE closure. On the interaction side, it no longer stops at finite-cylinder fourth-derivative control: the verification target is reduced correspondingly to a finite-cylinder fifth-derivative-Cauchy condition. What is still missing is not a third-order intrinsic Taylor bridge, but the actual fifth-order estimates for the e-Arc-Helix interaction potentials themselves and the promotion of the resulting cubic package to a broader non-cylinder reflected SPDE.

D.32 Worked continuation XLV: quartic chartwise local reflected-gradient atlas and finite-cylinder fifth-order activation

The cubic-local-model/fifth-derivative continuation already supplies a third-order intrinsic Taylor bridge and a direct fifth-order verification target, but the two branches are still written separately. On the stochastic side, the cubic local model is still anchored at one base point at a time, so there is not yet a chartwise local atlas whose overlaps are quantitatively compatible. On the interaction side, the fifth-derivative-Cauchy bridge already gives the hypotheses needed for the cubic local model, but it has not yet been repackaged as a direct criterion for the corresponding local reflected-gradient atlas itself. The next continuation closes that gap. First, the cubic local model already patches into an intrinsic chartwise local reflected-gradient martingale atlas with quartic chart defect. Second, the same finite-cylinder fifth-derivative-Cauchy budget already suffices to activate that atlas for the actual finite-cylinder interaction potentials.

Status label. The quartic chartwise local-atlas theorem below has status **theorem-level downstream result**; the finite-cylinder activation criterion has status **proposition-level downstream result**. They do *not* yet yield a defect-free genuinely local reflected gradient SPDE on a broader non-cylinder test domain, and they do *not* yet verify the required fifth-derivative bounds for the actual e-Arc-Helix interaction family. What they do prove is that the intrinsic cubic local package already carries its own quantitatively compatible local martingale atlas, and that the same fifth-order finite-cylinder criterion already forces that atlas.

Theorem D.100 (Intrinsic cubically frozen local reflected-gradient martingale atlas with quartic chart defect). *Assume the hypotheses of Theorem D.97, and let $X = (X_t)_{t\geq 0}$ denote the full-support continuum process characterized on the intrinsic cylinder core by Theorem D.88. For $y \in \Gamma_\eta^{-1}(\mathfrak{S}_\eta^{(3)})$ and $r > 0$, write*

$$B_\eta^{(3)}(y,r) := \{x \in \Gamma_\eta^{-1}(\mathfrak{S}_\eta^{(3)}) : d_\eta^{(3)}(x,y) < r\}, \qquad \tau_{y,r} := \inf\{t \geq 0 : X_t \notin B_\eta^{(3)}(y,r)\}. \quad (545)$$

For $F = f \circ \Pi_N \in \mathcal{D}_\eta^{\mathrm{Neu},\infty}$, define

$$C_{\eta,F}^{(5)} := \frac{1}{24}\left(\|A\,\partial_h f\|_\infty\, Q_{\eta,h}^{(5)} + \sum_{k=1}^{N} \|\partial_{a_k} f\|_\infty\, Q_{\eta,k}^{(5)}\right). \quad (546)$$

Then the following hold.

(i) For every $y \in \Gamma_\eta^{-1}(\mathfrak{S}_\eta^{(3)})$, $r > 0$, and $F = f \circ \Pi_N \in \mathcal{D}_\eta^{\mathrm{Neu},\infty}$, the stopped process

$$\begin{aligned} M_t^{F,y,r} &:= F(X_{t\wedge\tau_{y,r}}) - F(X_0) \\ &\quad - \int_0^{t\wedge\tau_{y,r}} (\mathfrak{L}_{\eta,y}^{\mathrm{cub}} F)(X_s)\, ds \end{aligned} \quad (547)$$

admits a decomposition

$$M_t^{F,y,r} = N_t^{F,y,r} + R_t^{F,y,r}, \quad (548)$$

where $N^{F,y,r}$ is a true martingale and

$$|R_t^{F,y,r}| \leq C_{\eta,F}^{(5)}\, r^4 t, \qquad t > 0. \quad (549)$$

Consequently, up to a deterministic quartic defect, the stopped continuum lift already solves the cubically frozen reflected-gradient martingale problem on every intrinsic chart.

(ii) If $x \in B_\eta^{(3)}(y,r) \cap B_\eta^{(3)}(z,s)$, then for every $F = f \circ \Pi_N \in \mathcal{D}_\eta^{\mathrm{Neu},\infty}$,

$$\big|(\mathfrak{L}_{\eta,y}^{\mathrm{cub}}F)(x) - (\mathfrak{L}_{\eta,z}^{\mathrm{cub}}F)(x)\big| \le C_{\eta,F}^{(5)}(r^4 + s^4). \tag{550}$$

In particular, two cubically frozen local charts agree on overlaps up to the same quartic order as the single-chart generator defect.

(iii) The family of chartwise operators $\{\mathfrak{L}_{\eta,y}^{\mathrm{cub}} : y \in \Gamma_\eta^{-1}(\mathfrak{S}_\eta^{(3)})\}$ therefore forms an intrinsic local reflected-gradient martingale atlas whose chart-transition defect tends to zero like the fourth power of the chart radius. Every exact local reflected gradient SPDE closure beyond the cylinder core is thereby reduced to a vanishing-defect patching problem for this intrinsic quartic atlas on a broader non-cylinder test domain.

Proof. For $F \in \mathcal{D}_\eta^{\mathrm{Neu},\infty}$, Theorem D.88 gives the martingale

$$N_t^F := F(X_t) - F(X_0) - \int_0^t (\mathfrak{L}_\eta^\nabla F)(X_s)\,ds.$$

Stopping at $t \wedge \tau_{y,r}$ preserves the martingale property, so

$$N_t^{F,y,r} := F(X_{t\wedge\tau_{y,r}}) - F(X_0) - \int_0^{t\wedge\tau_{y,r}} (\mathfrak{L}_\eta^\nabla F)(X_s)\,ds$$

is a true martingale. Subtracting (547) from this identity yields

$$R_t^{F,y,r} = \int_0^{t\wedge\tau_{y,r}} \big[(\mathfrak{L}_\eta^\nabla F)(X_s) - (\mathfrak{L}_{\eta,y}^{\mathrm{cub}}F)(X_s)\big]ds.$$

On $[0,\tau_{y,r}]$ one has $d_\eta^{(3)}(X_s,y) \le r$, so Theorem D.97 and (546) give

$$\big|(\mathfrak{L}_\eta^\nabla F)(X_s) - (\mathfrak{L}_{\eta,y}^{\mathrm{cub}}F)(X_s)\big| \le C_{\eta,F}^{(5)}r^4.$$

Integrating over $s \in [0, t\wedge\tau_{y,r}]$ gives (549), proving item (i).

For item (ii), apply item (iii) of Theorem D.97 twice, once with base point y and once with base point z. If $x \in B_\eta^{(3)}(y,r) \cap B_\eta^{(3)}(z,s)$, then

$$\begin{aligned}\big|(\mathfrak{L}_{\eta,y}^{\mathrm{cub}}F)(x) - (\mathfrak{L}_{\eta,z}^{\mathrm{cub}}F)(x)\big| &\le \big|(\mathfrak{L}_\eta^\nabla F)(x) - (\mathfrak{L}_{\eta,y}^{\mathrm{cub}}F)(x)\big| + \big|(\mathfrak{L}_\eta^\nabla F)(x) - (\mathfrak{L}_{\eta,z}^{\mathrm{cub}}F)(x)\big| \\ &\le C_{\eta,F}^{(5)}r^4 + C_{\eta,F}^{(5)}s^4,\end{aligned}$$

which is exactly (550). Item (iii) is simply the resulting reformulation of items (i)–(ii). □

Proposition D.101 (Finite-cylinder fifth-derivative criterion for the intrinsic quartic local reflected-gradient atlas)**.** *Assume the hypotheses of Proposition D.98. If the finite-cylinder fifth-derivative-Cauchy summability condition* (544) *holds, then all conclusions of Theorem D.100 apply. Equivalently, together with the lower-order first-, second-, third-, and fourth-derivative conditions already imported by Proposition D.95, the actual finite-cylinder fifth-derivative-Cauchy budget on $V_{\eta,N}$ is already sufficient to produce the intrinsic quartic-defect cubically frozen local reflected-gradient atlas for the full-support continuum process.*

Proof. Proposition D.98 yields the summable weighted C^5 increment budget required by Theorem D.97. Hence the cubic local-model estimate (537) is available. Theorem D.100 then applies directly and yields the claimed local-atlas conclusions. □

Integrated downstream conclusion after the quartic-local-atlas/fifth-order activation continuation. This continuation adds a forty-fifth downstream message. Positively, the intrinsic continuum lift no longer has only a cubic local model with quartic generator error around one frozen base point: it already carries an intrinsic chartwise local reflected-gradient martingale atlas built from those cubically frozen generators, with quartic defect on each chart and quartic mismatch on chart overlaps. Verification is correspondingly no longer split into separate stochastic and interaction branches: the same finite-cylinder fifth-derivative-Cauchy condition already activates that local atlas. Cautionarily, the manuscript still does not prove defect-free patching of this atlas to a genuinely local reflected gradient SPDE on a broader non-cylinder domain, and it still does not verify the fifth-derivative-Cauchy bounds for the actual e-Arc-Helix interaction family. What remains open is therefore exact vanishing-defect local reflected gradient SPDE closure and the concrete fifth-order estimates — or a sharp obstruction — for the real interaction potentials.

Remark D.102. The new forty-fifth message is again deliberately intermediate, but it finally binds the two remaining existence branches together. On the stochastic side, it no longer stops at a single cubic freezing rule: the continuum lift now comes with an intrinsic cubically frozen local atlas whose chart defect already decays like the fourth power of the chart radius. On the interaction side, it no longer says only that fifth derivatives would yield a cubic local model: the same finite-cylinder fifth-derivative-Cauchy budget already forces this local atlas itself. What is still missing is not an intrinsic local surrogate or a concrete fifth-order activation criterion, but defect-free passage from that surrogate atlas to a broader non-cylinder reflected gradient SPDE together with the actual fifth-order estimates for the e-Arc-Helix interaction potentials.

D.33 Worked continuation XLVI: vanishing-radius local exactness and concrete tensor-majorant fifth-order verification

The quartic-local-atlas/fifth-order activation continuation removes the separation between the stochastic and interaction branches at the level of existence, but two practical gaps remain. On the stochastic side, the atlas is still written with a finite chart defect, so one still has to explain whether the local generator is at least asymptotically unambiguous when the chart radius shrinks to zero. On the interaction side, the finite-cylinder fifth-derivative-Cauchy condition is already the right abstract activation criterion, but it is still tensorially cumbersome to verify directly on the actual interaction potentials. The next continuation addresses both points. First, the quartic atlas already has an exact vanishing-radius local generator germ on the intrinsic cylinder core: shrinking charts become asymptotically exact local reflected-gradient martingale problems, and the local generator limit is unique. Second, the fifth-order activation problem already reduces further to a concrete tensor-majorant mode-decay criterion that can be checked row-by-row on the finite-cylinder interaction family.

Status label. The vanishing-radius local-exactness statement below has status **theorem-level downstream result**; the tensor-majorant activation criterion has status **proposition-level downstream result**. They do *not* yet yield a defect-free genuinely local reflected gradient SPDE on a broader non-cylinder test domain, and they do *not* yet verify the required majorants for the actual e-Arc-Helix interaction family. What they do prove is that the quartic intrinsic atlas already determines an exact local generator germ on the cylinder core, and that concrete mode-decay tensor envelopes already suffice to activate the fifth-order local-atlas package.

Theorem D.103 (Vanishing-radius exactness and local generator germ for the intrinsic quartic atlas)**.** *Assume the hypotheses of Theorem D.100. Fix $F = f \circ \Pi_N \in \mathcal{D}_\eta^{\mathrm{Neu},\infty}$ and $x \in \Gamma_\eta^{-1}(\mathfrak{S}_\eta^{(3)})$. For $r > 0$, define the local atlas oscillation*

$$\Omega_{\eta,F}^{\mathrm{atlas}}(x,r) := \sup\Big\{\big|(\mathfrak{L}_{\eta,y}^{\mathrm{cub}}F)(u) - (\mathfrak{L}_{\eta,z}^{\mathrm{cub}}F)(u)\big| : y,z,u \in B_\eta^{(3)}(x,r)\Big\}, \tag{551}$$

and let

$$\sigma_{x,r} := \inf\{t \geq 0 : X_t \notin B_\eta^{(3)}(x,r)\}. \tag{552}$$

Then the following hold.

(i) For every $r > 0$,

$$\Omega_{\eta,F}^{\mathrm{atlas}}(x,r) \leq 32\, C_{\eta,F}^{(5)}\, r^4. \tag{553}$$

In particular, the quartic atlas is already uniformly Cauchy on shrinking intrinsic balls.

(ii) If $(y_r)_{r>0}$ is any family with $y_r \in B_\eta^{(3)}(x,r)$ for every $r > 0$, then

$$\left|(\mathfrak{L}_{\eta,y_r}^{\mathrm{cub}} F)(x) - (\mathfrak{L}_\eta^\nabla F)(x)\right| \leq C_{\eta,F}^{(5)}\, r^4. \tag{554}$$

Consequently the limit

$$\mathfrak{G}_{\eta,x}^{\mathrm{loc}} F := \lim_{r\downarrow 0} (\mathfrak{L}_{\eta,y_r}^{\mathrm{cub}} F)(x) \tag{555}$$

exists, is independent of the choice of $(y_r)_{r>0}$, and satisfies

$$\mathfrak{G}_{\eta,x}^{\mathrm{loc}} F = (\mathfrak{L}_\eta^\nabla F)(x). \tag{556}$$

Thus the intrinsic quartic atlas already determines a unique exact local generator germ on the cylinder core.

(iii) If $(y_r)_{r>0}$ is any family with $y_r \in B_\eta^{(3)}(x,r)$ and

$$\begin{aligned} M_t^{F,x,r;y_r} &:= F(X_{t\wedge\sigma_{x,r}}) - F(X_0) \\ &\quad - \int_0^{t\wedge\sigma_{x,r}} (\mathfrak{L}_{\eta,y_r}^{\mathrm{cub}} F)(X_s)\, ds, \end{aligned} \tag{557}$$

then there exists a true martingale $N^{F,x,r}$ such that

$$M_t^{F,x,r;y_r} = N_t^{F,x,r} + \widetilde{R}_t^{F,x,r;y_r} \tag{558}$$

and

$$\left|\widetilde{R}_t^{F,x,r;y_r}\right| \leq 16\, C_{\eta,F}^{(5)}\, r^4 t, \qquad t \geq 0. \tag{559}$$

Hence for every $T > 0$,

$$\sup_{0\leq t\leq T} \left|\widetilde{R}_t^{F,x,r;y_r}\right| \leq 16\, C_{\eta,F}^{(5)}\, r^4 T \xrightarrow[r\downarrow 0]{} 0, \tag{560}$$

uniformly over all choices of centers $y_r \in B_\eta^{(3)}(x,r)$. Thus the quartic atlas is already asymptotically exact on shrinking charts.

In particular, the remaining local-SPDE problem is no longer ambiguity of the intrinsic local generator on the cylinder core, but extension of the exact germ (555) from that core to a broader non-cylinder test domain together with a defect-free patching principle.

Proof. If $y, z, u \in B_\eta^{(3)}(x,r)$, then $u \in B_\eta^{(3)}(y,2r) \cap B_\eta^{(3)}(z,2r)$. Applying item (ii) of Theorem D.100 with radii $2r$ and $2r$ gives

$$\left|(\mathfrak{L}_{\eta,y}^{\mathrm{cub}} F)(u) - (\mathfrak{L}_{\eta,z}^{\mathrm{cub}} F)(u)\right| \leq C_{\eta,F}^{(5)}\big((2r)^4 + (2r)^4\big) = 32\, C_{\eta,F}^{(5)}\, r^4.$$

Taking the supremum yields (553), proving item (i).

For item (ii), if $y_r \in B_\eta^{(3)}(x,r)$ then $d_\eta^{(3)}(x,y_r) \le r$. Item (iii) of Theorem D.97 therefore yields (554). In particular,

$$\big|(\mathfrak{L}^{\mathrm{cub}}_{\eta,y_r}F)(x) - (\mathfrak{L}^{\nabla}_{\eta}F)(x)\big| \to 0 \qquad (r \downarrow 0),$$

so the limit (555) exists and equals $(\mathfrak{L}^{\nabla}_{\eta}F)(x)$. Independence of the chosen family $(y_r)_{r>0}$ follows either from this identity or directly from item (i) with $u = x$.

For item (iii), let

$$N_t^{F,x,r} := F(X_{t\wedge\sigma_{x,r}}) - F(X_0) - \int_0^{t\wedge\sigma_{x,r}} (\mathfrak{L}^{\nabla}_{\eta}F)(X_s)\,ds.$$

By Theorem D.88, this is a true martingale. Subtracting (557) gives

$$\widetilde{R}_t^{F,x,r;y_r} = \int_0^{t\wedge\sigma_{x,r}} \big[(\mathfrak{L}^{\nabla}_{\eta}F)(X_s) - (\mathfrak{L}^{\mathrm{cub}}_{\eta,y_r}F)(X_s)\big]\,ds.$$

On $[0,\sigma_{x,r}]$ one has $d_\eta^{(3)}(X_s,x) \le r$, while $d_\eta^{(3)}(x,y_r) \le r$, so $d_\eta^{(3)}(X_s,y_r) \le 2r$. Item (iii) of Theorem D.97 therefore gives

$$\big|(\mathfrak{L}^{\nabla}_{\eta}F)(X_s) - (\mathfrak{L}^{\mathrm{cub}}_{\eta,y_r}F)(X_s)\big| \le C^{(5)}_{\eta,F}(2r)^4 = 16\,C^{(5)}_{\eta,F}\,r^4.$$

Integrating over $s \in [0, t\wedge\sigma_{x,r}]$ yields (559), and (560) follows immediately. □

Proposition D.104 (Concrete tensor-majorant criterion for fifth-order atlas activation). *Assume the hypotheses of Proposition D.98. Suppose there exist nonnegative sequences $(\Xi^{(5)}_{\eta,n})_{n\ge1}$ and $(u_j)_{j\ge1}$ such that*

$$U_\eta^{(5)} := \sum_{j=1}^{\infty} \frac{u_j}{(2j+1)^3} < \infty, \qquad \sum_{n=1}^{\infty} \Xi^{(5)}_{\eta,n} < \infty, \tag{561}$$

and, for every $n \ge 1$ and every admissible choice of indices in $\{1,\dots,n\}$,

$$\Lambda^{hhhhh}_{\eta,n} \le \Xi^{(5)}_{\eta,n}, \tag{562}$$
$$\Lambda^{hhhha}_{\eta,n,k} \le \Xi^{(5)}_{\eta,n} u_k, \tag{563}$$
$$\Lambda^{hhhaa}_{\eta,n,k,\ell} \le \Xi^{(5)}_{\eta,n} u_k u_\ell, \tag{564}$$
$$\Lambda^{hhaaa}_{\eta,n,j,k,\ell} \le \Xi^{(5)}_{\eta,n} u_j u_k u_\ell, \tag{565}$$
$$\Lambda^{haaaa}_{\eta,n,i,j,k,\ell} \le \Xi^{(5)}_{\eta,n} u_i u_j u_k u_\ell, \tag{566}$$
$$\Lambda^{aaaaa}_{\eta,n,i,j,k,\ell,m} \le \Xi^{(5)}_{\eta,n} u_i u_j u_k u_\ell u_m. \tag{567}$$

Then the finite-cylinder fifth-derivative-Cauchy summability condition (544) holds, with the explicit bound

$$\begin{aligned}
&\sum_{n=1}^{\infty}\Bigg(\Lambda^{hhhhh}_{\eta,n} + 4\sum_{k=1}^{n}\frac{\Lambda^{hhhha}_{\eta,n,k}}{(2k+1)^3} + 6\sum_{k,\ell=1}^{n}\frac{\Lambda^{hhhaa}_{\eta,n,k,\ell}}{(2k+1)^3(2\ell+1)^3}\\
&\qquad + 6\sum_{j,k,\ell=1}^{n}\frac{\Lambda^{hhaaa}_{\eta,n,j,k,\ell}}{(2j+1)^3(2k+1)^3(2\ell+1)^3} + 4\sum_{i,j,k,\ell=1}^{n}\frac{\Lambda^{haaaa}_{\eta,n,i,j,k,\ell}}{(2i+1)^3(2j+1)^3(2k+1)^3(2\ell+1)^3}\\
&\qquad + \sum_{i,j,k,\ell,m=1}^{n}\frac{\Lambda^{aaaaa}_{\eta,n,i,j,k,\ell,m}}{(2i+1)^3(2j+1)^3(2k+1)^3(2\ell+1)^3(2m+1)^3}\Bigg)\\
&\le \big(1 + 4U_\eta^{(5)} + 6(U_\eta^{(5)})^2 + 6(U_\eta^{(5)})^3 + 4(U_\eta^{(5)})^4 + (U_\eta^{(5)})^5\big)\sum_{n=1}^{\infty}\Xi^{(5)}_{\eta,n} < \infty.
\end{aligned} \tag{568}$$

Consequently all conclusions of Proposition D.101 and Theorem D.103 apply. In particular, the fifth-order atlas-activation problem already reduces to finding one summable row-majorant sequence and one summable mode-decay envelope.

Proof. Insert the bounds (562)–(567) into the left-hand side of (544). The m-fold mixed terms factor into $\Xi^{(5)}_{\eta,n}$ times products of the mode envelope u_j divided by $(2j+1)^3$. Because all summands are nonnegative,

$$\sum_{k=1}^{n} \frac{u_k}{(2k+1)^3} \le U^{(5)}_\eta, \qquad \sum_{k,\ell=1}^{n} \frac{u_k u_\ell}{(2k+1)^3(2\ell+1)^3} \le (U^{(5)}_\eta)^2,$$

and similarly for the third-, fourth-, and fifth-order mixed sums. Therefore the entire weighted fifth-order budget is bounded by the right-hand side of (568). Finite summability follows from (561). Proposition D.101 then yields the quartic local atlas, and Theorem D.103 upgrades that atlas to the asserted vanishing-radius exactness and local generator germ. □

Integrated downstream conclusion after the vanishing-radius/tensor-majorant continuation. This continuation adds a forty-sixth downstream message. Positively, the quartic intrinsic local atlas no longer carries only a finite-radius defect estimate: it already determines a unique exact local reflected-gradient generator germ on the intrinsic cylinder core, and shrinking charts become asymptotically exact local martingale problems. Verification is also no longer posed only as the raw finite-cylinder fifth-derivative-Cauchy budget: it already reduces further to a concrete tensor-majorant mode-decay criterion. Cautionarily, the manuscript still does not extend that exact local generator germ to a broader non-cylinder test domain, and it still does not verify the required tensor-majorant or fifth-derivative bounds for the actual e-Arc-Helix interaction family. What remains open is therefore non-cylinder promotion of the exact local germ and concrete fifth-order mode-decay verification — or a sharp obstruction — for the real interaction potentials.

Remark D.105. The new forty-sixth message is deliberately asymptotic rather than definitive. On the stochastic side, it does not yet solve the local SPDE problem, but it does show that the quartic atlas already has no generator ambiguity in the vanishing-radius limit on the cylinder core. On the interaction side, it does not yet verify the actual fifth-order bounds, but it does replace the abstract fifth-derivative-Cauchy activation question by a concrete search for summable row-majorants and mode-decay envelopes. What is still missing is precisely the bridge from these exact local germs and explicit majorants to a broader non-cylinder reflected gradient SPDE for the real e-Arc-Helix interaction family.

D.34 Worked continuation XLVII: graph-closed local extension and row-sum kernel fifth-order activation

The vanishing-radius/tensor-majorant continuation isolates the two remaining difficulties much more sharply, but one stochastic gap and one verification gap still remain. On the stochastic side, Theorem D.103 already shows that the quartic intrinsic atlas determines a unique exact local generator germ on the intrinsic cylinder core, yet a genuinely local SPDE still requires a non-cylinder test domain on which that germ is both closed and exact after stopping. On the interaction side, Proposition D.104 already reduces fifth-order activation to one global mode-decay envelope, but actual finite-cylinder interaction families are more likely to present row-dependent mode profiles than one uniform sequence. The next continuation shows that both issues already admit one further intrinsic reduction. First, the exact local germ already extends canonically, by local graph closure, to a ballwise non-cylinder domain of asymptotically cylindrical observables, and on that domain the stopped local martingale problem is exact. Second, fifth-order atlas activation already follows from a weighted row-sum kernel criterion allowing the mode envelopes to vary with the released cylinder index.

Status label. The local graph-closure statement below has status **theorem-level downstream result**; the weighted row-sum activation criterion has status **proposition-level downstream result**. They do *not* yet identify a genuinely local reflected gradient SPDE with explicit coefficients on a natural differential test algebra, and they do *not* yet verify the required row-sum kernels for the actual e-Arc-Helix interaction family. What they do prove is that the exact local generator germ already has a canonical graph-closed non-cylinder extension on intrinsic balls, and that fifth-order activation already reduces to weighted row-sum kernel bounds that may be checked cylinder-by-cylinder rather than through one global tensor envelope.

Theorem D.106 (Graph-closed non-cylinder extension of the exact local generator germ). *Assume the hypotheses of Theorem D.103. Fix $x \in \Gamma_\eta^{-1}(\mathfrak{S}_\eta^{(3)})$ and $r > 0$, and write*

$$\mathbb{B}_\eta(x,r) := B_\eta^{(3)}(x,r) \cap \Gamma_\eta^{-1}(\mathfrak{S}_\eta^{(3)}). \tag{569}$$

Call a pair $(G,H) \in C(\mathbb{B}_\eta(x,r)) \times C(\mathbb{B}_\eta(x,r))$ locally graph-admissible *if there exists a sequence $(F_m)_{m\geq 1} \subset \mathcal{D}_\eta^{\mathrm{Neu},\infty}$ such that*

$$\sup_{u\in\mathbb{B}_\eta(x,r)} \left|F_m(u) - G(u)\right| + \sup_{u\in\mathbb{B}_\eta(x,r)} \left|\mathfrak{G}^{\mathrm{loc}}_{\eta,u} F_m - H(u)\right| \xrightarrow[m\to\infty]{} 0, \tag{570}$$

where $\mathfrak{G}^{\mathrm{loc}}_{\eta,u} F_m$ is the exact local generator germ from Theorem D.103. Let $\mathcal{D}^{\mathrm{gcl}}_{\eta,x,r}$ denote the collection of all G for which some such H exists. Then the following hold.

(i) If (G,H) and $(G,\widetilde{H})$ are both locally graph-admissible on $\mathbb{B}_\eta(x,r)$, then $H = \widetilde{H}$ on $\mathbb{B}_\eta(x,r)$. Consequently the operator

$$\overline{\mathfrak{G}}^{\mathrm{loc}}_{\eta,x,r} G := H, \tag{571}$$

is well defined on $\mathcal{D}^{\mathrm{gcl}}_{\eta,x,r}$.

(ii) Every cylinder-core observable $F \in \mathcal{D}_\eta^{\mathrm{Neu},\infty}$ restricts to an element of $\mathcal{D}^{\mathrm{gcl}}_{\eta,x,r}$, and for every $u \in \mathbb{B}_\eta(x,r)$,

$$\overline{\mathfrak{G}}^{\mathrm{loc}}_{\eta,x,r}\big(F|_{\mathbb{B}_\eta(x,r)}\big)(u) = \mathfrak{G}^{\mathrm{loc}}_{\eta,u} F = (\mathfrak{L}_\eta^{\nabla} F)(u). \tag{572}$$

(iii) Let $G \in \mathcal{D}^{\mathrm{gcl}}_{\eta,x,r}$, fix $0 < \rho < r$, and define

$$\sigma_{x,\rho} := \inf\{t \geq 0 : X_t \notin B_\eta^{(3)}(x,\rho)\}. \tag{573}$$

Then, whenever $X_0 \in \mathbb{B}_\eta(x,\rho)$, the process

$$\begin{aligned} M_t^{G,x,\rho} &:= G(X_{t\wedge\sigma_{x,\rho}}) - G(X_0) \\ &\quad - \int_0^{t\wedge\sigma_{x,\rho}} \big(\overline{\mathfrak{G}}^{\mathrm{loc}}_{\eta,x,r} G\big)(X_s)\, ds \end{aligned} \tag{574}$$

is a true martingale.

(iv) If $(G_m)_{m\geq1} \subset \mathcal{D}^{\mathrm{gcl}}_{\eta,x,r}$ satisfies

$$\sup_{u\in\mathbb{B}_\eta(x,r)} |G_m(u) - G(u)| + \sup_{u\in\mathbb{B}_\eta(x,r)} \left|\overline{\mathfrak{G}}^{\mathrm{loc}}_{\eta,x,r} G_m(u) - H(u)\right| \xrightarrow[m\to\infty]{} 0 \tag{575}$$

for some $G, H \in C(\mathbb{B}_\eta(x,r))$, then $G \in \mathcal{D}^{\mathrm{gcl}}_{\eta,x,r}$ and

$$\overline{\mathfrak{G}}^{\mathrm{loc}}_{\eta,x,r} G = H. \tag{576}$$

Thus the exact local generator germ already has a canonical graph-closed non-cylinder extension on every intrinsic ball.

Proof. For item (i), suppose $(F_m)_{m\ge1}$ and $(\widetilde{F}_m)_{m\ge1}$ witness local graph-admissibility of (G,H) and $(G,\widetilde{H})$, respectively. Fix $0<\rho<r$ and $u\in\mathbb{B}_\eta(x,\rho)$. For each m, Theorem D.88 applied to the cylinder observable $F_m-\widetilde{F}_m$ gives a true martingale

$$N_t^{(m)} := (F_m-\widetilde{F}_m)(X_{t\wedge\sigma_{x,\rho}}) - (F_m-\widetilde{F}_m)(u) - \int_0^{t\wedge\sigma_{x,\rho}} (\mathfrak{L}_\eta^{\nabla}(F_m-\widetilde{F}_m))(X_s)\,ds.$$

Taking expectations under $\mathbb{P}_u$ and using (570) together with (556) yields

$$\mathbb{E}_u\int_0^{t\wedge\sigma_{x,\rho}}(H-\widetilde{H})(X_s)\,ds = 0, \qquad t\ge 0.$$

Dividing by t and sending $t\downarrow 0$, continuity of $H-\widetilde{H}$ and continuity of the stopped paths imply $(H-\widetilde{H})(u)=0$. Since $u\in\mathbb{B}_\eta(x,\rho)$ and $0<\rho<r$ were arbitrary, $H=\widetilde{H}$ on $\mathbb{B}_\eta(x,r)$, proving item (i).

Item (ii) follows by taking the constant approximating sequence $F_m\equiv F$ and invoking (556).

For item (iii), let $(F_m)_{m\ge1}\subset\mathcal{D}_\eta^{\mathrm{Neu},\infty}$ witness local graph-admissibility of $(G,\overline{\mathfrak{G}}_{\eta,x,r}^{\mathrm{loc}}G)$. For each m, the stopped cylinder process

$$M_t^{(m)} := F_m(X_{t\wedge\sigma_{x,\rho}}) - F_m(X_0) - \int_0^{t\wedge\sigma_{x,\rho}}(\mathfrak{L}_\eta^{\nabla}F_m)(X_s)\,ds$$

is a true martingale by Theorem D.88. Because $X_{s\wedge\sigma_{x,\rho}}\in\mathbb{B}_\eta(x,r)$ for all $s\ge0$, the uniform convergences in (570) and (556) show that, for every $T>0$,

$$\sup_{0\le t\le T}\left|M_t^{(m)}-M_t^{G,x,\rho}\right| \le 2\sup_{\mathbb{B}_\eta(x,r)}|F_m-G| + T\sup_{\mathbb{B}_\eta(x,r)}\left|\mathfrak{G}_{\eta,\cdot}^{\mathrm{loc}}F_m-\overline{\mathfrak{G}}_{\eta,x,r}^{\mathrm{loc}}G\right| \xrightarrow[m\to\infty]{} 0.$$

Hence $M^{G,x,\rho}$ is the locally uniform limit of true martingales, and the martingale property passes to the limit by dominated convergence.

For item (iv), for each m choose a cylinder observable $F_m\in\mathcal{D}_\eta^{\mathrm{Neu},\infty}$ such that

$$\sup_{\mathbb{B}_\eta(x,r)}|F_m-G_m| + \sup_{\mathbb{B}_\eta(x,r)}\left|\mathfrak{G}_{\eta,\cdot}^{\mathrm{loc}}F_m-\overline{\mathfrak{G}}_{\eta,x,r}^{\mathrm{loc}}G_m\right| \le 2^{-m}.$$

Combining this with (575) gives

$$\sup_{\mathbb{B}_\eta(x,r)}|F_m-G| + \sup_{\mathbb{B}_\eta(x,r)}\left|\mathfrak{G}_{\eta,\cdot}^{\mathrm{loc}}F_m-H\right| \xrightarrow[m\to\infty]{} 0.$$

Thus $(F_m)_{m\ge1}$ witnesses local graph-admissibility of (G,H), proving (576). This is exactly the asserted graph-closure property. ☐

Proposition D.107 (Weighted row-sum kernel criterion for fifth-order atlas activation). *Assume the hypotheses of Proposition D.98. Suppose there exist nonnegative sequences $(\Xi_{\eta,n}^{(5)})_{n\ge1}$ and a nonnegative array $(A_{\eta,n,j}^{(5)})_{n,j\ge1}$ such that*

$$\rho_{\eta,n}^{(5)} := \sum_{j=1}^{\infty}\frac{A_{\eta,n,j}^{(5)}}{(2j+1)^3}<\infty, \qquad \sum_{n=1}^{\infty}\Xi_{\eta,n}^{(5)}\left(1+\rho_{\eta,n}^{(5)}\right)^5<\infty, \tag{577}$$

and, for every $n \geq 1$ and every admissible choice of indices in $\{1, \ldots, n\}$,

$$\Lambda^{hhhhh}_{\eta,n} \leq \Xi^{(5)}_{\eta,n}, \tag{578}$$

$$\Lambda^{hhhha}_{\eta,n,k} \leq \Xi^{(5)}_{\eta,n} A^{(5)}_{\eta,n,k}, \tag{579}$$

$$\Lambda^{hhhaa}_{\eta,n,k,\ell} \leq \Xi^{(5)}_{\eta,n} A^{(5)}_{\eta,n,k} A^{(5)}_{\eta,n,\ell}, \tag{580}$$

$$\Lambda^{hhaaa}_{\eta,n,j,k,\ell} \leq \Xi^{(5)}_{\eta,n} A^{(5)}_{\eta,n,j} A^{(5)}_{\eta,n,k} A^{(5)}_{\eta,n,\ell}, \tag{581}$$

$$\Lambda^{haaaa}_{\eta,n,i,j,k,\ell} \leq \Xi^{(5)}_{\eta,n} A^{(5)}_{\eta,n,i} A^{(5)}_{\eta,n,j} A^{(5)}_{\eta,n,k} A^{(5)}_{\eta,n,\ell}, \tag{582}$$

$$\Lambda^{aaaaa}_{\eta,n,i,j,k,\ell,m} \leq \Xi^{(5)}_{\eta,n} A^{(5)}_{\eta,n,i} A^{(5)}_{\eta,n,j} A^{(5)}_{\eta,n,k} A^{(5)}_{\eta,n,\ell} A^{(5)}_{\eta,n,m}. \tag{583}$$

Then the finite-cylinder fifth-derivative-Cauchy summability condition (544) holds, with the explicit bound

$$\begin{aligned}
\sum_{n=1}^{\infty} \Bigg(\Lambda^{hhhhh}_{\eta,n} &+ 4 \sum_{k=1}^{n} \frac{\Lambda^{hhhha}_{\eta,n,k}}{(2k+1)^3} + 6 \sum_{k,\ell=1}^{n} \frac{\Lambda^{hhhaa}_{\eta,n,k,\ell}}{(2k+1)^3(2\ell+1)^3} \\
&+ 6 \sum_{j,k,\ell=1}^{n} \frac{\Lambda^{hhaaa}_{\eta,n,j,k,\ell}}{(2j+1)^3(2k+1)^3(2\ell+1)^3} + 4 \sum_{i,j,k,\ell=1}^{n} \frac{\Lambda^{haaaa}_{\eta,n,i,j,k,\ell}}{(2i+1)^3(2j+1)^3(2k+1)^3(2\ell+1)^3} \\
&+ \sum_{i,j,k,\ell,m=1}^{n} \frac{\Lambda^{aaaaa}_{\eta,n,i,j,k,\ell,m}}{(2i+1)^3(2j+1)^3(2k+1)^3(2\ell+1)^3(2m+1)^3} \Bigg) \\
\leq \sum_{n=1}^{\infty} \Xi^{(5)}_{\eta,n} &\Big(1 + 4\rho^{(5)}_{\eta,n} + 6(\rho^{(5)}_{\eta,n})^2 + 6(\rho^{(5)}_{\eta,n})^3 + 4(\rho^{(5)}_{\eta,n})^4 + (\rho^{(5)}_{\eta,n})^5 \Big) \\
\leq \sum_{n=1}^{\infty} \Xi^{(5)}_{\eta,n} &\big(1 + \rho^{(5)}_{\eta,n}\big)^5 < \infty.
\end{aligned} \tag{584}$$

Consequently all conclusions of Proposition D.101, Theorem D.103, and Theorem D.106 apply. In particular, the fifth-order activation problem already reduces to finding one weighted row-sum mode profile for each released finite cylinder.

Proof. Insert the bounds (578)–(583) into the left-hand side of (544). For each fixed n, every mixed sum factors into $\Xi^{(5)}_{\eta,n}$ times a power of the weighted row-sum $\rho^{(5)}_{\eta,n}$. More precisely,

$$\sum_{k=1}^{n} \frac{A^{(5)}_{\eta,n,k}}{(2k+1)^3} \leq \rho^{(5)}_{\eta,n}, \qquad \sum_{k,\ell=1}^{n} \frac{A^{(5)}_{\eta,n,k} A^{(5)}_{\eta,n,\ell}}{(2k+1)^3(2\ell+1)^3} \leq (\rho^{(5)}_{\eta,n})^2,$$

and similarly for the third-, fourth-, and fifth-order mixed sums. This yields the first inequality in (584). The second inequality follows because, for every nonnegative ρ,

$$1 + 4\rho + 6\rho^2 + 6\rho^3 + 4\rho^4 + \rho^5 \leq (1+\rho)^5.$$

Finite summability therefore follows from (577). Proposition D.101 then yields the quartic local atlas, Theorem D.103 yields the exact local generator germ on the cylinder core, and Theorem D.106 upgrades that germ to the asserted graph-closed non-cylinder local extension. □

Integrated downstream conclusion after the graph-closure/row-sum continuation. This continuation adds a forty-seventh downstream message. Positively, the exact local reflected-gradient generator is no longer confined to the cylinder core: it already extends canonically, by graph closure, to a non-cylinder local domain on every intrinsic ball, and the stopped martingale problem is exact there. Verification is also no longer forced to use one global mode envelope:

weighted row-sum kernels depending on the released cylinder already suffice. Cautionarily, the manuscript still does not identify this graph-closed local extension with a genuinely differential reflected SPDE on a natural test algebra, and it still does not verify the required row-sum kernels for the actual e-Arc-Helix interaction potentials. What remains open is therefore natural-domain promotion of the graph-closed local generator and concrete weighted row-sum verification — or a sharp obstruction — for the real interaction family.

Remark D.108. The new forty-seventh message is again intentionally intermediate. On the stochastic side, it does not yet produce a defect-free local reflected gradient SPDE, but it does show that a canonical non-cylinder local generator domain already exists intrinsically and carries exact stopped martingale problems. On the interaction side, it does not yet verify the actual fifth-order bounds, but it does replace both the raw fifth-derivative-Cauchy budget and the global tensor-majorant envelope by a row-by-row weighted kernel search. What is still missing is the genuinely differential/SPDE identification of that graph-closed operator together with concrete row-sum estimates for the real e-Arc-Helix interaction potentials.

D.35 Worked continuation XLVIII: exact local-generator sheaf and banded-plus-tail fifth-order activation

The graph-closure/row-sum continuation leaves the downstream problem in a much sharper form, but one operator-level issue and one verification issue still remain. On the stochastic side, Theorem D.106 already gives an exact graph-closed local generator on every intrinsic ball, yet a credible local reflected SPDE would also like those ballwise operators to patch without ambiguity across arbitrary intrinsic open sets. On the interaction side, Proposition D.107 already reduces fifth-order activation to row-dependent weighted mode envelopes, but actual finite-cylinder interaction families may be more naturally described by a finite near-band together with one ambient tail profile shared across all released cylinders. The next continuation shows that both reductions are already intrinsic. First, the graph-closed local operators already glue into an exact local-generator sheaf on arbitrary intrinsic open sets. Second, weighted row-sum activation already follows from a banded-plus-tail kernel criterion reducing verification further to finite interaction windows plus one summable ambient tail profile.

Status label. The sheaf-level patching statement below has status **theorem-level downstream result**; the banded-plus-tail verification reduction has status **proposition-level downstream result**. They do *not* yet identify the resulting exact local-generator sheaf with a genuinely differential reflected SPDE carrying explicit local coefficients on a natural test algebra, and they do *not* yet verify the required banded-plus-tail bounds for the actual e-Arc-Helix interaction family. What they do prove is that the row-sum-activated graph-closed local operators already patch without defect across intrinsic open sets, and that the fifth-order activation problem already reduces to finitely many local interaction windows together with one ambient summable tail profile.

Theorem D.109 (Exact local-generator sheaf activated by weighted row-sum kernels). *Assume the hypotheses of Proposition D.107. For every intrinsic ball $\mathbb{B}_\eta(x,r)$, let $(\mathcal{D}^{\mathrm{gcl}}_{\eta,x,r}, \overline{\mathfrak{G}}^{\mathrm{loc}}_{\eta,x,r})$ be the graph-closed local domain/operator from Theorem D.106. Then the following hold.*

(i) If $\mathbb{B}_\eta(y,s) \subset \mathbb{B}_\eta(x,r)$ and $G \in \mathcal{D}^{\mathrm{gcl}}_{\eta,x,r}$, then

$$G|_{\mathbb{B}_\eta(y,s)} \in \mathcal{D}^{\mathrm{gcl}}_{\eta,y,s}, \qquad \overline{\mathfrak{G}}^{\mathrm{loc}}_{\eta,y,s}\big(G|_{\mathbb{B}_\eta(y,s)}\big) = \big(\overline{\mathfrak{G}}^{\mathrm{loc}}_{\eta,x,r}G\big)|_{\mathbb{B}_\eta(y,s)}. \tag{585}$$

(ii) For every open set $U \subset \Gamma_\eta^{-1}(\mathfrak{S}_\eta^{(3)})$, define

$$\mathcal{D}_\eta^{\mathrm{sh}}(U) := \Big\{ G \in C(U) : \textit{for every } u \in U \textit{ there exist } x \in \Gamma_\eta^{-1}(\mathfrak{S}_\eta^{(3)}) \textit{ and } r > 0 \\ \textit{with } u \in \mathbb{B}_\eta(x, r/2) \subset \mathbb{B}_\eta(x, r) \subset U \textit{ and } G|_{\mathbb{B}_\eta(x,r)} \in \mathcal{D}^{\mathrm{gcl}}_{\eta,x,r} \Big\}. \tag{586}$$

Then there is a unique pointwise-defined local operator $\mathfrak{G}^{\mathrm{sh}}_{\eta,U}$ on $\mathcal{D}_\eta^{\mathrm{sh}}(U)$ such that whenever $u \in \mathbb{B}_\eta(x, r/2) \subset \mathbb{B}_\eta(x, r) \subset U$ and $G|_{\mathbb{B}_\eta(x,r)} \in \mathcal{D}^{\mathrm{gcl}}_{\eta,x,r}$,

$$(\mathfrak{G}^{\mathrm{sh}}_{\eta,U} G)(u) = \big(\overline{\mathfrak{G}}^{\mathrm{loc}}_{\eta,x,r}(G|_{\mathbb{B}_\eta(x,r)})\big)(u). \tag{587}$$

(iii) If $U = \bigcup_{\alpha \in A} U_\alpha$ is an open cover and for each α one has $G_\alpha \in \mathcal{D}_\eta^{\mathrm{sh}}(U_\alpha)$ such that

$$G_\alpha = G_\beta \qquad \textit{on } U_\alpha \cap U_\beta \qquad (\alpha, \beta \in A), \tag{588}$$

then there exists a unique $G \in \mathcal{D}_\eta^{\mathrm{sh}}(U)$ with $G|_{U_\alpha} = G_\alpha$ for all α, and the local generators glue as well:

$$(\mathfrak{G}^{\mathrm{sh}}_{\eta,U} G)|_{U_\alpha} = \mathfrak{G}^{\mathrm{sh}}_{\eta,U_\alpha} G_\alpha \qquad (\alpha \in A). \tag{589}$$

(iv) Let $G \in \mathcal{D}_\eta^{\mathrm{sh}}(U)$ and choose $\mathbb{B}_\eta(x, r) \subset U$ such that $G|_{\mathbb{B}_\eta(x,r)} \in \mathcal{D}^{\mathrm{gcl}}_{\eta,x,r}$. Then for every $0 < \rho < r/2$, whenever $X_0 \in \mathbb{B}_\eta(x, \rho)$, the process

$$M_t^{G,U,x,\rho} := G(X_{t\wedge\sigma_{x,\rho}}) - G(X_0) \\ - \int_0^{t\wedge\sigma_{x,\rho}} (\mathfrak{G}^{\mathrm{sh}}_{\eta,U} G)(X_s)\, ds \tag{590}$$

is a true martingale. Moreover, on every such representing ball, G is the local graph-norm limit of restricted cylinder observables from $\mathcal{D}_\eta^{\mathrm{Neu},\infty}$.

Consequently the weighted row-sum fifth-order criterion already activates a defect-free operator-level patching on all intrinsic open sets; what remains is the genuinely differential/SPDE identification of that exact local-generator sheaf.

Proof. For item (i), let $(F_m)_{m\geq 1} \subset \mathcal{D}_\eta^{\mathrm{Neu},\infty}$ witness local graph-admissibility of G on $\mathbb{B}_\eta(x, r)$, with local generator $H := \overline{\mathfrak{G}}^{\mathrm{loc}}_{\eta,x,r} G$. Restricting the uniform convergences from (570) to the smaller ball $\mathbb{B}_\eta(y, s)$ shows that the same sequence witnesses local graph-admissibility of $G|_{\mathbb{B}_\eta(y,s)}$ there with candidate generator $H|_{\mathbb{B}_\eta(y,s)}$. By the uniqueness part of Theorem D.106, this candidate is exactly $\overline{\mathfrak{G}}^{\mathrm{loc}}_{\eta,y,s}(G|_{\mathbb{B}_\eta(y,s)})$, proving (585).

For item (ii), fix $u \in U$ and choose any representing ball $\mathbb{B}_\eta(x, r) \subset U$ with $u \in \mathbb{B}_\eta(x, r/2)$ and $G|_{\mathbb{B}_\eta(x,r)} \in \mathcal{D}^{\mathrm{gcl}}_{\eta,x,r}$. Define $(\mathfrak{G}^{\mathrm{sh}}_{\eta,U} G)(u)$ by (587). To check independence of the choice, choose another representing ball $\mathbb{B}_\eta(y, s) \subset U$ with $u \in \mathbb{B}_\eta(y, s/2)$ and $G|_{\mathbb{B}_\eta(y,s)} \in \mathcal{D}^{\mathrm{gcl}}_{\eta,y,s}$. Because both balls are open and contain u, there exists $\delta > 0$ such that $u \in \mathbb{B}_\eta(u, \delta) \subset \mathbb{B}_\eta(x, r) \cap \mathbb{B}_\eta(y, s)$. Applying item (i) twice shows that both restrictions of G belong to $\mathcal{D}^{\mathrm{gcl}}_{\eta,u,\delta}$, and the uniqueness statement in Theorem D.106 forces the two candidate generator values to agree on $\mathbb{B}_\eta(u, \delta)$, hence at u. Thus $\mathfrak{G}^{\mathrm{sh}}_{\eta,U}$ is well defined.

For item (iii), define G pointwise on U by the local formula $G|_{U_\alpha} := G_\alpha$, which is well defined because of (588). Let $u \in U$; choose α with $u \in U_\alpha$ and then choose a representing ball $\mathbb{B}_\eta(x, r) \subset U_\alpha$ with $u \in \mathbb{B}_\eta(x, r/2)$ and $G_\alpha|_{\mathbb{B}_\eta(x,r)} \in \mathcal{D}^{\mathrm{gcl}}_{\eta,x,r}$. Since $G = G_\alpha$ on that ball, it follows that $G \in \mathcal{D}_\eta^{\mathrm{sh}}(U)$. The compatibility of local generators on overlaps is exactly the well-definedness argument from item (ii), applied to the family $(G_\alpha)_{\alpha\in A}$. Uniqueness is immediate from pointwise uniqueness of the glued function.

For item (iv), apply Theorem D.106(iii) on the representing ball $\mathbb{B}_\eta(x,r)$. Because $\rho < r/2$, the stopped paths remain inside the ball on which the local representation is valid, and (587) identifies the sheaf generator with the graph-closed generator along those stopped paths. Therefore (590) is exactly the martingale from (574). The final approximation statement follows directly from the definition of $\mathcal{D}^{\mathrm{gcl}}_{\eta,x,r}$: each local representative is graph-admissible by cylinder observables, hence is a local graph-norm limit of restricted cylinder-core functions. □

Proposition D.110 (Banded-plus-tail criterion for weighted row-sum fifth-order activation). *Assume the hypotheses of Proposition D.98. Suppose there exist nonnegative sequences* $(\Xi^{(5)}_{\eta,n})_{n\ge1}$, $(\Theta^{(5)}_{\eta,n})_{n\ge1}$, *integers* $(b_{\eta,n})_{n\ge1}$, *and a nonnegative sequence* $(v^{(5)}_j)_{j\ge1}$ *such that*

$$V^{(5)}_\eta := \sum_{j=1}^{\infty} \frac{v^{(5)}_j}{(2j+1)^3} < \infty, \qquad B^{(5)}_{\eta,n} := \sum_{j=1}^{b_{\eta,n}} \frac{1}{(2j+1)^3}, \tag{591}$$

and

$$\sum_{n=1}^{\infty} \Xi^{(5)}_{\eta,n}\Big(1+\Theta^{(5)}_{\eta,n}B^{(5)}_{\eta,n}+V^{(5)}_\eta\Big)^5 < \infty. \tag{592}$$

Define

$$A^{(5)}_{\eta,n,j} := \Theta^{(5)}_{\eta,n}\mathbf{1}_{\{1\le j\le b_{\eta,n}\}} + v^{(5)}_j. \tag{593}$$

Assume further that for every $n \ge 1$ *and every admissible choice of indices in* $\{1,\dots,n\}$,

$$\Lambda^{hhhhh}_{\eta,n} \le \Xi^{(5)}_{\eta,n}, \tag{594}$$

$$\Lambda^{hhhha}_{\eta,n,k} \le \Xi^{(5)}_{\eta,n}A^{(5)}_{\eta,n,k}, \tag{595}$$

$$\Lambda^{hhhaa}_{\eta,n,k,\ell} \le \Xi^{(5)}_{\eta,n}A^{(5)}_{\eta,n,k}A^{(5)}_{\eta,n,\ell}, \tag{596}$$

$$\Lambda^{hhaaa}_{\eta,n,j,k,\ell} \le \Xi^{(5)}_{\eta,n}A^{(5)}_{\eta,n,j}A^{(5)}_{\eta,n,k}A^{(5)}_{\eta,n,\ell}, \tag{597}$$

$$\Lambda^{haaaa}_{\eta,n,i,j,k,\ell} \le \Xi^{(5)}_{\eta,n}A^{(5)}_{\eta,n,i}A^{(5)}_{\eta,n,j}A^{(5)}_{\eta,n,k}A^{(5)}_{\eta,n,\ell}, \tag{598}$$

$$\Lambda^{aaaaa}_{\eta,n,i,j,k,\ell,m} \le \Xi^{(5)}_{\eta,n}A^{(5)}_{\eta,n,i}A^{(5)}_{\eta,n,j}A^{(5)}_{\eta,n,k}A^{(5)}_{\eta,n,\ell}A^{(5)}_{\eta,n,m}. \tag{599}$$

Then the hypotheses of Proposition D.107 hold, with

$$\rho^{(5)}_{\eta,n} = \sum_{j=1}^{\infty} \frac{A^{(5)}_{\eta,n,j}}{(2j+1)^3} \le \Theta^{(5)}_{\eta,n}B^{(5)}_{\eta,n} + V^{(5)}_\eta. \tag{600}$$

Consequently all conclusions of Proposition D.107 and Theorem D.109 apply. In particular, the fifth-order activation problem already reduces to finitely many local interaction windows together with one ambient summable tail profile.

Proof. Set $A^{(5)}_{\eta,n,j}$ by (593). Then (600) follows immediately from (591). The derivative bounds (594)–(599) are exactly the row-sum factorization assumptions required in Proposition D.107. Moreover,

$$\sum_{n=1}^{\infty} \Xi^{(5)}_{\eta,n}(1+\rho^{(5)}_{\eta,n})^5 \le \sum_{n=1}^{\infty} \Xi^{(5)}_{\eta,n}\Big(1+\Theta^{(5)}_{\eta,n}B^{(5)}_{\eta,n}+V^{(5)}_\eta\Big)^5 < \infty$$

by (592). Proposition D.107 therefore applies and yields the fifth-order atlas activation, the exact local generator germ, the graph-closed non-cylinder extension, and the weighted row-sum local operator package. Theorem D.109 then upgrades those ballwise operators to the asserted exact local-generator sheaf on intrinsic open sets. □

Integrated downstream conclusion after the sheaf/banded-tail continuation. This continuation adds a forty-eighth downstream message. Positively, the exact local reflected-gradient generator is no longer only ballwise graph-closed: it already patches without defect into an exact local-generator sheaf on arbitrary intrinsic open sets, so operator-level overlap consistency is no longer an open issue. Verification is also no longer posed only as a raw weighted row-sum search: it already reduces further to finitely many local interaction windows plus one ambient summable tail profile. Cautionarily, the manuscript still does not identify this exact sheaf operator with a genuinely differential reflected SPDE on a natural non-cylinder test algebra, and it still does not verify the required banded-plus-tail bounds for the actual e-Arc-Helix interaction potentials. What remains open is therefore differential/SPDE identification of the exact local-generator sheaf together with concrete banded-plus-tail verification — or a sharp obstruction — for the real interaction family.

Remark D.111. The new forty-eighth message is the first one in this chain where the two former pure-existence directions genuinely meet. On the stochastic side, the issue is no longer patching or operator ambiguity: the local generator already glues exactly across intrinsic open sets. On the interaction side, the issue is no longer the existence of some abstract fifth-order activation budget: it already suffices to isolate a finite near-band and one ambient summable tail profile. What is still missing is the final differential identification of this exact sheaf operator as a reflected gradient SDE/SPDE for the actual e-Arc-Helix interaction family, together with concrete verification of the required banded-plus-tail estimates.

D.36 Worked continuation XLIX: chartwise differential realization and banded-plus-tail jet activation

The exact-sheaf/banded-tail continuation leaves one identification problem on the stochastic side and one practical reduction problem on the interaction side. On the stochastic side, Theorem D.109 already patches the ballwise graph-closed local operators into an exact local-generator sheaf on arbitrary intrinsic open sets, but one would still like a genuinely differential local formula on a natural local test sheaf rather than only an abstract operator-level gluing statement. On the interaction side, Proposition D.110 already reduces fifth-order activation to a finite near-band plus one ambient summable tail profile, but concrete identification of the resulting chartwise drift should ideally depend only on first and second finite-cylinder jets once that fifth-order activation has already been secured. The next continuation shows that both reductions are already intrinsic. First, the exact local-generator sheaf already admits a chartwise reflected-gradient differential realization on the local Neumann cylinder sheaf. Second, once the banded-plus-tail fifth-order activation is available, concrete identification of that chartwise differential drift already reduces further to a first/second banded-plus-tail jet criterion for the finite-cylinder interaction potentials themselves.

Status label. The chartwise differential realization below has status **theorem-level downstream result**; the jet-level verification reduction has status **proposition-level downstream result**. They do *not* yet identify one single global natural non-cylinder graph domain carrying a fully defect-free reflected gradient SPDE/SDE formulation for the whole continuum process, and they do *not* yet verify the required banded-plus-tail jet estimates for the actual e-Arc-Helix interaction family. What they do prove is that the exact local-generator sheaf is already differential on a natural local Neumann chart sheaf, and that concrete identification of its chartwise drift already reduces further to first/second finite-cylinder jet budgets once the fifth-order atlas/sheaf activation has been turned on.

Theorem D.112 (Chartwise differential realization of the exact local-generator sheaf). *Assume the hypotheses of Proposition D.110. For every intrinsic open set $U \subset \Gamma_\eta^{-1}(\mathfrak{S}_\eta^{(3)})$, define the* *local Neumann cylinder sheaf*

$$\mathcal{N}_\eta^{\mathrm{loc}}(U) := \Big\{ G \in \mathcal{D}_\eta^{\mathrm{sh}}(U) : \textit{for every } u \in U \textit{ there exist } x \in \Gamma_\eta^{-1}(\mathfrak{S}_\eta^{(3)}) \textit{ and } r > 0 \\ \textit{with } u \in \mathbb{B}_\eta(x, r/2) \subset \mathbb{B}_\eta(x,r) \subset U \textit{ and some } F \in \mathcal{D}_\eta^{\mathrm{Neu},\infty} \\ \textit{such that } G = F \textit{ on } \mathbb{B}_\eta(x,r) \Big\}. \tag{601}$$

Then the following hold.

(i) The assignment $U \mapsto \mathcal{N}_\eta^{\mathrm{loc}}(U)$ *is a sheaf of algebras on* $\Gamma_\eta^{-1}(\mathfrak{S}_\eta^{(3)})$*; in particular it is stable under restriction, finite sums, finite products, and exact gluing on open covers.*

(ii) There exist unique continuous coefficient fields $b_{\eta,h}^U \in C(U)$ *and* $b_{\eta,j}^U \in C(U)$ $(j \geq 1)$ *such that, for every compact* $K \Subset U$,

$$\sup_{u \in K} \left(|b_{\eta,h}^U(u)| + \sum_{j=1}^\infty \frac{|b_{\eta,j}^U(u)|}{(2j+1)^3} \right) < \infty. \tag{602}$$

(iii) If $G \in \mathcal{N}_\eta^{\mathrm{loc}}(U)$, $u \in U$, *and* $F \in \mathcal{D}_\eta^{\mathrm{Neu},\infty}$ *agrees with* G *on some intrinsic ball around* u*, then the cylinder derivatives of* F *at* u *depend only on* G*. Writing these well-defined local derivatives as* $\partial_h G(u)$, $\partial_{hh} G(u)$, $\partial_{a_j} G(u)$, *and* $\partial_{a_j a_j} G(u)$*, one has the exact chartwise differential formula*

$$\begin{aligned}(\mathfrak{G}_{\eta,U}^{\mathrm{sh}} G)(u) = \frac{1}{2} \partial_{hh} G(u) + \frac{1}{2} \sum_{j=1}^\infty \partial_{a_j a_j} G(u) \\ - b_{\eta,h}^U(u)\, \partial_h G(u) - \sum_{j=1}^\infty b_{\eta,j}^U(u)\, \partial_{a_j} G(u),\end{aligned} \tag{603}$$

where, on every local chart, only finitely many derivative terms are nonzero.

(iv) If $V \subset U$ *is intrinsic open, then*

$$b_{\eta,h}^V = b_{\eta,h}^U|_V, \qquad b_{\eta,j}^V = b_{\eta,j}^U|_V \quad (j \geq 1). \tag{604}$$

Consequently the exact local-generator sheaf already has a chartwise reflected-gradient differential realization on the natural local Neumann test sheaf $\mathcal{N}_\eta^{\mathrm{loc}}$.

(v) Let $G \in \mathcal{N}_\eta^{\mathrm{loc}}(U)$ *and choose* $\mathbb{B}_\eta(x,r) \subset U$ *on which* G *agrees with some* $F \in \mathcal{D}_\eta^{\mathrm{Neu},\infty}$*. Then for every* $0 < \rho < r/2$*, whenever* $X_0 \in \mathbb{B}_\eta(x,\rho)$*, the process*

$$\begin{aligned} M_t^{G,U,x,\rho,\mathrm{diff}} := G(X_{t\wedge\sigma_{x,\rho}}) - G(X_0) \\ - \int_0^{t\wedge\sigma_{x,\rho}} (\mathfrak{G}_{\eta,U}^{\mathrm{sh}} G)(X_s)\, ds \end{aligned} \tag{605}$$

is a true martingale, and the generator inside the integral is given exactly by (603).

Proof. Item (i) follows directly from the definition (601): restriction is immediate, finite sums and products preserve local representability by cylinder observables, and exact gluing follows from the gluing property already established in Theorem D.109.

For item (ii), let $H(u) := h(u)$ and $A_j(u) := a_j(u)$. Both are local cylinder observables, hence belong to $\mathcal{N}_\eta^{\mathrm{loc}}(U)$. Define

$$b_{\eta,h}^U := -\mathfrak{G}_{\eta,U}^{\mathrm{sh}} H, \qquad b_{\eta,j}^U := -\mathfrak{G}_{\eta,U}^{\mathrm{sh}} A_j \quad (j \geq 1). \tag{606}$$

Continuity follows from the continuity of the sheaf operator on each representing ball. The weighted bound (602) is inherited from the weighted C^1-gradient package of Theorem D.85 together with the banded-plus-tail activation provided by Proposition D.110, which already guarantees the required local summability on every compact intrinsic chart.

For item (iii), fix $u \in U$ and let $F \in \mathcal{D}_\eta^{\mathrm{Neu},\infty}$ coincide with G on some ball $\mathbb{B}_\eta(x,r) \subset U$ containing u. If $\widetilde{F} \in \mathcal{D}_\eta^{\mathrm{Neu},\infty}$ is another such representative on a possibly smaller ball around u, then $F - \widetilde{F}$ vanishes on a neighborhood of u; hence Theorem D.109 and the exactness of the local sheaf operator force

$$\mathfrak{G}^{\mathrm{sh}}_{\eta,U}(F-\widetilde{F})(u) = 0,$$

so the local cylinder derivatives of F and $\widetilde{F}$ agree at u. Thus the displayed derivatives are well defined. On the chosen chart, Theorem D.88 gives the explicit cylinder formula

$$\mathfrak{L}_\eta^{\nabla} F = \frac{1}{2}\partial_{hh}F + \frac{1}{2}\sum_{j=1}^{\infty}\partial_{a_ja_j}F - b^U_{\eta,h}\partial_h F - \sum_{j=1}^{\infty} b^U_{\eta,j}\partial_{a_j}F,$$

with only finitely many nonzero derivative terms because F is cylinder. Since $G = F$ near u, Theorem D.109 identifies $\mathfrak{G}^{\mathrm{sh}}_{\eta,U}G$ with $\mathfrak{L}_\eta^{\nabla}F$ there, proving (603).

Item (iv) is the restriction compatibility already built into Theorem D.109: if $V \subset U$, then the local coordinate observables H and A_j restrict identically, so (606) together with (585) yields (604).

For item (v), choose a representing cylinder observable F on $\mathbb{B}_\eta(x,r)$. The stopped martingale property for F follows from Theorem D.88, while Theorem D.109 upgrades the same stopped process to the abstract sheaf operator. Since the operator inside that stopped martingale is exactly (603) by item (iii), the process (605) is a true martingale with the claimed chartwise differential generator. □

Proposition D.113 (Banded-plus-tail jet criterion for chartwise differential activation). *Assume the hypotheses of Proposition D.110. For each released finite-cylinder increment set*

$$\Delta_{\eta,n} := V_{\eta,n} - V_{\eta,n-1} \qquad (n \geq 1), \tag{607}$$

with the convention $V_{\eta,0} \equiv 0$. Suppose there exist nonnegative sequences

$$\left(\Upsilon^{(1)}_{\eta,n}\right)_{n\geq 1}, \quad \left(\Upsilon^{(2)}_{\eta,n}\right)_{n\geq 1}, \quad \left(\beta_{\eta,n}\right)_{n\geq 1}, \quad \left(\Theta^{(1)}_{\eta,n}\right)_{n\geq 1}, \quad \left(\Theta^{(2)}_{\eta,n}\right)_{n\geq 1},$$

and ambient tails $(v_j^{(1)})_{j\geq 1}$, $(v_j^{(2)})_{j\geq 1}$ satisfying

$$V_\eta^{(1)} := \sum_{j=1}^{\infty}\frac{v_j^{(1)}}{(2j+1)^3} < \infty, \qquad V_\eta^{(2)} := \sum_{j=1}^{\infty}\frac{v_j^{(2)}}{(2j+1)^3} < \infty, \tag{608}$$

for which, uniformly on every intrinsic compact chart and for all admissible indices $j, k \geq 1$,

$$|\partial_h \Delta_{\eta,n}| \leq \Upsilon^{(1)}_{\eta,n}, \tag{609}$$

$$|\partial_{a_j}\Delta_{\eta,n}| \leq \Upsilon^{(1)}_{\eta,n}\left(\Theta^{(1)}_{\eta,n}\mathbf{1}_{\{j\leq\beta_{\eta,n}\}} + v_j^{(1)}\right), \tag{610}$$

$$|\partial_{hh}\Delta_{\eta,n}| \leq \Upsilon^{(2)}_{\eta,n}, \tag{611}$$

$$|\partial_{ha_j}\Delta_{\eta,n}| \leq \Upsilon^{(2)}_{\eta,n}\left(\Theta^{(2)}_{\eta,n}\mathbf{1}_{\{j\leq\beta_{\eta,n}\}} + v_j^{(2)}\right), \tag{612}$$

$$|\partial_{a_ja_k}\Delta_{\eta,n}| \leq \Upsilon^{(2)}_{\eta,n}\left(\Theta^{(2)}_{\eta,n}\mathbf{1}_{\{|j-k|\leq\beta_{\eta,n}\}} + v_j^{(2)}v_k^{(2)}\right). \tag{613}$$

Assume also that

$$\sum_{n=1}^{\infty} \Upsilon_{\eta,n}^{(1)}\left(1+\Theta_{\eta,n}^{(1)}\beta_{\eta,n}+V_\eta^{(1)}\right)<\infty, \qquad \sum_{n=1}^{\infty} \Upsilon_{\eta,n}^{(2)}\left(1+\Theta_{\eta,n}^{(2)}\beta_{\eta,n}+(V_\eta^{(2)})^2\right)<\infty. \tag{614}$$

For every compact $K \Subset U$, define the local weighted jet seminorm on $\mathcal{N}_\eta^{\mathrm{loc}}(U)$ by

$$\begin{aligned} \|G\|_{K,\mathrm{jet}} := {} & \sup_{u\in K}|G(u)| + \sup_{u\in K}|\partial_h G(u)| + \sup_{u\in K}\sum_{j=1}^{\infty}\frac{|\partial_{a_j}G(u)|}{(2j+1)^3} \\ & + \sup_{u\in K}|\partial_{hh}G(u)| + \sup_{u\in K}\sum_{j=1}^{\infty}\frac{|\partial_{ha_j}G(u)|}{(2j+1)^3} + \sup_{u\in K}\sum_{j,k=1}^{\infty}\frac{|\partial_{a_j a_k}G(u)|}{(2j+1)^3(2k+1)^3}. \end{aligned} \tag{615}$$

Then the following hold.

(i) For every intrinsic open U and compact $K \Subset U$, the finite-cylinder drift jets converge locally uniformly to the chartwise differential coefficients from Theorem D.112:

$$\sup_{u\in K}\left(\left|b_{\eta,h}^U(u)-\partial_h V_{\eta,N}(u)\right| + \sum_{j=1}^{\infty}\frac{\left|b_{\eta,j}^U(u)-\partial_{a_j}V_{\eta,N}(u)\right|}{(2j+1)^3}\right) \xrightarrow[N\to\infty]{} 0. \tag{616}$$

(ii) The operator $\mathfrak{G}_{\eta,U}^{\mathrm{sh}}$ is continuous on $\mathcal{N}_\eta^{\mathrm{loc}}(U)$ with respect to the family of seminorms (615)*, hence extends uniquely to the local weighted C^2-jet completion of $\mathcal{N}_\eta^{\mathrm{loc}}(U)$.*

(iii) On that completed local test class, the exact chartwise differential formula (603) *remains valid, and the stopped local martingale problem from* (605) *remains exact.*

In particular, once fifth-order atlas/sheaf activation is available, concrete identification of the chartwise differential drift already reduces to a first/second banded-plus-tail jet verification for the finite-cylinder interaction potentials themselves.

Proof. The bounds (609)–(613) imply, after summing the telescoping series $V_{\eta,N}=\sum_{n=1}^{N}\Delta_{\eta,n}$, that the first and second finite-cylinder jets are Cauchy in the weighted compact-open norms appearing in (615). Indeed, for $M>N$,

$$\partial_h(V_{\eta,M}-V_{\eta,N}) = \sum_{n=N+1}^{M}\partial_h\Delta_{\eta,n}, \qquad \partial_{a_j}(V_{\eta,M}-V_{\eta,N}) = \sum_{n=N+1}^{M}\partial_{a_j}\Delta_{\eta,n},$$

and the weighted sums are controlled by (614). The same argument applied to (611)–(613) yields local uniform Cauchy control of the second jets. Hence the limits defining the coefficients in Theorem D.112 are precisely the locally uniform limits of the finite-cylinder gradients, proving item (i).

For item (ii), the differential formula (603) together with the compact weighted bounds on the limiting coefficients shows that, on every compact $K \Subset U$,

$$\sup_{u\in K}|\mathfrak{G}_{\eta,U}^{\mathrm{sh}}G(u)| \le C_K\,\|G\|_{K,\mathrm{jet}} \qquad (G\in\mathcal{N}_\eta^{\mathrm{loc}}(U))$$

for some finite constant C_K. Thus the operator is continuous in the jet topology and extends uniquely to the corresponding local weighted C^2-jet completion.

Item (iii) follows by approximating every element of the completion by local cylinder observables in the seminorms (615). The differential formula passes to the limit by the continuity established in item (ii), while the stopped local martingale identity passes to the limit by dominated convergence exactly as in the proofs of Theorems D.106 and D.109. This yields the asserted exact martingale problem on the completed local test class. □

Integrated downstream conclusion after the chartwise-differential/jet continuation. This continuation adds a forty-ninth downstream message. Positively, the exact local-generator sheaf is no longer only an abstract operator-level object: it already has a chartwise reflected-gradient differential realization on a natural local Neumann test sheaf, and concrete recovery of its local drift is no longer forced to proceed through raw fifth-order budgets once the atlas is activated — it already reduces further to first/second banded-plus-tail jet bounds for the finite-cylinder interaction potentials. Cautionarily, the manuscript still does not identify one single global natural non-cylinder graph domain carrying a fully defect-free reflected gradient SPDE/SDE for the whole continuum process, and it still does not verify the required jet bounds for the actual e-Arc-Helix interaction family. What remains open is therefore single-domain promotion of the chartwise differential operator together with concrete banded-plus-tail jet verification — or a sharp obstruction — for the real interaction family.

Remark D.114. The new forty-ninth message is the first one in this chain where the exact local-generator sheaf stops being merely abstract and becomes differential on a natural local test sheaf. At the same time, verification no longer has to begin with the highest-order activation budget once that activation is available; it may begin with first and second concrete banded-plus-tail jets of the finite-cylinder interaction potentials. What is still missing is a single global natural domain on which these local differential representatives glue into one reflected gradient SPDE/SDE object for the full continuum process, together with actual jet estimates for the genuine e-Arc-Helix interaction family.

D.37 Worked continuation L: single natural-domain closure and direct finite-cylinder C^2 stabilization

The chartwise-differential/jet continuation leaves one final chart-dependence issue on the stochastic side and one remaining increment-versus-potential issue on the interaction side. On the stochastic side, Theorem D.112 and Proposition D.113 already give exact chartwise differential operators on the local weighted C^2-jet completion of the Neumann cylinder sheaf, but one would still like one single global natural non-cylinder graph domain rather than a family of chartwise completions. On the interaction side, Proposition D.113 already reduces concrete drift identification to first/second banded-plus-tail jets of the released increments, but one would still like a direct verification criterion stated on the finite-cylinder interaction potentials $V_{\eta,N}$ themselves. The next continuation shows that both reductions are already intrinsic. First, the chartwise completed differential operators close canonically to one single natural non-cylinder local martingale domain. Second, activation of that single-domain package already reduces further to a direct finite-cylinder local-window C^2-Cauchy criterion on the interaction potentials themselves.

Status label. The single-domain closure below has status **theorem-level downstream result**; the direct finite-cylinder verification reduction has status **proposition-level downstream result**. They do *not* yet identify an explicit noise/reflection decomposition for a classical reflected SPDE/SDE, and they do *not* yet verify the required C^2 stabilization bounds for the actual e-Arc-Helix interaction family. What they do prove is that the chartwise differential reflected-gradient representatives already close to one single natural non-cylinder local martingale domain, and that activation of this single-domain package already reduces further to direct first/second derivative stabilization of the finite-cylinder interaction potentials themselves.

Theorem D.115 (Single natural-domain closure of the chartwise differential reflected-gradient operator). *Assume the hypotheses of Proposition D.113. Write*

$$\Omega_\eta := \Gamma_\eta^{-1}(\mathfrak{S}_\eta^{(3)}).$$

For every intrinsic open set $U \subset \Omega_\eta$ and compact $K \Subset U$, define the local graph seminorm on $\mathcal{N}_\eta^{\mathrm{loc}}(U)$ by

$$\|G\|_{K,\mathrm{graph}} := \|G\|_{K,\mathrm{jet}} + \sup_{u \in K} \big|(\mathfrak{G}^{\mathrm{sh}}_{\eta,U} G)(u)\big|. \tag{617}$$

Let $\mathcal{D}_\eta^{\mathrm{nat}}(U)$ be the completion of $\mathcal{N}_\eta^{\mathrm{loc}}(U)$ for the family of seminorms (617). Then the following hold.

(i) The assignment $U \mapsto \mathcal{D}_\eta^{\mathrm{nat}}(U)$ is a sheaf of graph-closed algebras on Ω_η; in particular it is stable under restriction, finite sums, finite products, and exact gluing on intrinsic open covers.

(ii) For every intrinsic open $U \subset \Omega_\eta$ there exists a unique continuous operator

$$\mathfrak{G}^{\mathrm{nat}}_{\eta,U} : \mathcal{D}_\eta^{\mathrm{nat}}(U) \longrightarrow C(U)$$

extending $\mathfrak{G}^{\mathrm{sh}}_{\eta,U}$ such that, whenever $G_m \in \mathcal{N}_\eta^{\mathrm{loc}}(U)$ converges to $G \in \mathcal{D}_\eta^{\mathrm{nat}}(U)$ in the graph seminorms on compact subsets of U,

$$\mathfrak{G}^{\mathrm{nat}}_{\eta,U} G = \lim_{m \to \infty} \mathfrak{G}^{\mathrm{sh}}_{\eta,U} G_m \qquad \text{locally uniformly on } U. \tag{618}$$

(iii) If $V \subset U$ are intrinsic open, then

$$(\mathfrak{G}^{\mathrm{nat}}_{\eta,U} G)|_V = \mathfrak{G}^{\mathrm{nat}}_{\eta,V}(G|_V) \qquad \big(G \in \mathcal{D}_\eta^{\mathrm{nat}}(U)\big). \tag{619}$$

Consequently the local operators patch uniquely to a chart-independent global operator on the full intrinsic support.

(iv) For every intrinsic open $U \subset \Omega_\eta$ and every $G \in \mathcal{D}_\eta^{\mathrm{nat}}(U)$, the exact chartwise differential formula (603) remains valid with the derivatives interpreted in the local weighted C^2-jet completion supplied by Proposition D.113. In particular the coefficient fields $b^U_{\eta,h}$ and $b^U_{\eta,j}$ continue to represent the exact local reflected-gradient drift on $\mathcal{D}_\eta^{\mathrm{nat}}(U)$.

(v) For the full support domain

$$\mathcal{D}_\eta^{\mathrm{nat}} := \mathcal{D}_\eta^{\mathrm{nat}}(\Omega_\eta), \qquad \mathfrak{G}^{\mathrm{nat}}_\eta := \mathfrak{G}^{\mathrm{nat}}_{\eta,\Omega_\eta},$$

the cylinder core $\mathcal{D}_\eta^{\mathrm{Neu},\infty}$ is dense in the local graph topology. If $G \in \mathcal{D}_\eta^{\mathrm{nat}}$, $x \in \Omega_\eta$, and $\rho > 0$ satisfy $\mathbb{B}_\eta(x,\rho) \Subset \Omega_\eta$, then

$$\begin{aligned} M_t^{G,\mathrm{nat},x,\rho} := {} & G(X_{t\wedge\sigma_{x,\rho}}) - G(X_0) \\ & - \int_0^{t\wedge\sigma_{x,\rho}} \big(\mathfrak{G}^{\mathrm{nat}}_\eta G\big)(X_s)\, ds \end{aligned} \tag{620}$$

is a true martingale.

(vi) Consequently the full intrinsic continuum process already carries a single chart-independent exact local reflected-gradient martingale problem on the natural non-cylinder domain $\mathcal{D}_\eta^{\mathrm{nat}}$.

Proof. By Proposition D.113, on every intrinsic open set U the operator $\mathfrak{G}^{\mathrm{sh}}_{\eta,U}$ is continuous for the local weighted C^2-jet seminorms (615), and the exact chartwise differential formula already extends to the corresponding local completion. Hence the graph seminorms (617) are equivalent on compact subsets to the jet seminorms of Proposition D.113; the completion $\mathcal{D}_\eta^{\mathrm{nat}}(U)$ is therefore the same completed local test class, now equipped with its graph interpretation. This gives item (ii) and the differential formula in item (iv) immediately.

For item (i), restriction is inherited from the restriction maps on $\mathcal{N}_\eta^{\mathrm{loc}}$, while stability under finite sums and products follows because the graph seminorm is controlled by the jet seminorm together with the Leibniz rule for the chartwise differential formula. Exact gluing on open covers follows by taking compatible local Cauchy sequences on the cover, gluing them first in the sheaf $\mathcal{N}_\eta^{\mathrm{loc}}$ via Theorem D.112, and then passing to the graph limits on compact subsets.

Item (iii) is a direct consequence of uniqueness of graph limits: if G_m approximates G on U, then $G_m|_V$ approximates $G|_V$ on V, and the operator limits agree on overlaps because the prelimit sheaf operators already agree there by Theorem D.109.

For item (v), density of $\mathcal{D}_\eta^{\mathrm{Neu},\infty}$ is built into the definition of the completion. Choose $G_m \in \mathcal{D}_\eta^{\mathrm{Neu},\infty}$ converging to G in the local graph seminorms on a ball containing $\mathbb{B}_\eta(x,\rho)$. For each m, the stopped process

$$G_m(X_{t\wedge\sigma_{x,\rho}}) - G_m(X_0) - \int_0^{t\wedge\sigma_{x,\rho}} (\mathfrak{L}_\eta^\nabla G_m)(X_s)\,ds$$

is a true martingale by Theorem D.88. Since $\mathfrak{L}_\eta^\nabla G_m = \mathfrak{G}_\eta^{\mathrm{nat}} G_m$ on the cylinder core and both G_m and the generators converge uniformly on the stopped chart, dominated convergence passes the martingale identity to the limit and yields (620). Item (vi) is merely the restatement of item (v) in local martingale-problem language. □

Proposition D.116 (Finite-cylinder local-window C^2-Cauchy criterion for the single natural–domain package). *Assume the hypotheses of Proposition D.110. Suppose there exist nonnegative sequences*

$$\left(\Lambda_{\eta,N}^{(1)}\right)_{N\geq 0}, \quad \left(\Lambda_{\eta,N}^{(2)}\right)_{N\geq 0}, \quad \left(\rho_{\eta,N}\right)_{N\geq 0}, \quad \left(\Xi_{\eta,N}^{(1)}\right)_{N\geq 0}, \quad \left(\Xi_{\eta,N}^{(2)}\right)_{N\geq 0},$$

and common tails $(w_j^{(1)})_{j\geq 1}$, $(w_j^{(2)})_{j\geq 1}$ satisfying

$$W_\eta^{(1)} := \sum_{j=1}^\infty \frac{w_j^{(1)}}{(2j+1)^3} < \infty, \qquad W_\eta^{(2)} := \sum_{j=1}^\infty \frac{w_j^{(2)}}{(2j+1)^3} < \infty, \tag{621}$$

for which, uniformly on every intrinsic compact chart, for all admissible $M > N \geq 0$ and all $j,k \geq 1$,

$$|\partial_h(V_{\eta,M} - V_{\eta,N})| \leq \Lambda_{\eta,N}^{(1)}, \tag{622}$$

$$|\partial_{a_j}(V_{\eta,M} - V_{\eta,N})| \leq \Lambda_{\eta,N}^{(1)}\left(\Xi_{\eta,N}^{(1)}\mathbf{1}_{\{j\leq\rho_{\eta,N}\}} + w_j^{(1)}\right), \tag{623}$$

$$|\partial_{hh}(V_{\eta,M} - V_{\eta,N})| \leq \Lambda_{\eta,N}^{(2)}, \tag{624}$$

$$|\partial_{ha_j}(V_{\eta,M} - V_{\eta,N})| \leq \Lambda_{\eta,N}^{(2)}\left(\Xi_{\eta,N}^{(2)}\mathbf{1}_{\{j\leq\rho_{\eta,N}\}} + w_j^{(2)}\right), \tag{625}$$

$$|\partial_{a_ja_k}(V_{\eta,M} - V_{\eta,N})| \leq \Lambda_{\eta,N}^{(2)}\left(\Xi_{\eta,N}^{(2)}\mathbf{1}_{\{|j-k|\leq\rho_{\eta,N}\}} + w_j^{(2)}w_k^{(2)}\right). \tag{626}$$

Assume also that

$$\sum_{N=0}^\infty \Lambda_{\eta,N}^{(1)}\left(1 + \Xi_{\eta,N}^{(1)}\rho_{\eta,N} + W_\eta^{(1)}\right) < \infty, \qquad \sum_{N=0}^\infty \Lambda_{\eta,N}^{(2)}\left(1 + \Xi_{\eta,N}^{(2)}\rho_{\eta,N} + (W_\eta^{(2)})^2\right) < \infty. \tag{627}$$

Then the following hold.

(i) *The hypotheses of Proposition D.113 hold for the released increments $\Delta_{\eta,n} = V_{\eta,n} - V_{\eta,n-1}$ with*

$$\Upsilon_{\eta,n}^{(r)} := \Lambda_{\eta,n-1}^{(r)}, \qquad \beta_{\eta,n} := \rho_{\eta,n-1}, \qquad \Theta_{\eta,n}^{(r)} := \Xi_{\eta,n-1}^{(r)} \quad (r = 1,2).$$

(ii) Consequently Proposition D.113 and Theorem D.115 apply. In particular the chartwise drift coefficients and the single natural-domain operator are determined by the locally uniform first/second derivative limits of the finite-cylinder interaction potentials.

(iii) For every intrinsic compact chart K and every cylinder observable $F \in \mathcal{D}_\eta^{\mathrm{Neu},\infty}$, the finite-cylinder reflected-gradient approximants determined by $V_{\eta,N}$ are Cauchy in the graph seminorms (617), *and their local graph limit is exactly the natural-domain operator $\mathfrak{G}_\eta^{\mathrm{nat}} F$.*

In particular, activation of the single natural-domain package already reduces to a direct finite-cylinder local-window C^2-Cauchy verification on the interaction potentials themselves.

Proof. Taking $M = n$ and $N = n - 1$ in (622)–(626) yields the increment bounds required in Proposition D.113, with the displayed identification of the budget sequences. The summability condition (627) is exactly the translated version of (614). This proves item (i).

Item (ii) is therefore immediate from Proposition D.113 and Theorem D.115. For item (iii), the same estimates show that for every fixed cylinder observable F the corresponding finite-cylinder drift coefficients form a Cauchy family in the compact graph seminorms. Their limit agrees with the limiting drift fields identified in Proposition D.113, hence by uniqueness of the graph closure it equals $\mathfrak{G}_\eta^{\mathrm{nat}} F$. □

Integrated downstream conclusion after the natural-domain/C^2-stabilization continuation. This continuation adds a fiftieth downstream message. Positively, the chartwise differential reflected-gradient representatives are no longer only a compatible local family: they already close to one single natural non-cylinder local martingale domain for the intrinsic continuum process. Verification is also no longer forced to pass through released increments even at the first/second-jet stage: direct finite-cylinder local-window C^2 stabilization already suffices to activate the same single-domain package. Cautionarily, the manuscript still does not identify an explicit Brownian/reflection decomposition for a classical reflected gradient SPDE/SDE on that natural domain, and it still does not verify the required C^2 stabilization bounds for the actual e-Arc-Helix interaction potentials. What remains open is therefore explicit SPDE/SDE identification of the single natural-domain operator together with concrete finite-cylinder C^2 verification — or a sharp obstruction — for the genuine interaction family.

Remark D.117. The new fiftieth message is the first one in this chain where the stochastic and interaction reductions have both crossed the same threshold. On the stochastic side, there is already one single natural non-cylinder local martingale domain rather than only a chartwise sheaf of operators. On the interaction side, there is already a direct finite-cylinder C^2 route to that same domain rather than only an increment-level jet criterion. What is still missing is explicit identification of the resulting exact local operator with a classical reflected gradient SPDE/SDE and actual C^2 estimates for the genuine e-Arc-Helix interaction family.

D.38 Worked continuation LI: chart-free local Itô characteristics and one-profile finite-cylinder C^2 activation

The previous continuation produced a single natural non-cylinder local martingale domain together with a direct finite-cylinder local-window C^2 route to that domain. The next intrinsic sharpening again has two complementary parts. On the stochastic side, once one works on the same natural domain but restricts to a canonical square-gradient subdomain, the exact local martingales already carry chart-free Itô characteristics whose predictable brackets are given by the intrinsic continuum carré du champ and whose martingale parts are represented by the same countable Brownian family that already drives the full-support strong continuum realization. On the verification side, the direct local-window C^2 criterion can be compressed further to a one-profile condition involving only one radius sequence, one tail profile, and one amplitude budget. This still does not identify a chart-free Skorokhod reflection measure or a classical reflected gradient SPDE/SDE on the whole natural domain, and it still does not verify the new one-profile bounds for the genuine e-Arc-Helix interaction family. But it moves the remaining gap almost entirely to the reflection side and to one especially concrete finite-cylinder verification problem.

Status label. The Itô-characteristic theorem below has status **theorem-level downstream result**. It does *not* yet identify the exact local reflected-gradient operator with a classical reflected gradient SPDE/SDE carrying an explicit chart-free Skorokhod term, but it does show that on a canonical square-gradient subdomain the local martingale/bracket characteristics are already exact and are already driven by the common Brownian family from the strong continuum lift. The verification reduction below is also **proposition-level downstream result**. It does *not* verify the required bounds for the actual e-Arc-Helix interaction potentials, but it reduces the activation problem further to one especially concrete one-profile local-window/tail C^2 majorant.

For every intrinsic open set $U \subset \Omega_\eta$ and compact $K \Subset U$, define the local Itô seminorm on the square-gradient subspace of $\mathcal{N}_\eta^{\mathrm{loc}}(U)$ by

$$\|F\|_{K,\mathrm{Ito}} := \|F\|_{K,\mathrm{graph}} + \sup_{u=(h,\mathbf{a})\in K} \left(A(h)\,|\partial_h F(u)|^2 + \sum_{j=1}^{\infty} |\partial_{a_j} F(u)|^2 \right)^{1/2}. \tag{628}$$

Let $\mathcal{I}_\eta^{\mathrm{loc}}(U)$ denote the collection of all local test functions for which the seminorms (628) are finite on every compact $K \Subset U$, and let $\mathcal{D}_\eta^{\mathrm{Ito}}(U)$ be the completion of $\mathcal{I}_\eta^{\mathrm{loc}}(U)$ for that family of seminorms.

Theorem D.118 (Chart-free local Itô characteristics on the single natural domain). *Assume the hypotheses of Proposition D.116, and let*

$$Z^x = (H^x, A_1^x, A_2^x, \dots)$$

be the full-support strong continuum lift from Theorem D.73, started from $x \in \Omega_\eta$ and driven by the common Brownian family $(\beta, B_1, B_2, \dots)$. Then the following hold.

(i) The assignment $U \mapsto \mathcal{D}_\eta^{\mathrm{Ito}}(U)$ is a sheaf of local square-gradient domains contained in $\mathcal{D}_\eta^{\mathrm{nat}}(U)$. Moreover, there is a unique jointly continuous bilinear operator

$$\Gamma_{\eta,U}^{\mathrm{Ito}} : \mathcal{D}_\eta^{\mathrm{Ito}}(U) \times \mathcal{D}_\eta^{\mathrm{Ito}}(U) \longrightarrow C(U)$$

extending the cylinder carré-du-champ formula

$$\Gamma_\eta^{\mathrm{cyl}}(F,G)(h,\mathbf{a}) := A(h)\,\partial_h F(h,\mathbf{a})\,\partial_h G(h,\mathbf{a}) + \sum_{j=1}^{\infty} \partial_{a_j} F(h,\mathbf{a})\,\partial_{a_j} G(h,\mathbf{a}), \tag{629}$$

where the sum is finite on cylinder observables.

(ii) If $U \subset \Omega_\eta$ is intrinsic open, $\mathbb{B}_\eta(x,\rho) \Subset U$, and $F, G \in \mathcal{D}_\eta^{\mathrm{Ito}}(U)$, then the stopped process

$$M_t^{F,\mathrm{Ito},x,\rho} := F(Z^x_{t\wedge\sigma_{x,\rho}}) - F(Z^x_0) - \int_0^{t\wedge\sigma_{x,\rho}} (\mathfrak{G}_\eta^{\mathrm{nat}} F)(Z^x_s)\,ds \tag{630}$$

is a continuous square-integrable martingale, and its predictable quadratic covariation with the analogous martingale for G is

$$\left\langle M^{F,\mathrm{Ito},x,\rho}, M^{G,\mathrm{Ito},x,\rho} \right\rangle_t = 2 \int_0^{t\wedge\sigma_{x,\rho}} \left(\Gamma_{\eta,U}^{\mathrm{Ito}}(F,G)\right)(Z^x_s)\,ds. \tag{631}$$

(iii) On the filtered probability space of Theorem D.73, every $F \in \mathcal{D}_\eta^{\mathrm{Ito}}(U)$ admits the exact local Brownian representation

$$\begin{aligned} M_t^{F,\mathrm{Ito},x,\rho} &= \int_0^{t\wedge\sigma_{x,\rho}} \sqrt{2A(H_s^x)}\,\partial_h F(Z_s^x)\,d\beta_s \\ &\quad + \lim_{J\to\infty}\sum_{j=1}^{J}\int_0^{t\wedge\sigma_{x,\rho}} \sqrt{2}\,\partial_{a_j}F(Z_s^x)\,dB_{j,s}, \end{aligned} \tag{632}$$

where the limit is taken in $L^2(\mathbb{P})$ uniformly on compact time intervals.

(iv) The coordinate maps $H(h,\mathbf{a}) := h$ and $A_n(h,\mathbf{a}) := a_n$ belong to $\mathcal{D}_\eta^{\mathrm{Ito}}(\Omega_\eta)$. Under the identification $\Gamma_\eta(Z^x) = Y^x$, the local brackets furnished by (631) agree with the coordinate Fukushima brackets of Theorem D.50. Consequently, what remains open is not the chart-free local martingale/bracket package itself, but explicit identification of the accompanying finite-variation/reflection term as a classical reflected gradient Skorokhod mechanism on the same natural domain.

Proof. For cylinder observables, Theorem D.88 gives the local martingale property together with the bracket formula (473), while Theorem D.73 and the triangular system of Theorem D.76 realize those same martingales on the common Brownian family $(\beta, B_1, B_2, \dots)$. In particular the cylinder carré-du-champ formula (629) is jointly continuous with respect to the seminorms (628) on compact subsets, because those seminorms control both the graph part and the local square-gradient part. Completing the local cylinder test classes therefore yields item (i) and the unique extension $\Gamma_{\eta,U}^{\mathrm{Ito}}$.

For item (ii), choose cylinder approximants $F_m, G_m \in \mathcal{I}_\eta^{\mathrm{loc}}(U)$ converging to F, G in the Itô seminorms on compact subsets of U. By Theorem D.115, the graph component of the seminorm already gives locally uniform convergence of F_m and of $\mathfrak{G}_\eta^{\mathrm{nat}} F_m$ on every stopped chart. Hence the stopped martingales for F_m converge in L^2 on compact time intervals to the process (630). At the same time, the square-gradient component of (628) implies local uniform convergence of the cylinder carré-du-champ fields $\Gamma_\eta^{\mathrm{cyl}}(F_m, G_m)$ to a continuous limit, necessarily $\Gamma_{\eta,U}^{\mathrm{Ito}}(F,G)$. Passing to the limit in the bracket identity for the cylinder approximants gives (631).

For item (iii), the same cylinder approximants satisfy the stochastic-integral representation furnished by the common Brownian family. The Itô isometry together with convergence in (628) shows that these stochastic integrals form a Cauchy sequence in L^2, uniformly on compact time intervals. Their limit must coincide with the martingale from item (ii), yielding (632). Item (iv) is immediate because the coordinate functions belong to the cylinder core, and their brackets were already computed in Theorem D.50; transporting through the strong realization $\Gamma_\eta(Z^x) = Y^x$ identifies the two descriptions. □

Proposition D.119 (One-profile finite-cylinder local-window/tail C^2 criterion for the chart-free Itô package). *Assume the hypotheses of Proposition D.110. Suppose there exist nonnegative sequences*

$$(\Lambda_{\eta,N})_{N\ge 0}, \qquad (\rho_{\eta,N})_{N\ge 0},$$

and a single common tail profile $(w_j)_{j\ge 1}$ satisfying

$$W_\eta := \sum_{j=1}^{\infty} \frac{w_j}{(2j+1)^3} < \infty, \tag{633}$$

for which, uniformly on every intrinsic compact chart, for all admissible $M > N \ge 0$ and all

$j, k \geq 1$,

$$|\partial_h(V_{\eta,M} - V_{\eta,N})| + |\partial_{hh}(V_{\eta,M} - V_{\eta,N})| \leq \Lambda_{\eta,N}, \tag{634}$$
$$|\partial_{a_j}(V_{\eta,M} - V_{\eta,N})| + |\partial_{ha_j}(V_{\eta,M} - V_{\eta,N})| \leq \Lambda_{\eta,N}(\mathbf{1}_{\{j \leq \rho_{\eta,N}\}} + w_j), \tag{635}$$
$$|\partial_{a_j a_k}(V_{\eta,M} - V_{\eta,N})| \leq \Lambda_{\eta,N}(\mathbf{1}_{\{|j-k| \leq \rho_{\eta,N}\}} + w_j w_k). \tag{636}$$

Assume also that

$$\sum_{N=0}^{\infty} \Lambda_{\eta,N}(1 + \rho_{\eta,N} + W_\eta + (W_\eta)^2) < \infty. \tag{637}$$

Then the hypotheses of Proposition D.116 hold with the identifications

$$\Lambda^{(1)}_{\eta,N} = \Lambda^{(2)}_{\eta,N} = \Lambda_{\eta,N}, \qquad \Xi^{(1)}_{\eta,N} = \Xi^{(2)}_{\eta,N} = 1, \qquad w^{(1)}_j = w^{(2)}_j = w_j.$$

Consequently Proposition D.116, Theorem D.115, and Theorem D.118 all apply. In particular, activation of the chart-free local Itô package already reduces to one direct finite-cylinder local-window/tail C^2 verification involving only one radius sequence, one tail profile, and one amplitude budget.

Proof. The estimates (634)–(636) are exactly the special case of (622)–(626) in which the first- and second-derivative amplitudes are identified and the two tail profiles coincide. The summability condition (637) is stronger than (627) under the same identifications. Proposition D.116 therefore applies directly, and the remaining conclusions are immediate from that proposition together with Theorems D.115 and D.118. □

Integrated downstream conclusion after the natural-Itô/one-profile continuation. This continuation adds a fifty-first downstream message. Positively, the single natural-domain reflected-gradient operator is no longer only an exact local martingale generator: on a canonical square-gradient subdomain it already carries an exact chart-free local Itô/carré-du-champ package driven by the same Brownian family as the full-support strong continuum lift. Verification is also no longer forced to track separate first/second derivative amplitudes and separate tail profiles: one local-window radius sequence, one common tail profile, and one amplitude budget already suffice. Cautionarily, the reflection term is still present only implicitly through the exact local operator and its cylinder approximants, and the new one-profile C^2 bounds are still not verified for the genuine e-Arc-Helix interaction potentials. What remains open is therefore explicit chart-free Skorokhod/SPDE identification of the reflection mechanism together with concrete one-profile local-window/tail C^2 verification — or a sharp obstruction — for the actual interaction family.

Remark D.120. The new fifty-first message is the first point in the chain where the single natural-domain operator already has exact chart-free local Itô characteristics on its own canonical square-gradient subdomain. The remaining stochastic gap is therefore no longer the absence of Brownian characteristics, but only the absence of a chart-free identification of the associated finite-variation part as a classical Skorokhod reflection term. On the interaction side, the remaining verification gap is likewise concentrated: it is enough to prove — or sharply fail to prove — one-profile local-window/tail C^2 bounds for the actual finite-cylinder interaction potentials.

D.39 Worked continuation LII: dual reflection-functional closure and one-step shell C^2 activation

The previous continuation left the two downstream gaps in a particularly sharp form. On the stochastic side, the natural Itô package already furnished exact chart-free local martingale and bracket data on a canonical square-gradient domain, but the reflection contribution was still present only implicitly through the operator and the coordinate local times of the strong continuum lift. On the interaction side, the activation problem had already been compressed to a one-profile local-window/tail C^2 budget, but that budget was still stated directly on all finite-cylinder differences $V_{\eta,M} - V_{\eta,N}$. The next intrinsic sharpening again pushes both directions at once. First, the coordinate reflected local times already patch into a single chart-free dual local reflection functional on the intrinsic finitely supported test space, still driven by the same Brownian family as the full-support strong continuum lift. Second, the one-profile C^2 package already reduces further to one-step released-shell local-window/tail bounds.

Status label. The reflection-functional statement below has status **theorem-level downstream result**. It does *not* yet extend the reflected local additive functional from finitely supported directions to the full natural Itô domain, and it does *not* yet identify its support/positivity as a classical chart-free Skorokhod normal term for a reflected gradient SPDE/SDE. What it does show is that the missing reflection object is no longer absent even on the continuum side: it already exists canonically as a chart-free dual local additive functional on the intrinsic finitely supported test space. The one-step activation reduction below is also a **proposition-level downstream result**. It does *not* verify the required bounds for the actual e-Arc-Helix interaction family, but it reduces the remaining C^2 verification problem further to released-shell one-step estimates.

Write

$$\mathcal{T}_\eta^{\mathrm{fs}} := \{\varphi = (\varphi_h, \varphi_1, \varphi_2, \dots) : \ \varphi_j = 0 \text{ for all but finitely many } j\},$$

for the intrinsic finitely supported test space, and for $\varphi \in \mathcal{T}_\eta^{\mathrm{fs}}$ define the associated linear cylinder observable

$$\ell_\varphi(h, \mathbf{a}) := \varphi_h h + \sum_{j=1}^{\infty} \varphi_j a_j, \tag{638}$$

where the sum is finite by construction.

Theorem D.121 (Chart-free dual local reflection functional on the intrinsic finitely supported test space)**.** *Assume the hypotheses of Proposition D.119, let*

$$Z^x = (H^x, A_1^x, A_2^x, \dots)$$

be the full-support strong continuum lift from Theorem D.73, and fix $x \in \Omega_\eta$ together with $\rho > 0$ such that $\mathbb{B}_\eta(x, \rho) \Subset \Omega_\eta$. Then there exists a unique family

$$\mathfrak{R}^{x,\rho} = \left(\mathfrak{R}_t^{x,\rho}\right)_{t \ge 0}$$

of linear maps

$$\mathfrak{R}_t^{x,\rho} : \mathcal{T}_\eta^{\mathrm{fs}} \longrightarrow \mathbb{R}$$

with the following properties.

(i) For every $\varphi \in \mathcal{T}_\eta^{\mathrm{fs}}$, the process $t \mapsto \mathfrak{R}_t^{x,\rho}[\varphi]$ is a continuous finite-variation additive functional on $[0, \infty)$, stopped at $\sigma_{x,\rho}$.

(ii) For every $\varphi = (\varphi_h, \varphi_1, \varphi_2, \dots) \in \mathcal{T}_\eta^{\mathrm{fs}}$, one has the exact stopped semimartingale decomposi-

tion

$$
\begin{aligned}
\ell_\varphi(Z^x_{t\wedge\sigma_{x,\rho}}) - \ell_\varphi(Z^x_0) = &\int_0^{t\wedge\sigma_{x,\rho}} \sqrt{2A(H^x_s)}\,\varphi_h\,d\beta_s \\
&+ \sum_{j=1}^{\infty}\int_0^{t\wedge\sigma_{x,\rho}} \sqrt{2}\,\varphi_j\,dB_{j,s} \\
&- \int_0^{t\wedge\sigma_{x,\rho}} \Big(\varphi_h b_{\eta,h}(Z^x_s) + \sum_{j=1}^{\infty}\varphi_j b_{\eta,j}(Z^x_s)\Big)\,ds \\
&+ \mathfrak{R}^{x,\rho}_t[\varphi],
\end{aligned}
\tag{639}
$$

where both infinite sums are in fact finite because φ has finite support.

(iii) Writing $e_h := (1,0,0,\dots)$ and $e_n := (0,\dots,0,1,0,\dots)$ for the basis vectors of $\mathcal{T}^{\mathrm{fs}}_\eta$, the dual reflection functional agrees with the explicit coordinate reflected local times from Theorem D.76:

$$
\mathfrak{R}^{x,\rho}_t[e_h] = L^{0,x}_{t\wedge\sigma_{x,\rho}} - L^{h_*,x}_{t\wedge\sigma_{x,\rho}}, \tag{640}
$$

$$
\mathfrak{R}^{x,\rho}_t[e_n] = K^{+,x}_{n,t\wedge\sigma_{x,\rho}} - K^{-,x}_{n,t\wedge\sigma_{x,\rho}} \qquad (n \ge 1). \tag{641}
$$

Consequently, for every $\varphi \in \mathcal{T}^{\mathrm{fs}}_\eta$,

$$
\begin{aligned}
\mathfrak{R}^{x,\rho}_t[\varphi] = &\varphi_h\big(L^{0,x}_{t\wedge\sigma_{x,\rho}} - L^{h_*,x}_{t\wedge\sigma_{x,\rho}}\big) \\
&+ \sum_{j=1}^{\infty}\varphi_j\big(K^{+,x}_{j,t\wedge\sigma_{x,\rho}} - K^{-,x}_{j,t\wedge\sigma_{x,\rho}}\big),
\end{aligned}
\tag{642}
$$

again with only finitely many nonzero terms.

(iv) For every $\varphi \in \mathcal{T}^{\mathrm{fs}}_\eta$, the finite-variation process $\mathfrak{R}^{x,\rho}[\varphi]$ has zero quadratic variation. Hence the remaining stochastic gap is no longer the absence of any chart-free reflection candidate on continuum test directions, but only extension of this dual local additive functional beyond finitely supported directions together with its identification as a classical Skorokhod normal/reflection mechanism on the full natural Itô domain.

Proof. Take first $\varphi = e_h$. The H-equation from Theorem D.76 gives

$$
H^x_{t\wedge\sigma_{x,\rho}} - H^x_0 = \int_0^{t\wedge\sigma_{x,\rho}} \sqrt{2A(H^x_s)}\,d\beta_s - \int_0^{t\wedge\sigma_{x,\rho}} b_{\eta,h}(Z^x_s)\,ds + L^{0,x}_{t\wedge\sigma_{x,\rho}} - L^{h_*,x}_{t\wedge\sigma_{x,\rho}}.
$$

This is exactly (639) with the choice (640). For $\varphi = e_n$, the n-th coordinate equation from the same theorem gives

$$
A^x_n(t\wedge\sigma_{x,\rho}) - A^x_n(0) = \int_0^{t\wedge\sigma_{x,\rho}} \sqrt{2}\,dB_{n,s} - \int_0^{t\wedge\sigma_{x,\rho}} b_{\eta,n}(Z^x_s)\,ds + K^{+,x}_{n,t\wedge\sigma_{x,\rho}} - K^{-,x}_{n,t\wedge\sigma_{x,\rho}},
$$

which is exactly (639) with (641).

Now let $\varphi \in \mathcal{T}^{\mathrm{fs}}_\eta$ be arbitrary. Since ℓ_φ is a finite linear combination of the coordinate maps $H, A_1, A_2, \dots$, summing the already established basis decompositions yields (639) together with the explicit formula (642). Uniqueness follows because the basis vectors $e_h, e_1, e_2, \dots$ span $\mathcal{T}^{\mathrm{fs}}_\eta$. Finally, every process (642) is a finite linear combination of continuous increasing local times, hence is continuous, of finite variation, and has zero quadratic variation. This proves all items. □

Proposition D.122 (One-step released-shell criterion for the one-profile local-window/tail C^2 package). *Assume the hypotheses of Proposition D.110. For each $n \geq 1$ write*

$$\Delta_{\eta,n} := V_{\eta,n} - V_{\eta,n-1}, \tag{643}$$

with the convention $V_{\eta,0} \equiv 0$. Suppose there exist nonnegative sequences

$$(\lambda_{\eta,n})_{n\geq 1}, \qquad (r_{\eta,n})_{n\geq 1},$$

with $r_{\eta,n+1} \leq r_{\eta,n}$ for all n, and a single common tail profile $(w_j)_{j\geq 1}$ satisfying

$$W_\eta := \sum_{j=1}^{\infty} \frac{w_j}{(2j+1)^3} < \infty, \tag{644}$$

for which, uniformly on every intrinsic compact chart, for all $n \geq 1$ and all $j, k \geq 1$,

$$|\partial_h \Delta_{\eta,n}| + |\partial_{hh} \Delta_{\eta,n}| \leq \lambda_{\eta,n}, \tag{645}$$

$$|\partial_{a_j} \Delta_{\eta,n}| + |\partial_{ha_j} \Delta_{\eta,n}| \leq \lambda_{\eta,n}\big(\mathbf{1}_{\{j\leq r_{\eta,n}\}} + w_j\big), \tag{646}$$

$$|\partial_{a_j a_k} \Delta_{\eta,n}| \leq \lambda_{\eta,n}\big(\mathbf{1}_{\{|j-k|\leq r_{\eta,n}\}} + w_j w_k\big). \tag{647}$$

Assume also that

$$\sum_{n=1}^{\infty} n\, \lambda_{\eta,n}\big(1 + r_{\eta,n} + W_\eta + (W_\eta)^2\big) < \infty. \tag{648}$$

Then Proposition D.119 applies with

$$\Lambda_{\eta,N} := \sum_{n=N+1}^{\infty} \lambda_{\eta,n}, \qquad \rho_{\eta,N} := r_{\eta,N+1} \qquad (N \geq 0), \tag{649}$$

and the same tail profile $(w_j)_{j\geq 1}$. Consequently Theorem D.118, Theorem D.121, and Proposition D.119 all apply. In particular, activation of the chart-free natural Itô/dual-reflection package already reduces to one-step released-shell local-window/tail C^2 estimates.

Proof. Fix $M > N \geq 0$. By telescoping,

$$V_{\eta,M} - V_{\eta,N} = \sum_{n=N+1}^{M} \Delta_{\eta,n}.$$

Summing (645) gives

$$|\partial_h(V_{\eta,M} - V_{\eta,N})| + |\partial_{hh}(V_{\eta,M} - V_{\eta,N})| \leq \sum_{n=N+1}^{M} \lambda_{\eta,n} \leq \Lambda_{\eta,N}.$$

Because the window radii are nonincreasing, $r_{\eta,n} \leq r_{\eta,N+1} = \rho_{\eta,N}$ for every $n \geq N+1$. Therefore

$$\mathbf{1}_{\{j\leq r_{\eta,n}\}} \leq \mathbf{1}_{\{j\leq \rho_{\eta,N}\}}, \qquad \mathbf{1}_{\{|j-k|\leq r_{\eta,n}\}} \leq \mathbf{1}_{\{|j-k|\leq \rho_{\eta,N}\}},$$

for all $n \geq N+1$, and summing (646)–(647) yields exactly the one-profile bounds (635)–(636) with the identifications (649).

It remains to verify the summability condition (637). Since each $\lambda_{\eta,n}$ contributes to $\Lambda_{\eta,N}$ only for $0 \leq N \leq n-1$, one has

$$\sum_{N=0}^{\infty} \Lambda_{\eta,N}\big(1 + \rho_{\eta,N} + W_\eta + (W_\eta)^2\big) \leq \sum_{n=1}^{\infty} n\, \lambda_{\eta,n}\big(1 + r_{\eta,n} + W_\eta + (W_\eta)^2\big),$$

which is finite by (648). Proposition D.119 therefore applies, and the remaining conclusions follow immediately. ∎

Integrated downstream conclusion after the dual-reflection/one-step-shell continuation. This continuation adds a fifty-second downstream message. Positively, the natural Itô package no longer leaves the reflection side only as an unnamed missing finite-variation term: on the intrinsic finitely supported test space it already carries a chart-free dual local reflection functional compatible with the same Brownian family and the explicit coordinate reflected local times of the strong continuum lift. Verification is also no longer posed at the level of arbitrary finite-cylinder differences: one-step released-shell local-window/tail C^2 bounds already suffice to activate the same package. Cautionarily, this dual reflection functional is still only defined on finitely supported directions and is not yet identified with a full normal-cone/Skorokhod field on the whole natural Itô domain, and the one-step shell bounds are still not verified for the genuine e-Arc-Helix interaction family. What remains open is therefore chart-free extension and Skorokhod identification of the dual reflection functional together with concrete one-step shell C^2 verification — or a sharp obstruction — for the actual interaction family.

Remark D.123. The new fifty-second message is the first point in the chain where the continuum process already carries a chart-free reflected local additive functional on its intrinsic finitely supported test space, rather than only an implicit finite-variation gap. At the same time, the remaining verification problem is no longer phrased in terms of all pairwise cylinder differences: it is enough to prove — or sharply fail to prove — one-step released-shell local-window/tail C^2 bounds for the actual finite-cylinder interaction increments. What is still missing is extension of that dual reflection functional to the whole natural Itô domain together with a genuinely classical Skorokhod/SPDE interpretation.

D.40 Worked continuation LIII: graph-closed reflection closure and one-seminorm shell C^2 activation

The previous continuation already produced a chart-free dual reflection functional on the intrinsic finitely supported test space and reduced activation of the natural Itô package to one-step released-shell local-window/tail C^2 bounds. The next intrinsic sharpening again advances both remaining gaps together. On the stochastic side, the finitely supported dual reflection package is already graph-closable against the same local Itô topology that governs the natural square-gradient domain. Thus one already obtains a chart-free graph-closed reflection operator on a canonical linear subspace of the natural Itô domain, even though a full Skorokhod or normal-cone interpretation on the whole natural domain is still missing. On the verification side, the released-shell C^2 bounds can be compressed further to one scalar shell seminorm per released mode. So the remaining concrete interaction problem is no longer to track three separate families of first/second derivative inequalities, but to prove — or sharply fail to prove — one summable shell budget.

Status label. The graph-closure theorem below has status **theorem-level downstream result**. It does *not* yet identify the resulting closed reflection graph with a positive chart-free normal-cone or Skorokhod mechanism on the full natural Itô domain, but it shows that the finitely supported dual reflection functional is no longer an isolated core object: it already closes canonically inside the same local Itô topology as the natural square-gradient martingale package. The shell-seminorm reduction below is also a **proposition-level downstream result**. It does *not* verify the required bounds for the genuine e-Arc-Helix interaction family, but it compresses the remaining one-step C^2 verification to one scalar seminorm per released shell.

For every intrinsic open set $U \subset \Omega_\eta$ define the finitely supported linear core

$$\mathcal{L}_\eta^{\mathrm{fs}}(U) := \{\ell_\varphi|_U : \varphi \in \mathcal{T}_\eta^{\mathrm{fs}}\} \subset \mathcal{D}_\eta^{\mathrm{Ito}}(U),$$

and for $F = \ell_\varphi|_U \in \mathcal{L}_\eta^{\mathrm{fs}}(U)$ write

$$\mathfrak{R}^{x,\rho}[F] := \mathfrak{R}^{x,\rho}[\varphi]$$

for the finitely supported dual reflection functional from Theorem D.121. Let $\mathcal{A}_{\text{fv}}^{x,\rho}$ denote the vector space of continuous finite-variation additive functionals on $[0,\infty)$ stopped at $\sigma_{x,\rho}$, endowed with local uniform convergence in probability on compact time intervals.

Theorem D.124 (Graph-closed chart-free reflection operator on the linear natural Itô core). *Assume the hypotheses of Proposition D.119. Fix an intrinsic open set $U \subset \Omega_\eta$, a point $x \in U$, and $\rho > 0$ such that $\mathbb{B}_\eta(x,\rho) \Subset U$. Then the graph of the map*

$$\mathcal{L}_\eta^{\text{fs}}(U) \ni F \longmapsto \mathfrak{R}^{x,\rho}[F] \in \mathcal{A}_{\text{fv}}^{x,\rho}$$

is closable in $\mathcal{D}_\eta^{\text{Ito}}(U) \times \mathcal{A}_{\text{fv}}^{x,\rho}$ for the local Itô seminorms (628) on the first factor and local uniform convergence in probability on the second factor. Equivalently, there exists a unique graph-closed linear operator

$$\mathfrak{R}_{\eta,U}^{x,\rho} : \mathcal{D}_\eta^{\text{ref}}(U;x,\rho) \longrightarrow \mathcal{A}_{\text{fv}}^{x,\rho},$$

with domain equal to the graph closure of $\mathcal{L}_\eta^{\text{fs}}(U)$, such that the following hold.

(i) Every $F \in \mathcal{L}_\eta^{\text{fs}}(U)$ belongs to $\mathcal{D}_\eta^{\text{ref}}(U;x,\rho)$ and

$$\mathfrak{R}_{\eta,U}^{x,\rho} F = \mathfrak{R}^{x,\rho}[F].$$

(ii) If $F_m \in \mathcal{L}_\eta^{\text{fs}}(U)$ converges to 0 in the local Itô seminorms on compact subsets of U and if $\mathfrak{R}^{x,\rho}[F_m]$ converges in $\mathcal{A}_{\text{fv}}^{x,\rho}$ to some additive functional A, then $A \equiv 0$. Hence the closed graph above is unique.

(iii) For every $F \in \mathcal{D}_\eta^{\text{ref}}(U;x,\rho)$ there exists a sequence $F_m \in \mathcal{L}_\eta^{\text{fs}}(U)$ converging to F in the local Itô seminorms such that the finitely supported decompositions from Theorem D.121 converge to a unique stopped semimartingale decomposition

$$F(Z^x_{t\wedge\sigma_{x,\rho}}) - F(Z^x_0) = M_t^{F,x,\rho} + A_t^{F,x,\rho} + \left(\mathfrak{R}_{\eta,U}^{x,\rho} F\right)_t, \tag{650}$$

where $M^{F,x,\rho}$ is the local Brownian martingale limit determined by the common Brownian family and $A^{F,x,\rho}$ is the locally uniform finite-variation drift limit of the same approximants.

(iv) What remains open is therefore no longer the existence of a chart-free closed reflection graph on a natural square-gradient domain, but only identification of that closed graph as a classical Skorokhod normal/reflection mechanism on the full natural Itô domain, together with positivity/support information beyond the linear core.

Proof. For $F = \ell_\varphi|_U \in \mathcal{L}_\eta^{\text{fs}}(U)$, Theorem D.121 gives an exact stopped semimartingale decomposition of the form

$$F(Z^x_{t\wedge\sigma_{x,\rho}}) - F(Z^x_0) = M_t^{F,x,\rho} + A_t^{F,x,\rho} + \mathfrak{R}^{x,\rho}[F]_t,$$

where $M^{F,x,\rho}$ is the finitely supported Brownian martingale part and $A^{F,x,\rho}$ is the corresponding absolutely continuous drift term. Now suppose $F_m \in \mathcal{L}_\eta^{\text{fs}}(U)$ converges to 0 in the local Itô seminorms on compact subsets and that $\mathfrak{R}^{x,\rho}[F_m] \to A$ in $\mathcal{A}_{\text{fv}}^{x,\rho}$. The observable terms $F_m(Z^x_{t\wedge\sigma_{x,\rho}}) - F_m(Z^x_0)$ converge to 0 locally uniformly in probability because the graph part of the Itô seminorm controls locally uniform convergence on stopped compact charts. The Brownian martingale parts converge to 0 in L^2, uniformly on compact time intervals, by the same Itô-isometry argument used in Theorem D.118. The drift terms $A^{F_m,x,\rho}$ converge to 0 locally uniformly because the same graph control forces the corresponding finite-cylinder drifts to be Cauchy and then vanishing in the compact graph seminorms. Passing to the limit in the displayed decomposition therefore yields $A \equiv 0$, which proves closability and item (ii).

Items (i) and (iii) are then simply the standard graph-closure consequences: the closed graph determines a unique domain $\mathcal{D}_\eta^{\text{ref}}(U;x,\rho)$ and a unique closed operator $\mathfrak{R}_{\eta,U}^{x,\rho}$ extending the finitely supported dual core. Any two approximating sequences have the same limit because their difference satisfies the already proved zero-limit property. Item (iv) is exactly the interpretation of this graph-closure statement. ∎

For every $n \geq 1$ and every choice of nonnegative window radius $r_{\eta,n}$ together with a tail profile $(w_j)_{j\geq 1}$, write

$$\omega_{\eta,n}(j) := \mathbf{1}_{\{j \leq r_{\eta,n}\}} + w_j, \qquad \Omega_{\eta,n}(j,k) := \mathbf{1}_{\{|j-k| \leq r_{\eta,n}\}} + w_j w_k.$$

Given the released shell $\Delta_{\eta,n} = V_{\eta,n} - V_{\eta,n-1}$, define its local-window/tail shell seminorm to be the least number $\mathcal{S}^{(2)}_{\eta,n} \geq 0$ for which, uniformly on every intrinsic compact chart and for all $j, k \geq 1$,

$$|\partial_h \Delta_{\eta,n}| + |\partial_{hh} \Delta_{\eta,n}| \leq \mathcal{S}^{(2)}_{\eta,n}, \tag{651}$$

$$|\partial_{a_j} \Delta_{\eta,n}| + |\partial_{ha_j} \Delta_{\eta,n}| \leq \mathcal{S}^{(2)}_{\eta,n}\, \omega_{\eta,n}(j), \tag{652}$$

$$|\partial_{a_j a_k} \Delta_{\eta,n}| \leq \mathcal{S}^{(2)}_{\eta,n}\, \Omega_{\eta,n}(j,k). \tag{653}$$

Proposition D.125 (One-seminorm released-shell criterion for the chart-free natural Itô/reflection package). *Assume the hypotheses of Proposition D.110. Suppose there exist nonnegative sequences $(r_{\eta,n})_{n\geq 1}$ and $(\mathcal{S}^{(2)}_{\eta,n})_{n\geq 1}$ with $r_{\eta,n+1} \leq r_{\eta,n}$ for all n, together with a single common tail profile $(w_j)_{j\geq 1}$ satisfying*

$$W_\eta := \sum_{j=1}^{\infty} \frac{w_j}{(2j+1)^3} < \infty, \tag{654}$$

and, uniformly on every intrinsic compact chart, the shell-seminorm bounds (651)–(653) *for all $n \geq 1$. Assume also that*

$$\sum_{n=1}^{\infty} n\, \mathcal{S}^{(2)}_{\eta,n} \big(1 + r_{\eta,n} + W_\eta + (W_\eta)^2\big) < \infty. \tag{655}$$

Then Proposition D.122 applies with $\lambda_{\eta,n} := \mathcal{S}^{(2)}_{\eta,n}$. Consequently Proposition D.119, Theorem D.118, Theorem D.121, and Theorem D.124 all apply. In particular, activation of the chart-free natural Itô / dual-reflection / closed-reflection-graph package already reduces to one scalar released-shell seminorm budget per mode.

Proof. The shell-seminorm inequalities (651)–(653) are exactly the hypotheses (645)–(647) of Proposition D.122 with the identification $\lambda_{\eta,n} = \mathcal{S}^{(2)}_{\eta,n}$. The summability assumption (655) is likewise exactly the shell summability condition (648) under the same identification. Proposition D.122 therefore applies immediately, and the remaining conclusions follow by chaining with Proposition D.119, Theorem D.118, Theorem D.121, and Theorem D.124. □

Integrated downstream conclusion after the graph-closed reflection/shell-seminorm continuation. This continuation adds a fifty-third downstream message. Positively, the reflection side is no longer confined to a finitely supported dual test core: that dual reflection package already closes canonically to a chart-free reflection graph inside the same local Itô topology as the natural square-gradient martingale domain. Verification is also no longer forced to track three separate released-shell derivative families: one scalar shell seminorm per released mode already suffices. Cautionarily, the closed reflection graph is still not identified with a genuine Skorokhod normal-cone mechanism on the full natural Itô domain, and the new shell seminorm bounds are still not verified for the genuine e-Arc-Helix interaction family. What remains open is therefore chart-free Skorokhod identification of that closed reflection graph together with concrete one-seminorm released-shell C^2 verification — or a sharp obstruction — for the actual interaction potentials.

Remark D.126. The new fifty-third message is the first point in the chain where the reflection side already has a graph-closed chart-free continuation inside the natural Itô topology, rather than only a finitely supported dual core. At the same time, the interaction side has been compressed to one scalar shell budget per released mode. The remaining gap is therefore concentrated even more sharply than before: identify the closed reflection graph as a genuine Skorokhod mechanism, and prove — or sharply fail to prove — the one-seminorm shell estimates for the actual e-Arc-Helix interaction family.

D.41 Worked continuation LIV: first-order reflection calculus and fixed-profile shell C^2 activation

The previous continuation left the two downstream gaps in an even sharper form. On the stochastic side, the chart-free reflection package already closed canonically on a linear natural Itô core, but the remaining Skorokhod question still lacked any equally chart-free first-order calculus beyond that linear level. On the interaction side, the released-shell activation problem had already been compressed to one scalar shell seminorm per mode, but the allowed local windows and tail profiles could still vary from shell to shell. The next intrinsic sharpening again pushes both directions at once. First, the closed reflection package already extends from the linear core to a canonical multiplicative cylinder-generated natural-Itô core on which it obeys exact first-order product and scalar chain rules. Second, the same released-shell activation problem already reduces further to one fixed local window, one common tail profile, and one amplitude sequence.

Status label. The reflection-calculus statement below has status **theorem-level downstream result**. It does *not* yet identify the closed reflection graph with a classical chart-free Skorokhod normal-cone term for the full continuum process, and it does *not* yet show that the resulting signed finite-variation part is supported exactly on the geometric contact set of a global reflected gradient SPDE/SDE. What it does show is that the closed reflection package is no longer merely linear: on a canonical multiplicative natural-Itô core it already behaves like a first-order derivation under products and scalar compositions. The fixed-profile activation statement is likewise a **proposition-level downstream result**. It does *not* verify the required bounds for the actual e-Arc-Helix interaction family, but it reduces the remaining released-shell C^2 verification problem further to one especially concrete frozen local-window/tail template.

Let $\mathcal{A}_\eta^{\mathrm{cyl}}(U)$ denote the bounded smooth cylinder algebra on an intrinsic open set $U \subset \Omega_\eta$, i.e.

$$\mathcal{A}_\eta^{\mathrm{cyl}}(U) := \Big\{ \Phi\big(\ell_{\varphi^{(1)}}, \dots, \ell_{\varphi^{(m)}}\big)\big|_U : m \geq 1,\ \Phi \in C_b^2(\mathbb{R}^m),\\ \varphi^{(1)}, \dots, \varphi^{(m)} \in \mathcal{T}_\eta^{\mathrm{fs}} \Big\}, \tag{656}$$

where $C_b^2(\mathbb{R}^m)$ denotes the bounded twice continuously differentiable functions with bounded first and second derivatives. For fixed $x \in \Omega_\eta$ and $\rho > 0$ with $\mathbb{B}_\eta(x,\rho) \Subset U$, let $\mathcal{D}_\eta^{\mathrm{alg}}(U; x, \rho)$ be the collection of all $F \in \mathcal{D}_\eta^{\mathrm{Ito}}(U) \cap \mathcal{D}_\eta^{\mathrm{ref}}(U; x, \rho)$ for which there exists a sequence $F_m \in \mathcal{A}_\eta^{\mathrm{cyl}}(U)$ converging to F in the local Itô seminorms on compact subsets of U and whose closed reflection terms converge in $\mathcal{A}_{\mathrm{fv}}^{x,\rho}$.

Theorem D.127 (Chart-free first-order reflection calculus on a canonical multiplicative natural-Itô core). *Assume the hypotheses of Proposition D.125. Fix an intrinsic open set $U \subset \Omega_\eta$, $x \in \Omega_\eta$, and $\rho > 0$ with $\mathbb{B}_\eta(x,\rho) \Subset U$. Then the following hold.*

(i) The class $\mathcal{D}_\eta^{\mathrm{alg}}(U; x, \rho)$ is a multiplicative vector subspace of $\mathcal{D}_\eta^{\mathrm{Ito}}(U) \cap \mathcal{D}_\eta^{\mathrm{ref}}(U; x, \rho)$ containing the bounded cylinder algebra $\mathcal{A}_\eta^{\mathrm{cyl}}(U)$. It is stable under bounded C^2 scalar calculus: if $F \in \mathcal{D}_\eta^{\mathrm{alg}}(U; x, \rho)$ and $\Psi \in C_b^2(\mathbb{R})$, then $\Psi \circ F \in \mathcal{D}_\eta^{\mathrm{alg}}(U; x, \rho)$.

(ii) If $F, G \in \mathcal{D}^{\mathrm{alg}}_{\eta}(U; x, \rho)$, then the graph-closed reflection operator satisfies the exact stopped first-order product rule

$$\left(\mathfrak{R}^{x,\rho}_{\eta,U}(FG)\right)_t = \int_0^t F\left(Z^x_{s\wedge\sigma_{x,\rho}}\right) d\left(\mathfrak{R}^{x,\rho}_{\eta,U}G\right)_s + \int_0^t G\left(Z^x_{s\wedge\sigma_{x,\rho}}\right) d\left(\mathfrak{R}^{x,\rho}_{\eta,U}F\right)_s. \tag{657}$$

No additional quadratic-variation term appears in the reflection part.

(iii) If $F \in \mathcal{D}^{\mathrm{alg}}_{\eta}(U; x, \rho)$ and $\Psi \in C^2_b(\mathbb{R})$, then the graph-closed reflection operator satisfies the exact stopped scalar chain rule

$$\left(\mathfrak{R}^{x,\rho}_{\eta,U}(\Psi \circ F)\right)_t = \int_0^t \Psi'\left(F(Z^x_{s\wedge\sigma_{x,\rho}})\right) d\left(\mathfrak{R}^{x,\rho}_{\eta,U}F\right)_s. \tag{658}$$

(iv) Consequently, what remains open is no longer the absence of any chart-free first-order calculus for the closed reflection package on the natural Itô side, but only identification of that first-order closed graph as a classical chart-free Skorokhod normal-cone/reflected-gradient mechanism for the full continuum process.

Proof. For $F \in \mathcal{A}^{\mathrm{cyl}}_{\eta}(U)$ there exist $m \geq 1$, $\Phi \in C^2_b(\mathbb{R}^m)$, and finitely supported directions $\varphi^{(1)}, \ldots, \varphi^{(m)} \in \mathcal{T}^{\mathrm{fs}}_{\eta}$ such that

$$F = \Phi\left(\ell_{\varphi^{(1)}}, \ldots, \ell_{\varphi^{(m)}}\right)\big|_U.$$

Applying the multidimensional Itô formula to the exact stopped semimartingale decompositions furnished by Theorem D.121 yields a stopped decomposition for $F(Z^x_{t\wedge\sigma_{x,\rho}}) - F(Z^x_0)$ whose martingale and drift parts are exactly those dictated by Theorem D.118 and the natural generator, while the reflection contribution is

$$d\left(\mathfrak{R}^{x,\rho}_{\eta,U}F\right)_t = \sum_{i=1}^m \partial_i\Phi\left(\ell_{\varphi^{(1)}}(Z^x_{t\wedge\sigma_{x,\rho}}), \ldots, \ell_{\varphi^{(m)}}(Z^x_{t\wedge\sigma_{x,\rho}})\right) d\left(\mathfrak{R}^{x,\rho}[\varphi^{(i)}]\right)_t.$$

Because every $\mathfrak{R}^{x,\rho}[\varphi^{(i)}]$ has zero quadratic variation by Theorem D.121, no additional quadratic-variation term can enter the reflection part. Taking $\Phi(u,v) = uv$ gives (657) on the cylinder algebra, and taking $\Phi = \Psi$ in one variable gives (658) there.

Now let $F, G \in \mathcal{D}^{\mathrm{alg}}_{\eta}(U; x, \rho)$ and choose cylinder-algebra approximants $F_m, G_m \in \mathcal{A}^{\mathrm{cyl}}_{\eta}(U)$ converging to F, G in the local Itô seminorms on compact subsets with closed reflection terms converging in $\mathcal{A}^{x,\rho}_{\mathrm{fv}}$. Since the Itô seminorms control locally uniform convergence on stopped compact charts, the products $F_m G_m$ converge to FG in the same seminorms, and the product-rule identities for F_m, G_m pass to the limit in $\mathcal{A}^{x,\rho}_{\mathrm{fv}}$. This proves that $FG \in \mathcal{D}^{\mathrm{alg}}_{\eta}(U; x, \rho)$ and yields (657). The same argument with the scalar cylinder approximants $\Psi \circ F_m$ proves closure under bounded C^2 scalar calculus and gives (658). Item (iv) is exactly the interpretation of these first-order identities. □

Fix once and for all a nonnegative integer r_η and a nonnegative tail profile $(w_j)_{j\geq 1}$. Set

$$\omega_\eta(j) := \mathbf{1}_{\{j \leq r_\eta\}} + w_j, \qquad \Omega_\eta(j,k) := \mathbf{1}_{\{|j-k| \leq r_\eta\}} + w_j w_k.$$

For each released shell $\Delta_{\eta,n} = V_{\eta,n} - V_{\eta,n-1}$, let $\Lambda^{(2),\mathrm{fp}}_{\eta,n} \geq 0$ denote the least constant for which, uniformly on every intrinsic compact chart and for all $j, k \geq 1$,

$$|\partial_h \Delta_{\eta,n}| + |\partial_{hh}\Delta_{\eta,n}| \leq \Lambda^{(2),\mathrm{fp}}_{\eta,n}, \tag{659}$$

$$|\partial_{a_j}\Delta_{\eta,n}| + |\partial_{ha_j}\Delta_{\eta,n}| \leq \Lambda^{(2),\mathrm{fp}}_{\eta,n}\, \omega_\eta(j), \tag{660}$$

$$|\partial_{a_j a_k}\Delta_{\eta,n}| \leq \Lambda^{(2),\mathrm{fp}}_{\eta,n}\, \Omega_\eta(j,k). \tag{661}$$

Proposition D.128 (Fixed-profile released-shell criterion for the closed reflection/first-order natural-Itô package). *Assume the hypotheses of Proposition D.110. Suppose the fixed tail profile satisfies*

$$W_\eta^{\mathrm{fp}} := \sum_{j=1}^{\infty} \frac{w_j}{(2j+1)^3} < \infty, \tag{662}$$

and, uniformly on every intrinsic compact chart, the bounds (659)–(661) *hold for all* $n \geq 1$. *Assume also that*

$$\sum_{n=1}^{\infty} n\, \Lambda_{\eta,n}^{(2),\mathrm{fp}} \big(1 + r_\eta + W_\eta^{\mathrm{fp}} + (W_\eta^{\mathrm{fp}})^2\big) < \infty. \tag{663}$$

Then Proposition D.125 applies with the identifications

$$r_{\eta,n} = r_\eta, \qquad \mathcal{S}_{\eta,n}^{(2)} = \Lambda_{\eta,n}^{(2),\mathrm{fp}}, \qquad W_\eta = W_\eta^{\mathrm{fp}},$$

for all $n \geq 1$. *Consequently Theorem D.118, Theorem D.121, Theorem D.124, Theorem D.127, and Proposition D.125 all apply. In particular, activation of the chart-free natural Itô / dual-reflection / closed-reflection / first-order-reflection package already reduces to one fixed local window, one common tail profile, and one amplitude sequence for the released shells.*

Proof. The fixed-profile bounds (659)–(661) are exactly the shell-seminorm hypotheses (651)–(653) of Proposition D.125 in the special case where the local window and tail profile do not depend on n. The summability condition (663) is likewise exactly (655) under the same identifications. Proposition D.125 therefore applies immediately, and the remaining conclusions follow by chaining with Theorems D.118, D.121, D.124, and D.127. □

Integrated downstream conclusion after the first-order-reflection/fixed-profile continuation. This continuation adds a fifty-fourth downstream message. Positively, the chart-free reflection package is no longer only a closed additive operator on a linear natural Itô core: on a canonical cylinder-generated multiplicative core it already satisfies exact first-order product and scalar chain rules. Verification is also no longer forced to allow shell-dependent windows and shell-dependent tail profiles: one fixed local window, one common tail profile, and one amplitude sequence already suffice. Cautionarily, the new first-order reflection calculus still does not identify the actual supporting contact measure or its signed normal-cone meaning on the full natural Itô domain, and the frozen-profile shell bounds are still not verified for the genuine e-Arc-Helix interaction family. What remains open is therefore chart-free Skorokhod/normal-cone identification of that closed first-order reflection calculus together with concrete fixed-profile released-shell C^2 verification — or a sharp obstruction — for the actual interaction family.

Remark D.129. The new fifty-fourth message is the first point in the chain where the closed reflection package already has a genuinely first-order chart-free calculus on a nontrivial multiplicative natural-Itô core rather than only on linear observables. At the same time, the interaction side has been frozen to one common local window/tail geometry. The remaining gap is therefore now even more concentrated: identify the first-order closed reflection calculus as a genuine Skorokhod normal-cone mechanism, and prove — or sharply fail to prove — one fixed-profile released-shell C^2 estimate for the actual e-Arc-Helix interaction family.

D.42 Worked continuation LV: common contact-clock factorization and fixed-template telescoping C^2 activation

The previous continuation concentrated the remaining stochastic gap into a chart-free first-order reflection calculus on a canonical multiplicative natural Itô core, and it concentrated the remaining verification gap into one fixed released-shell geometry with one amplitude sequence.

One further intrinsic sharpening is available on both sides before any genuinely external input is needed. First, the closed first-order reflection calculus already factors through a *single* chart-free increasing contact clock on the same natural Itô domain: all closed reflection terms on the multiplicative core become densities with respect to one common finite-variation carrier. Second, the fixed-profile shell verification problem can be read directly on the actual telescoping finite-cylinder interaction family $(V_{\eta,N})_{N\geq 0}$ through one fixed template seminorm, rather than by tracking shellwise derivative arrays separately.

Status label. The contact-clock statement below has status **theorem-level downstream result**. It does *not* yet identify the common contact clock with the true geometric contact measure of a global Skorokhod reflected gradient SPDE/SDE, and it does *not* yet prove that its density field is the actual normal-cone selector of the full continuum process. What it does show is that the remaining reflection ambiguity is no longer the absence of a single chart-free carrier: the entire closed first-order reflection calculus already factors through one common increasing additive functional on the natural Itô side. The telescoping-criterion statement is likewise a **proposition-level downstream result**. It does *not* verify the needed estimates for the genuine e-Arc-Helix interaction family, but it reduces the fixed-profile activation problem further to one fixed template seminorm applied directly to the actual finite-cylinder interaction differences.

Theorem D.130 (Chart-free common-contact-clock factorization of the closed first-order reflection calculus). *Assume the hypotheses of Proposition D.128. Fix an intrinsic open set $U \subset \Omega_\eta$, $x \in \Omega_\eta$, and $\rho > 0$ with $\mathbb{B}_\eta(x,\rho) \Subset U$. Then there exist a continuous increasing additive functional*

$$\mathfrak{L}^{x,\rho}_{\eta,U} \in \mathcal{A}^{x,\rho}_{\mathrm{fv}}$$

and a linear assignment

$$F \longmapsto \mathfrak{n}^{x,\rho}_{\eta,U}[F]$$

from $\mathcal{D}^{\mathrm{alg}}_\eta(U; x, \rho)$ into predictable locally integrable processes on $[0, \sigma_{x,\rho}]$ such that:

(i) for every $F \in \mathcal{D}^{\mathrm{alg}}_\eta(U; x, \rho)$,

$$\left(\mathfrak{R}^{x,\rho}_{\eta,U} F\right)_t = \int_0^t \mathfrak{n}^{x,\rho}_{\eta,U}[F]_s \, d\mathfrak{L}^{x,\rho}_{\eta,U}(s), \qquad t \geq 0; \tag{664}$$

(ii) if $F = \ell_\varphi|_U$ with $\varphi \in \mathcal{T}^{\mathrm{fs}}_\eta$, then $\mathfrak{n}^{x,\rho}_{\eta,U}[F]$ is the Radon–Nikodým density of $d\big(\mathfrak{R}^{x,\rho}[\varphi]\big)$ with respect to $d\mathfrak{L}^{x,\rho}_{\eta,U}$; in particular, every closed reflection term on the multiplicative natural-Itô core is absolutely continuous with respect to one common chart-free contact clock;

(iii) if $F, G \in \mathcal{D}^{\mathrm{alg}}_\eta(U; x, \rho)$ and $\Psi \in C^2_b(\mathbb{R})$, then the density assignment satisfies the stopped first-order identities

$$\mathfrak{n}^{x,\rho}_{\eta,U}[FG]_s = F\big(Z^x_{s\wedge\sigma_{x,\rho}}\big)\, \mathfrak{n}^{x,\rho}_{\eta,U}[G]_s + G\big(Z^x_{s\wedge\sigma_{x,\rho}}\big)\, \mathfrak{n}^{x,\rho}_{\eta,U}[F]_s, \tag{665}$$

$$\mathfrak{n}^{x,\rho}_{\eta,U}[\Psi \circ F]_s = \Psi'\big(F(Z^x_{s\wedge\sigma_{x,\rho}})\big)\, \mathfrak{n}^{x,\rho}_{\eta,U}[F]_s, \tag{666}$$

for $d\mathfrak{L}^{x,\rho}_{\eta,U}$-almost every s;

(iv) consequently, what remains open is no longer the existence of a chart-free first-order reflection density carried by one common contact clock, but only identification of $\mathfrak{L}^{x,\rho}_{\eta,U}$ as the genuine Skorokhod/contact measure and of $\mathfrak{n}^{x,\rho}_{\eta,U}$ as the corresponding geometric normal-cone selector for the full continuum reflected gradient dynamics.

Proof. Choose once and for all a countable family $\{\varphi^{(m)}\}_{m\geq 1} \subset \mathcal{T}_\eta^{\mathrm{fs}}$ whose cylinder linear observables generate the bounded cylinder algebra. For each m, let

$$A_m := \mathfrak{R}^{x,\rho}[\varphi^{(m)}] \in \mathcal{A}_{\mathrm{fv}}^{x,\rho}.$$

By Theorem D.121, each A_m is a continuous finite-variation additive functional. By standard countable domination of finite-variation measures on a fixed stopped chart, there exists a continuous increasing additive functional $\mathfrak{L}_{\eta,U}^{x,\rho} \in \mathcal{A}_{\mathrm{fv}}^{x,\rho}$ such that

$$dA_m \ll d\mathfrak{L}_{\eta,U}^{x,\rho} \qquad \text{for every } m \geq 1.$$

Let g_m denote the corresponding Radon–Nikodým density.

If $F \in \mathcal{A}_\eta^{\mathrm{cyl}}(U)$ is written as

$$F = \Phi\big(\ell_{\varphi^{(m_1)}}, \dots, \ell_{\varphi^{(m_r)}}\big)\big|_U, \qquad \Phi \in C_b^2(\mathbb{R}^r),$$

then the product and scalar chain rules from Theorem D.127 show that its closed reflection term is exactly

$$d\big(\mathfrak{R}_{\eta,U}^{x,\rho} F\big)_s = \sum_{i=1}^r \partial_i \Phi\big(\ell_{\varphi^{(m_1)}}(Z_{s\wedge\sigma_{x,\rho}}^x), \dots, \ell_{\varphi^{(m_r)}}(Z_{s\wedge\sigma_{x,\rho}}^x)\big)\, dA_{m_i}(s).$$

Hence $d\big(\mathfrak{R}_{\eta,U}^{x,\rho} F\big) \ll d\mathfrak{L}_{\eta,U}^{x,\rho}$ and one may define

$$\mathfrak{n}_{\eta,U}^{x,\rho}[F]_s := \sum_{i=1}^r \partial_i \Phi\big(\ell_{\varphi^{(m_1)}}(Z_{s\wedge\sigma_{x,\rho}}^x), \dots, \ell_{\varphi^{(m_r)}}(Z_{s\wedge\sigma_{x,\rho}}^x)\big) g_{m_i}(s)$$

for $d\mathfrak{L}_{\eta,U}^{x,\rho}$-almost every s. Because the closed reflection term itself is representation-independent, uniqueness of Radon–Nikodým densities shows that this definition is independent of the chosen cylinder presentation. The identities (665)–(666) are then simply the density versions of (657)–(658).

Now let $F \in \mathcal{D}_\eta^{\mathrm{alg}}(U; x, \rho)$ and choose cylinder-algebra approximants $F_m \in \mathcal{A}_\eta^{\mathrm{cyl}}(U)$ converging to F in the local Itô seminorms with closed reflection terms converging in $\mathcal{A}_{\mathrm{fv}}^{x,\rho}$. Since all those closed reflection terms are already absolutely continuous with respect to the same clock $\mathfrak{L}_{\eta,U}^{x,\rho}$, their densities form a Cauchy family in local $L^1(d\mathfrak{L}_{\eta,U}^{x,\rho})$ and therefore converge to a unique limit, which we declare to be $\mathfrak{n}_{\eta,U}^{x,\rho}[F]$. This yields (664) for every $F \in \mathcal{D}_\eta^{\mathrm{alg}}(U; x, \rho)$, preserves linearity by construction, and passes the density-level product/chain rules to the limit. Item (iv) is exactly the interpretation of the resulting factorization. □

Keeping the fixed profile r_η, the weights ω_η, Ω_η, and the tail sum W_η^{fp} from Section D.41, define for each intrinsic compact chart $K \Subset U$ and each cylinder interaction observable G the fixed-template seminorm

$$\begin{aligned} \|G\|_{\eta,\mathrm{tpl},2;K} &:= \|\partial_h G\|_{L^\infty(K)} + \|\partial_{hh} G\|_{L^\infty(K)} \\ &\quad + \sup_{j\geq 1} \omega_\eta(j)^{-1} \Big(\|\partial_{a_j} G\|_{L^\infty(K)} + \|\partial_{ha_j} G\|_{L^\infty(K)}\Big) \\ &\quad + \sup_{j,k\geq 1} \Omega_\eta(j,k)^{-1} \|\partial_{a_j a_k} G\|_{L^\infty(K)}. \end{aligned} \tag{667}$$

Proposition D.131 (Fixed-template telescoping C^2 criterion on the finite-cylinder interaction family)**.** *Assume the hypotheses of Proposition D.110. Keep the fixed profile r_η, ω_η, Ω_η, and W_η^{fp} from Section D.41. Suppose there exists a sequence $(\Gamma_{\eta,n}^{(2),\mathrm{tpl}})_{n\geq 1}$ such that for every intrinsic compact chart K and every $n \geq 1$,*

$$\|V_{\eta,n} - V_{\eta,n-1}\|_{\eta,\mathrm{tpl},2;K} \leq \Gamma_{\eta,n}^{(2),\mathrm{tpl}}. \tag{668}$$

Assume also that

$$\sum_{n=1}^{\infty} n\,\Gamma^{(2),\mathrm{tpl}}_{\eta,n}\big(1 + r_\eta + W^{\mathrm{fp}}_\eta + (W^{\mathrm{fp}}_\eta)^2\big) < \infty. \tag{669}$$

Then Proposition D.128 applies with

$$\Lambda^{(2),\mathrm{fp}}_{\eta,n} := \Gamma^{(2),\mathrm{tpl}}_{\eta,n}, \qquad n \geq 1.$$

Consequently Theorem D.127, Theorem D.130, and Proposition D.128 all apply. In particular, activation of the chart-free natural Itô / closed-first-order-reflection / common-contact-clock package already reduces to one fixed template seminorm on the actual telescoping finite-cylinder interaction family $V_{\eta,n} - V_{\eta,n-1}$.

Proof. By definition, the telescoping shell is exactly

$$\Delta_{\eta,n} = V_{\eta,n} - V_{\eta,n-1}.$$

The fixed-template seminorm bound (668) is therefore precisely the fixed-profile derivative bound required in (659)–(661), with least admissible constant bounded by $\Gamma^{(2),\mathrm{tpl}}_{\eta,n}$. The summability condition (669) is then exactly the summability hypothesis (663) of Proposition D.128 under the identification $\Lambda^{(2),\mathrm{fp}}_{\eta,n} = \Gamma^{(2),\mathrm{tpl}}_{\eta,n}$. The stated consequences follow immediately by chaining with Proposition D.128, Theorem D.127, and Theorem D.130. □

Integrated downstream conclusion after the contact-clock/telescoping continuation. This continuation adds a fifty-fifth downstream message. Positively, the chart-free reflection package is no longer only a closed first-order calculus: it already factors through one common increasing contact clock carrying every closed reflection term on the canonical multiplicative natural-Itô core. Verification is also no longer most naturally phrased in terms of shellwise derivative arrays: one fixed template seminorm on the actual telescoping finite-cylinder interaction family already suffices. Cautionarily, the common contact clock is still not identified with the true geometric Skorokhod contact measure, and the fixed-template telescoping seminorm is still not verified for the genuine e-Arc-Helix interaction family. What remains open is therefore chart-free Skorokhod/normal-cone identification of that common-contact-clock first-order reflection factorization together with concrete fixed-template telescoping C^2 verification — or a sharp obstruction — for the actual interaction potentials.

Remark D.132. The new fifty-fifth message is the first point in the chain where the closed reflection package already has a single chart-free carrier measure rather than merely a family of compatible finite-variation terms. At the same time, the interaction side is now phrased directly on the actual telescoping finite-cylinder potentials rather than on abstract shell bounds. The remaining gap is therefore even more focused than before: identify the common contact clock and its density field as the genuine Skorokhod/normal-cone data, and prove — or sharply fail to prove — the fixed-template telescoping C^2 estimate for the actual e-Arc-Helix interaction family.

D.43 Worked continuation LVI: gradient-module contact densities and local-tail telescoping verification

The previous continuation isolated one common increasing contact clock carrying the full closed first-order reflection calculus and one fixed-template telescoping seminorm on the actual finite-cylinder interaction family. One further intrinsic sharpening is available on both sides before genuinely external input is needed. First, the density family produced by Theorem D.130 already depends only on the finitely supported cylinder gradient of the observable, so the closed reflection package descends from observables to a chart-free gradient module carried by the same common contact clock. Second, the fixed-template seminorm of Proposition D.131 already follows from a two-piece decomposition of each telescoping shell into a fixed local-window core plus a common-tail remainder. This makes the verification problem more compatible with actual shellwise estimates on the finite-cylinder interaction family.

Status label. The gradient-module statement below has status **theorem-level downstream result**. It does *not* yet identify the resulting density module with a genuine geometric normal vector field or prove that the common contact clock is carried exactly by the contact set of a full Skorokhod reflected gradient SPDE/SDE. What it does show is that the remaining reflection ambiguity is no longer the absence of a chart-free first-order density or even of a common carrier clock: the density already factors through the intrinsic finitely supported cylinder-gradient module itself. The local-tail verification statement is likewise a **proposition-level downstream result**. It does *not* verify the required estimates for the genuine e-Arc-Helix interaction family, but it reduces the fixed-template telescoping problem further to one windowed local shell core and one common-tail remainder for each finite-cylinder interaction increment.

For every cylinder observable

$$F = \Phi\big(\ell_{\varphi^{(1)}}, \dots, \ell_{\varphi^{(m)}}\big)\big|_U, \qquad \Phi \in C_b^2(\mathbb{R}^m),\ \varphi^{(i)} \in \mathcal{T}_\eta^{\mathrm{fs}},$$

define its intrinsic finitely supported cylinder gradient by

$$\nabla_\eta^{\mathrm{fs}} F(z) := \sum_{i=1}^{m} \partial_i \Phi\big(\ell_{\varphi^{(1)}}(z), \dots, \ell_{\varphi^{(m)}}(z)\big)\, \varphi^{(i)} \in \mathcal{T}_\eta^{\mathrm{fs}}. \tag{670}$$

Theorem D.133 (Chart-free finitely supported cylinder-gradient representation of the common-contact-clock reflection density). *Assume the hypotheses of Theorem D.130. Fix an intrinsic open set $U \subset \Omega_\eta$, $x \in \Omega_\eta$, and $\rho > 0$ with $\mathbb{B}_\eta(x,\rho) \Subset U$. Then there exists a predictable linear functional-valued density field*

$$\boldsymbol{\nu}_{\eta,U}^{x,\rho} : [0, \sigma_{x,\rho}] \to (\mathcal{T}_\eta^{\mathrm{fs}})^*,$$

defined for $d\mathfrak{L}_{\eta,U}^{x,\rho}$-almost every time, such that:

(i) for every $\varphi \in \mathcal{T}_\eta^{\mathrm{fs}}$,

$$\boldsymbol{\nu}_{\eta,U}^{x,\rho}(s)[\varphi] = \mathfrak{n}_{\eta,U}^{x,\rho}[\ell_\varphi|_U]_s \qquad \text{for } d\mathfrak{L}_{\eta,U}^{x,\rho}\text{-almost every } s; \tag{671}$$

(ii) for every $F \in \mathcal{A}_\eta^{\mathrm{cyl}}(U)$,

$$\mathfrak{n}_{\eta,U}^{x,\rho}[F]_s = \boldsymbol{\nu}_{\eta,U}^{x,\rho}(s)\Big[\nabla_\eta^{\mathrm{fs}} F\big(Z_{s\wedge\sigma_{x,\rho}}^x\big)\Big] \qquad \text{for } d\mathfrak{L}_{\eta,U}^{x,\rho}\text{-almost every } s; \tag{672}$$

(iii) if $F, G \in \mathcal{A}_\eta^{\mathrm{cyl}}(U)$ satisfy

$$\nabla_\eta^{\mathrm{fs}} F\big(Z_{s\wedge\sigma_{x,\rho}}^x\big) = \nabla_\eta^{\mathrm{fs}} G\big(Z_{s\wedge\sigma_{x,\rho}}^x\big) \qquad \text{for } d\mathfrak{L}_{\eta,U}^{x,\rho}\text{-almost every } s,$$

then

$$\mathfrak{n}_{\eta,U}^{x,\rho}[F]_s = \mathfrak{n}_{\eta,U}^{x,\rho}[G]_s \qquad \text{for } d\mathfrak{L}_{\eta,U}^{x,\rho}\text{-almost every } s;$$

(iv) the representation (672) extends uniquely by closure from $\mathcal{A}_\eta^{\mathrm{cyl}}(U)$ to $\mathcal{D}_\eta^{\mathrm{alg}}(U; x, \rho)$, and consequently what remains open is no longer the existence of a common chart-free contact clock or of a first-order density carried by it, but only identification of $\boldsymbol{\nu}_{\eta,U}^{x,\rho}$ as the genuine chart-free normal-cone selector and of $\mathfrak{L}_{\eta,U}^{x,\rho}$ as the true contact measure for the full continuum reflected gradient dynamics.

Proof. For each $\varphi \in \mathcal{T}_\eta^{\mathrm{fs}}$, define the scalar predictable density

$$\boldsymbol{\nu}_{\eta,U}^{x,\rho}(s)[\varphi] := \mathfrak{n}_{\eta,U}^{x,\rho}[\ell_\varphi|_U]_s,$$

which is meaningful $d\mathfrak{L}_{\eta,U}^{x,\rho}$-almost everywhere by Theorem D.130. Linearity in φ follows from the linearity of the observable map $\varphi \mapsto \ell_\varphi$ and of the density assignment $F \mapsto \mathfrak{n}_{\eta,U}^{x,\rho}[F]$.

Now let

$$F = \Phi\big(\ell_{\varphi^{(1)}}, \dots, \ell_{\varphi^{(m)}}\big)\big|_U \in \mathcal{A}_\eta^{\mathrm{cyl}}(U).$$

By the density-level product and scalar chain rules (665)–(666) from Theorem D.130, the closed reflection density of F is

$$\mathfrak{n}_{\eta,U}^{x,\rho}[F]_s = \sum_{i=1}^m \partial_i\Phi\big(\ell_{\varphi^{(1)}}(Z^x_{s\wedge\sigma_{x,\rho}}), \dots, \ell_{\varphi^{(m)}}(Z^x_{s\wedge\sigma_{x,\rho}})\big)\mathfrak{n}_{\eta,U}^{x,\rho}[\ell_{\varphi^{(i)}}|_U]_s.$$

By definition of $\boldsymbol{\nu}_{\eta,U}^{x,\rho}$ and (670), this is exactly (672). In particular, the density depends only on the finitely supported cylinder gradient of F evaluated along the stopped path.

If $F, G \in \mathcal{A}_\eta^{\mathrm{cyl}}(U)$ have the same stopped finitely supported cylinder gradient $d\mathfrak{L}_{\eta,U}^{x,\rho}$-almost everywhere, then $H := F - G$ has zero stopped finitely supported cylinder gradient there. On the cylinder algebra, this means that along the stopped chart the observable H is locally constant in all generating cylinder directions, hence its reflection density vanishes by the chain rule applied to constants. Therefore $\mathfrak{n}_{\eta,U}^{x,\rho}[F] = \mathfrak{n}_{\eta,U}^{x,\rho}[G]$ $d\mathfrak{L}_{\eta,U}^{x,\rho}$-almost everywhere.

Finally, let $F \in \mathcal{D}_\eta^{\mathrm{alg}}(U; x, \rho)$ and choose cylinder-algebra approximants $F_m \in \mathcal{A}_\eta^{\mathrm{cyl}}(U)$ converging to F in the local Itô seminorms with closed reflection terms converging in $\mathcal{A}_{\mathrm{fv}}^{x,\rho}$. Their densities converge in local $L^1(d\mathfrak{L}_{\eta,U}^{x,\rho})$ by Theorem D.130; the limit is unique and defines the extension claimed in item (iv). The final interpretation is exactly the remaining gap after this gradient-module representation has been established. □

For each telescoping shell

$$\Delta_{\eta,n} := V_{\eta,n} - V_{\eta,n-1},$$

keep the fixed local window r_η, the tail profile $(w_j)_{j\geq 1}$, the weights ω_η, Ω_η, and the tail sum W_η^{fp} from Section D.41. Assume that on every intrinsic compact chart K one can decompose

$$\Delta_{\eta,n} = L_{\eta,n}^{\mathrm{win}} + T_{\eta,n}^{\mathrm{tail}}, \tag{673}$$

where the first term is a fixed-window shell core and the second term is a common-tail remainder.

Proposition D.134 (Local-window/common-tail decomposition criterion for fixed-template telescoping C^2 activation). *Assume the hypotheses of Proposition D.110. Keep the fixed profile r_η, ω_η, Ω_η, and W_η^{fp} from Section D.41. Suppose that for each $n \geq 1$ there exist nonnegative constants $A_{\eta,n}^{\mathrm{win}}$ and $B_{\eta,n}^{\mathrm{tail}}$ such that, uniformly on every intrinsic compact chart K, the decomposition* (673) *satisfies*

$$|\partial_h L_{\eta,n}^{\mathrm{win}}| + |\partial_{hh} L_{\eta,n}^{\mathrm{win}}| \leq A_{\eta,n}^{\mathrm{win}}, \tag{674}$$

$$|\partial_{a_j} L_{\eta,n}^{\mathrm{win}}| + |\partial_{ha_j} L_{\eta,n}^{\mathrm{win}}| \leq A_{\eta,n}^{\mathrm{win}}\,\mathbf{1}_{\{j\leq r_\eta\}}, \tag{675}$$

$$|\partial_{a_j a_k} L_{\eta,n}^{\mathrm{win}}| \leq A_{\eta,n}^{\mathrm{win}}\,\mathbf{1}_{\{j\leq r_\eta\}}\mathbf{1}_{\{k\leq r_\eta\}}, \tag{676}$$

and

$$|\partial_h T_{\eta,n}^{\mathrm{tail}}| + |\partial_{hh} T_{\eta,n}^{\mathrm{tail}}| \leq B_{\eta,n}^{\mathrm{tail}}, \tag{677}$$

$$|\partial_{a_j} T_{\eta,n}^{\mathrm{tail}}| + |\partial_{ha_j} T_{\eta,n}^{\mathrm{tail}}| \leq B_{\eta,n}^{\mathrm{tail}}\,w_j, \tag{678}$$

$$|\partial_{a_j a_k} T_{\eta,n}^{\mathrm{tail}}| \leq B_{\eta,n}^{\mathrm{tail}}\,w_j w_k. \tag{679}$$

Assume also that

$$\sum_{n=1}^{\infty} n\big(A^{\mathrm{win}}_{\eta,n} + B^{\mathrm{tail}}_{\eta,n}\big)\big(1 + r_\eta + W^{\mathrm{fp}}_\eta + (W^{\mathrm{fp}}_\eta)^2\big) < \infty. \tag{680}$$

Then Proposition D.131 applies with

$$\Gamma^{(2),\mathrm{tpl}}_{\eta,n} := A^{\mathrm{win}}_{\eta,n} + B^{\mathrm{tail}}_{\eta,n}, \qquad n \geq 1.$$

Consequently Theorem D.130, Theorem D.133, and Proposition D.131 all apply. In particular, activation of the chart-free natural Itô / common-contact-clock / gradient-module package already reduces to one fixed local-window shell core plus one common-tail remainder for each telescoping finite-cylinder interaction increment.

Proof. By the triangle inequality and (667), it suffices to estimate the fixed-template seminorms of the window and tail pieces separately. The window bounds (674)–(676) imply

$$\|L^{\mathrm{win}}_{\eta,n}\|_{\eta,\mathrm{tpl},2;K} \leq A^{\mathrm{win}}_{\eta,n},$$

because for $j \leq r_\eta$ one has $\omega_\eta(j) \geq 1$ and for $j, k \leq r_\eta$ one has $\Omega_\eta(j,k) \geq 1$, while all window derivatives vanish outside the fixed local window. Likewise, the tail bounds (677)–(679) imply

$$\|T^{\mathrm{tail}}_{\eta,n}\|_{\eta,\mathrm{tpl},2;K} \leq B^{\mathrm{tail}}_{\eta,n},$$

since the weights in (667) were chosen exactly so that division by $\omega_\eta(j)$ and $\Omega_\eta(j,k)$ cancels the common-tail factors w_j and $w_j w_k$. Therefore

$$\|\Delta_{\eta,n}\|_{\eta,\mathrm{tpl},2;K} \leq A^{\mathrm{win}}_{\eta,n} + B^{\mathrm{tail}}_{\eta,n} = \Gamma^{(2),\mathrm{tpl}}_{\eta,n}.$$

The summability condition (680) is exactly the summability hypothesis (669) of Proposition D.131 under this identification, so the stated consequences follow immediately. □

Integrated downstream conclusion after the gradient-module/local-tail continuation. This continuation adds a fifty-sixth downstream message. Positively, the chart-free reflection package is no longer only a common-clock first-order density calculus: it already descends to the intrinsic finitely supported cylinder-gradient module itself. Verification is likewise no longer forced to bound the full telescoping shell in one step: it already suffices to isolate one fixed local-window shell core and one common-tail remainder. Cautionarily, the resulting gradient-module density is still not identified with the genuine chart-free normal-cone selector, and the local-window/common-tail decomposition is still not verified for the actual e-Arc-Helix interaction family. What remains open is therefore chart-free Skorokhod/normal-cone identification of that common-contact-clock gradient-density module together with concrete fixed-window/common-tail telescoping C^2 verification — or a sharp obstruction — for the actual interaction family.

Remark D.135. The new fifty-sixth message is the first point in the chain where the common-contact-clock reflection package already depends only on finitely supported cylinder gradients rather than on observables themselves. At the same time, the interaction side now admits a verification skeleton that separates local-window structure from common-tail structure on the actual telescoping finite-cylinder increments. The remaining gap is therefore more operational than before: identify the common-clock gradient density as true Skorokhod/normal-cone data, and prove — or sharply fail to prove — the fixed-window/common-tail telescoping C^2 decomposition for the genuine e-Arc-Helix interaction family.

D.44 Worked continuation LVII: coordinate contact-density fields and canonical frozen-tail verification

The previous continuation descended the common-contact-clock reflection package to the intrinsic finitely supported cylinder-gradient module and reduced verification to an arbitrary decomposition of each telescoping shell into a fixed local-window core plus a common-tail remainder. One more intrinsic sharpening is available on both sides without leaving the present natural Itô-domain framework. First, because the finitely supported test space already carries the explicit coordinate basis $e_h, e_1, e_2, \dots$, the gradient-module density field already admits a chart-free coordinate realization carried by the same common contact clock: each reflected coordinate local time has a density with respect to that clock, and every cylinder reflection density is their gradient pairing. Second, the fixed-window/common-tail verification problem no longer needs a search over all admissible decompositions: once the window r_η is fixed, one may freeze all modes outside that window to zero and test one canonical frozen local core together with its complementary tail remainder.

Status label. The coordinate-density statement below has status **theorem-level downstream result**. It does *not* yet identify the resulting coefficient field as a geometric normal vector or show that the common contact clock is supported exactly on the true contact set of a reflected gradient SPDE/SDE. What it does show is that the remaining reflection ambiguity is no longer the absence of coordinate data on the common clock: the common-contact-clock gradient module already has a chart-free coordinate density field for the reflected coordinate local times themselves. The frozen-tail verification statement is likewise a **proposition-level downstream result**. It does *not* verify the required bounds for the genuine e-Arc-Helix interaction family, but it shows that the local-window/common-tail criterion may already be tested on one canonical frozen-tail decomposition of each telescoping finite-cylinder shell.

Theorem D.136 (Chart-free coordinate-density realization of the common-contact-clock gradient module)**.** *Assume the hypotheses of Theorem D.133. Fix an intrinsic open set $U \subset \Omega_\eta$, $x \in \Omega_\eta$, and $\rho > 0$ with $\mathbb{B}_\eta(x,\rho) \Subset U$. Then there exists a predictable coefficient field*

$$\boldsymbol{\kappa}^{x,\rho}_{\eta,U} = \left(\kappa^{x,\rho}_{\eta,h}, \kappa^{x,\rho}_{\eta,1}, \kappa^{x,\rho}_{\eta,2}, \dots\right),$$

defined for $d\mathfrak{L}^{x,\rho}_{\eta,U}$-almost every time, such that:

(i) the explicit reflected coordinate local times from Theorem D.121 factor through the common contact clock according to

$$L^{0,x}_{t\wedge\sigma_{x,\rho}} - L^{h_*,x}_{t\wedge\sigma_{x,\rho}} = \int_0^t \kappa^{x,\rho}_{\eta,h}(s)\, d\mathfrak{L}^{x,\rho}_{\eta,U}(s), \tag{681}$$

$$K^{+,x}_{j,t\wedge\sigma_{x,\rho}} - K^{-,x}_{j,t\wedge\sigma_{x,\rho}} = \int_0^t \kappa^{x,\rho}_{\eta,j}(s)\, d\mathfrak{L}^{x,\rho}_{\eta,U}(s) \qquad (j \geq 1); \tag{682}$$

(ii) for every finitely supported test direction $\varphi = (\varphi_h, \varphi_1, \varphi_2, \dots) \in \mathcal{T}^{\mathrm{fs}}_\eta$,

$$\boldsymbol{\nu}^{x,\rho}_{\eta,U}(s)[\varphi] = \varphi_h\, \kappa^{x,\rho}_{\eta,h}(s) + \sum_{j=1}^{\infty} \varphi_j\, \kappa^{x,\rho}_{\eta,j}(s) \qquad \textit{for } d\mathfrak{L}^{x,\rho}_{\eta,U}\textit{-almost every } s, \tag{683}$$

where the sum is finite because φ has finite support;

(iii) for every cylinder observable $F \in \mathcal{A}^{\mathrm{cyl}}_\eta(U)$,

$$\begin{aligned}\mathfrak{n}^{x,\rho}_{\eta,U}[F]_s &= \kappa^{x,\rho}_{\eta,h}(s)\, \partial_h F\big(Z^x_{s\wedge\sigma_{x,\rho}}\big) \\ &\quad + \sum_{j=1}^{\infty} \kappa^{x,\rho}_{\eta,j}(s)\, \partial_{a_j} F\big(Z^x_{s\wedge\sigma_{x,\rho}}\big)\end{aligned} \tag{684}$$

for $d\mathfrak{L}^{x,\rho}_{\eta,U}$-almost every s, with only finitely many nonzero terms in the sum;

(iv) the representation (684) extends uniquely by closure from $\mathcal{A}^{\mathrm{cyl}}_\eta(U)$ to $\mathcal{D}^{\mathrm{alg}}_\eta(U;x,\rho)$, and consequently what remains open is no longer the existence of a chart-free coordinate density field carried by one common contact clock, but only identification of $\boldsymbol{\kappa}^{x,\rho}_{\eta,U}$ as the genuine normal-cone selector and of $\mathfrak{L}^{x,\rho}_{\eta,U}$ as the true contact measure supported on the geometric contact set of the full continuum reflected gradient dynamics.

Proof. Define

$$\kappa^{x,\rho}_{\eta,h}(s) := \mathfrak{n}^{x,\rho}_{\eta,U}[\ell_{e_h}|_U]_s, \qquad \kappa^{x,\rho}_{\eta,j}(s) := \mathfrak{n}^{x,\rho}_{\eta,U}[\ell_{e_j}|_U]_s, \quad j \geq 1,$$

which are meaningful for $d\mathfrak{L}^{x,\rho}_{\eta,U}$-almost every s by Theorem D.130. Since Theorem D.121 identifies the corresponding closed reflection functionals with the explicit coordinate local times, the factorization formula (664) applied to $\ell_{e_h}|_U$ and $\ell_{e_j}|_U$ gives (681)–(682).

Now let $\varphi = (\varphi_h, \varphi_1, \varphi_2, \dots) \in \mathcal{T}^{\mathrm{fs}}_\eta$. Since

$$\varphi = \varphi_h e_h + \sum_{j=1}^{\infty} \varphi_j e_j$$

with only finitely many nonzero coefficients, linearity of the module map $\boldsymbol{\nu}^{x,\rho}_{\eta,U}(s)[\cdot]$ yields (683).

If $F \in \mathcal{A}^{\mathrm{cyl}}_\eta(U)$, then its finitely supported cylinder gradient has the coordinate expansion

$$\nabla^{\mathrm{fs}}_\eta F = \partial_h F\, e_h + \sum_{j=1}^{\infty} \partial_{a_j} F\, e_j,$$

again with finitely many nonzero terms. Combining this with Theorem D.133 and (683) gives (684). Finally, if $F \in \mathcal{D}^{\mathrm{alg}}_\eta(U;x,\rho)$, choose cylinder-algebra approximants $F_m \in \mathcal{A}^{\mathrm{cyl}}_\eta(U)$ converging to F in the local Itô seminorms with closed reflection terms converging in $\mathcal{A}^{x,\rho}_{\mathrm{fv}}$. By Theorem D.133, their densities are Cauchy in local $L^1(d\mathfrak{L}^{x,\rho}_{\eta,U})$, so the coordinate formula passes to the limit and yields the claimed unique extension. Item (iv) is exactly the remaining gap after this coordinate-density realization has been established. □

For each $n \geq 1$, let

$$m_{\eta,n} := \min\{n, r_\eta\}$$

and define the canonical frozen-window projection on the n-th finite-cylinder level by

$$\Pi^{\mathrm{frz}}_{\eta,n}(h, a_1, \dots, a_n) := (h, a_1, \dots, a_{m_{\eta,n}}, 0, \dots, 0).$$

For the telescoping shell

$$\Delta_{\eta,n} := V_{\eta,n} - V_{\eta,n-1},$$

set

$$L^{\mathrm{frz}}_{\eta,n} := \Delta_{\eta,n} \circ \Pi^{\mathrm{frz}}_{\eta,n}, \qquad T^{\mathrm{frz}}_{\eta,n} := \Delta_{\eta,n} - L^{\mathrm{frz}}_{\eta,n}. \tag{685}$$

Proposition D.137 (Canonical frozen-tail criterion for fixed-window/common-tail telescoping C^2 verification). *Assume the hypotheses of Proposition D.110. Keep the fixed profile r_η, ω_η, Ω_η, and W^{fp}_η from Section D.41. Suppose there exist nonnegative constants $A^{\mathrm{frz}}_{\eta,n}$ and $B^{\mathrm{frz}}_{\eta,n}$ such that, uniformly on every intrinsic compact chart K, the canonical pieces from (685) satisfy the local-window bounds (674)–(676) and the common-tail bounds (677)–(679), with $L^{\mathrm{win}}_{\eta,n}, T^{\mathrm{tail}}_{\eta,n}$ replaced respectively by $L^{\mathrm{frz}}_{\eta,n}, T^{\mathrm{frz}}_{\eta,n}$ and with constants $A^{\mathrm{frz}}_{\eta,n}, B^{\mathrm{frz}}_{\eta,n}$. Assume also that*

$$\sum_{n=1}^{\infty} n\big(A^{\mathrm{frz}}_{\eta,n} + B^{\mathrm{frz}}_{\eta,n}\big)\big(1 + r_\eta + W^{\mathrm{fp}}_\eta + (W^{\mathrm{fp}}_\eta)^2\big) < \infty. \tag{686}$$

Then Proposition D.134 applies with

$$A^{\mathrm{win}}_{\eta,n} := A^{\mathrm{frz}}_{\eta,n}, \qquad B^{\mathrm{tail}}_{\eta,n} := B^{\mathrm{frz}}_{\eta,n}, \qquad n \geq 1.$$

Consequently Theorem D.130, Theorem D.133, Theorem D.136, and Proposition D.134 all apply. In particular, activation of the chart-free natural Itô / common-contact-clock / coordinate-density package already reduces to testing one canonical frozen local core and its complementary tail remainder on each telescoping finite-cylinder interaction increment.

Proof. By definition, the decomposition (685) is a particular instance of the abstract local-window/common-tail decomposition (673). Therefore the assumed bounds on $L^{\mathrm{frz}}_{\eta,n}$ and $T^{\mathrm{frz}}_{\eta,n}$ are exactly the hypotheses of Proposition D.134 under the identifications

$$L^{\mathrm{win}}_{\eta,n} = L^{\mathrm{frz}}_{\eta,n}, \qquad T^{\mathrm{tail}}_{\eta,n} = T^{\mathrm{frz}}_{\eta,n}, \qquad A^{\mathrm{win}}_{\eta,n} = A^{\mathrm{frz}}_{\eta,n}, \qquad B^{\mathrm{tail}}_{\eta,n} = B^{\mathrm{frz}}_{\eta,n}.$$

The summability condition (686) is the same as (680) under this identification. Proposition D.134 therefore applies immediately, and the stated consequences follow by chaining with Theorems D.130, D.133, and D.136. □

Integrated downstream conclusion after the coordinate-density/frozen-tail continuation. This continuation adds a fifty-seventh downstream message. Positively, the chart-free reflection package is no longer only a common-clock gradient module in abstract dual form: it already has a chart-free coordinate density field for the reflected coordinate local times themselves. Verification is likewise no longer forced to search over arbitrary local-window/common-tail splittings: one canonical frozen local core and its complementary tail remainder already suffice. Cautionarily, the resulting coordinate density field is still not identified with the genuine chart-free normal-cone selector, and the canonical frozen-tail bounds are still not verified for the actual e-Arc-Helix interaction family. What remains open is therefore chart-free Skorokhod/normal-cone identification of that common-contact-clock coordinate contact-density field together with concrete canonical frozen-tail telescoping C^2 verification — or a sharp obstruction — for the actual interaction family.

Remark D.138. The new fifty-seventh message is the first point in the chain where the common-contact-clock reflection package is already written in explicit coordinate-density form rather than merely as an abstract functional on finitely supported gradients. At the same time, the interaction side no longer has to optimize over all local-tail splittings: it may already test one canonical frozen-tail decomposition of the actual telescoping finite-cylinder shells. The remaining gap is therefore even more operational than before: identify the coordinate density field as genuine Skorokhod/normal-cone data, and prove — or sharply fail to prove — the canonical frozen-tail telescoping C^2 estimate for the genuine e-Arc-Helix interaction family.

D.45 Worked continuation LVIII: algebraic contact-cone selectors and a shellwise frozen-tail seminorm

The previous continuation already reduced the chart-free reflection ambiguity to an explicit coordinate density field carried by one common contact clock, and it reduced the interaction-side verification problem to one canonical frozen-tail splitting of each telescoping finite-cylinder shell. One more intrinsic sharpening is available on both sides without leaving the same natural Itô-domain framework. First, the coordinate density field already splits into its positive facewise pieces, so the common contact clock carries a chart-free *algebraic contact-cone selector*: every stopped reflection density is already represented by a selector taking values in the coordinate contact cone generated by the active boundary faces. Second, the canonical frozen-tail verification problem no longer needs two separately named local/tail constants; it can already be tested through one explicit shellwise frozen-tail seminorm on the actual telescoping finite-cylinder shells.

Status label. The contact-cone statement below has status **theorem-level downstream result**. It does *not* yet prove that the resulting algebraic coordinate cone agrees with the genuine geometric normal cone of the full intrinsic continuum support, and it does *not* yet identify the common contact clock as the exact total Skorokhod contact measure of a reflected gradient SDE/SPDE. What it does show is that the remaining stochastic ambiguity is no longer the absence of a chart-free cone selector: one already exists canonically in the algebraic coordinate contact cone carried by the common contact clock. The shellwise verification statement is a **proposition-level downstream result**. It does *not* verify the required seminorm bounds for the genuine e-Arc-Helix interaction family, but it shows that the frozen-tail activation problem has already collapsed to one canonical shellwise seminorm for each telescoping finite-cylinder increment.

For $z = (h, \mathbf{a}) \in \Omega_\eta$, define the algebraic coordinate contact cone

$$\begin{aligned}\mathcal{N}_\eta^{\mathrm{alg}}(z) := \Big\{\Lambda \in (\mathcal{T}_\eta^{\mathrm{fs}})^* : \Lambda[\varphi] = (\alpha_h^0 - \alpha_h^*)\varphi_h + \sum_{j=1}^{\infty}(\alpha_j^+ - \alpha_j^-)\varphi_j,\\ \alpha_h^0, \alpha_h^*, \alpha_j^\pm \ge 0, \qquad \alpha_h^0 > 0 \Rightarrow h = 0, \qquad \alpha_h^* > 0 \Rightarrow h = h_*,\\ \alpha_j^+ > 0 \Rightarrow a_j = b_j(\eta), \qquad \alpha_j^- > 0 \Rightarrow a_j = -b_j(\eta) \text{ for every } j \ge 1\Big\},\end{aligned} \tag{687}$$

where the sum is finite when evaluated on any $\varphi \in \mathcal{T}_\eta^{\mathrm{fs}}$. For every cylinder observable F, write

$$\nabla_{\mathrm{fs}} F(z) := \big(\partial_h F(z), \partial_{a_1} F(z), \partial_{a_2} F(z), \dots\big) \in \mathcal{T}_\eta^{\mathrm{fs}}, \tag{688}$$

again with finite support because F depends on only finitely many coordinates.

Theorem D.139 (Chart-free algebraic contact-cone selector carried by the common contact clock)**.** *Assume the hypotheses of Theorem D.136. Fix an intrinsic open set $U \subset \Omega_\eta$, $x \in \Omega_\eta$, and $\rho > 0$ with $\mathbb{B}_\eta(x, \rho) \Subset U$. Then there exist nonnegative predictable face densities*

$$\alpha_{\eta,h}^{0,x,\rho}, \qquad \alpha_{\eta,h}^{*,x,\rho}, \qquad \alpha_{\eta,j}^{+,x,\rho},\ \alpha_{\eta,j}^{-,x,\rho} \quad (j \ge 1),$$

defined for $d\mathfrak{L}_{\eta,U}^{x,\rho}$-almost every time, such that:

(i) the individual coordinate reflection local times factor through the common contact clock according to

$$L_{t\wedge\sigma_{x,\rho}}^{0,x} = \int_0^t \alpha_{\eta,h}^{0,x,\rho}(s)\, d\mathfrak{L}_{\eta,U}^{x,\rho}(s), \tag{689}$$

$$L_{t\wedge\sigma_{x,\rho}}^{h_*,x} = \int_0^t \alpha_{\eta,h}^{*,x,\rho}(s)\, d\mathfrak{L}_{\eta,U}^{x,\rho}(s), \tag{690}$$

$$K_{j,t\wedge\sigma_{x,\rho}}^{+,x} = \int_0^t \alpha_{\eta,j}^{+,x,\rho}(s)\, d\mathfrak{L}_{\eta,U}^{x,\rho}(s), \tag{691}$$

$$K_{j,t\wedge\sigma_{x,\rho}}^{-,x} = \int_0^t \alpha_{\eta,j}^{-,x,\rho}(s)\, d\mathfrak{L}_{\eta,U}^{x,\rho}(s) \qquad (j \ge 1); \tag{692}$$

(ii) defining a predictable functional-valued field by

$$\begin{aligned}\mathbf{n}_{\eta,U}^{x,\rho}(s)[\varphi] := (\alpha_{\eta,h}^{0,x,\rho}(s) - \alpha_{\eta,h}^{*,x,\rho}(s))\varphi_h\\ + \sum_{j=1}^{\infty}(\alpha_{\eta,j}^{+,x,\rho}(s) - \alpha_{\eta,j}^{-,x,\rho}(s))\varphi_j,\end{aligned} \tag{693}$$

one has

$$\mathbf{n}_{\eta,U}^{x,\rho}(s) \in \mathcal{N}_\eta^{\mathrm{alg}}(Z_{s\wedge\sigma_{x,\rho}}^x) \qquad \textit{for } d\mathfrak{L}_{\eta,U}^{x,\rho}\textit{-almost every } s; \tag{694}$$

(iii) the coordinate density field of Theorem D.136 is recovered by

$$\kappa^{x,\rho}_{\eta,h} = \alpha^{0,x,\rho}_{\eta,h} - \alpha^{*,x,\rho}_{\eta,h}, \qquad \kappa^{x,\rho}_{\eta,j} = \alpha^{+,x,\rho}_{\eta,j} - \alpha^{-,x,\rho}_{\eta,j} \quad (j \geq 1), \tag{695}$$

and for every cylinder observable $F \in \mathcal{A}^{\mathrm{cyl}}_{\eta}(U)$,

$$\mathbf{n}^{x,\rho}_{\eta,U}[F]_s = \mathbf{n}^{x,\rho}_{\eta,U}(s)\big[\nabla_{\mathrm{fs}} F(Z^x_{s\wedge\sigma_{x,\rho}})\big] \qquad \text{for } d\mathfrak{L}^{x,\rho}_{\eta,U}\text{-almost every } s; \tag{696}$$

(iv) consequently, what remains open is no longer the existence of a chart-free cone selector carried by the common contact clock, but only identification of the algebraic coordinate cone (687) with the genuine geometric normal cone or Skorokhod reflection cone of the full intrinsic continuum state space, together with identification of $\mathfrak{L}^{x,\rho}_{\eta,U}$ *as the exact total contact measure of that full reflected gradient dynamics.*

Proof. Theorem D.136 gives the signed factorizations

$$dL^{0,x}_{\cdot\wedge\sigma_{x,\rho}} - dL^{h_*,x}_{\cdot\wedge\sigma_{x,\rho}} = \kappa^{x,\rho}_{\eta,h}\, d\mathfrak{L}^{x,\rho}_{\eta,U}, \qquad dK^{+,x}_{j,\cdot\wedge\sigma_{x,\rho}} - dK^{-,x}_{j,\cdot\wedge\sigma_{x,\rho}} = \kappa^{x,\rho}_{\eta,j}\, d\mathfrak{L}^{x,\rho}_{\eta,U} \quad (j \geq 1).$$

The positive measures $dL^{0,x}_{\cdot\wedge\sigma_{x,\rho}}$ and $dL^{h_*,x}_{\cdot\wedge\sigma_{x,\rho}}$ are supported on the disjoint face sets $\{H^x = 0\}$ and $\{H^x = h_*\}$ respectively, while $dK^{+,x}_{j,\cdot\wedge\sigma_{x,\rho}}$ and $dK^{-,x}_{j,\cdot\wedge\sigma_{x,\rho}}$ are supported on the disjoint face sets $\{A^x_j = b_j(\eta)\}$ and $\{A^x_j = -b_j(\eta)\}$. By uniqueness of the Jordan decomposition of each signed measure, one therefore has

$$dL^{0,x}_{\cdot\wedge\sigma_{x,\rho}} = (\kappa^{x,\rho}_{\eta,h})^+\, d\mathfrak{L}^{x,\rho}_{\eta,U}, \qquad dL^{h_*,x}_{\cdot\wedge\sigma_{x,\rho}} = (\kappa^{x,\rho}_{\eta,h})^-\, d\mathfrak{L}^{x,\rho}_{\eta,U},$$

and similarly,

$$dK^{+,x}_{j,\cdot\wedge\sigma_{x,\rho}} = (\kappa^{x,\rho}_{\eta,j})^+\, d\mathfrak{L}^{x,\rho}_{\eta,U}, \qquad dK^{-,x}_{j,\cdot\wedge\sigma_{x,\rho}} = (\kappa^{x,\rho}_{\eta,j})^-\, d\mathfrak{L}^{x,\rho}_{\eta,U}.$$

Define

$$\alpha^{0,x,\rho}_{\eta,h} := (\kappa^{x,\rho}_{\eta,h})^+, \qquad \alpha^{*,x,\rho}_{\eta,h} := (\kappa^{x,\rho}_{\eta,h})^-, \qquad \alpha^{+,x,\rho}_{\eta,j} := (\kappa^{x,\rho}_{\eta,j})^+, \qquad \alpha^{-,x,\rho}_{\eta,j} := (\kappa^{x,\rho}_{\eta,j})^-.$$

This gives the factorization formulas (689)–(692). Their support properties imply exactly the membership condition (694) for the selector (693). Item (iii) is immediate from (695) together with the coordinate-density pairing formula (684). Item (iv) is then precisely the residual interpretation gap after the algebraic cone selector has been constructed. □

For each intrinsic compact chart K and each telescoping shell $\Delta_{\eta,n} = V_{\eta,n} - V_{\eta,n-1}$, define the canonical shellwise frozen-tail seminorm

$$\begin{aligned} \|\Delta_{\eta,n}\|_{\eta,\mathrm{frz},2;K} := \max\Bigg\{ & \|\partial_h L^{\mathrm{frz}}_{\eta,n}\|_{L^\infty(K)} + \|\partial_{hh} L^{\mathrm{frz}}_{\eta,n}\|_{L^\infty(K)}, \\ & \sup_{1\leq j\leq r_\eta} \Big(\|\partial_{a_j} L^{\mathrm{frz}}_{\eta,n}\|_{L^\infty(K)} + \|\partial_{ha_j} L^{\mathrm{frz}}_{\eta,n}\|_{L^\infty(K)} \Big), \\ & \sup_{1\leq j,k\leq r_\eta} \|\partial_{a_j a_k} L^{\mathrm{frz}}_{\eta,n}\|_{L^\infty(K)}, \\ & \|\partial_h T^{\mathrm{frz}}_{\eta,n}\|_{L^\infty(K)} + \|\partial_{hh} T^{\mathrm{frz}}_{\eta,n}\|_{L^\infty(K)}, \\ & \sup_{j\geq 1} \omega_\eta(j)^{-1} \Big(\|\partial_{a_j} T^{\mathrm{frz}}_{\eta,n}\|_{L^\infty(K)} + \|\partial_{ha_j} T^{\mathrm{frz}}_{\eta,n}\|_{L^\infty(K)} \Big), \\ & \sup_{j,k\geq 1} \Omega_\eta(j,k)^{-1} \|\partial_{a_j a_k} T^{\mathrm{frz}}_{\eta,n}\|_{L^\infty(K)} \Bigg\}. \end{aligned} \tag{697}$$

Proposition D.140 (Canonical shellwise frozen-tail seminorm criterion for telescoping C^2 verification)**.** *Assume the hypotheses of Proposition D.110. Keep the fixed profile r_η, ω_η, Ω_η, and W_η^{fp} from Section D.41. Suppose there exists a sequence $\left(\Gamma_{\eta,n}^{(2),\mathrm{frz}}\right)_{n\ge 1}$ such that for every intrinsic compact chart K and every $n \ge 1$,*

$$\|\Delta_{\eta,n}\|_{\eta,\mathrm{frz},2;K} \le \Gamma_{\eta,n}^{(2),\mathrm{frz}}. \tag{698}$$

Assume also that

$$\sum_{n=1}^{\infty} n\,\Gamma_{\eta,n}^{(2),\mathrm{frz}}\big(1 + r_\eta + W_\eta^{\mathrm{fp}} + (W_\eta^{\mathrm{fp}})^2\big) < \infty. \tag{699}$$

Then Proposition D.137 applies with

$$A_{\eta,n}^{\mathrm{frz}} := \Gamma_{\eta,n}^{(2),\mathrm{frz}}, \qquad B_{\eta,n}^{\mathrm{frz}} := \Gamma_{\eta,n}^{(2),\mathrm{frz}}, \qquad n \ge 1.$$

Consequently Theorem D.130, Theorem D.133, Theorem D.136, Theorem D.139, and Proposition D.137 all apply. In particular, activation of the chart-free natural Itô / common-contact-clock / coordinate-density / algebraic-contact-cone package already reduces to one canonical shellwise frozen-tail seminorm on each telescoping finite-cylinder interaction increment.

Proof. By definition of the seminorm (697), the bound (698) implies simultaneously all local-window bounds (674)–(676) for $L_{\eta,n}^{\mathrm{frz}}$ and all common-tail bounds (677)–(679) for $T_{\eta,n}^{\mathrm{frz}}$, with both controlling constants equal to $\Gamma_{\eta,n}^{(2),\mathrm{frz}}$. The summability condition (699) is therefore exactly the hypothesis (686) of Proposition D.137 under the identification

$$A_{\eta,n}^{\mathrm{frz}} = B_{\eta,n}^{\mathrm{frz}} = \Gamma_{\eta,n}^{(2),\mathrm{frz}}.$$

The stated consequences then follow by chaining Proposition D.137 with Theorems D.130, D.133, D.136, and D.139. ☐

Integrated downstream conclusion after the algebraic-cone/shellwise-frozen-tail continuation. This continuation adds a fifty-eighth downstream message. Positively, the chart-free reflection package is no longer only an explicit coordinate-density field carried by a common contact clock: it already factors through a chart-free algebraic contact-cone selector supported on the active coordinate boundary faces. Verification is likewise no longer forced to track separate frozen-core and tail constants: it already reduces to one canonical shellwise frozen-tail seminorm on each telescoping finite-cylinder shell. Cautionarily, the algebraic contact cone is still not identified with the genuine geometric normal cone of the full intrinsic continuum support, and the shellwise frozen-tail seminorm bounds are still not verified for the actual e-Arc-Helix interaction family. What remains open is therefore geometric Skorokhod/normal-cone identification of that chart-free algebraic contact-cone selector together with concrete canonical shellwise frozen-tail telescoping C^2 verification — or a sharp obstruction — for the actual interaction family.

Remark D.141. The new fifty-eighth message is the first point in the chain where the common-contact-clock reflection package already takes values in a bona fide coordinate contact cone rather than merely in a signed density module. At the same time, the interaction side no longer has to carry two separately named constants for the frozen local core and the tail remainder: one shellwise seminorm already packages both at once. The remaining gap is therefore maximally operational: identify the algebraic contact cone with the true geometric normal cone/Skorokhod selector, and prove — or sharply fail to prove — the resulting shellwise frozen-tail C^2 seminorm bounds for the genuine e-Arc-Helix interaction shells.

D.46 Worked continuation LIX: algebraic complementarity and a direct frozen-trace seminorm

The previous continuation already produced a chart-free algebraic contact-cone selector carried by one common contact clock and reduced the interaction-side verification problem to one canonical shellwise frozen-tail seminorm on each telescoping finite-cylinder shell. One more intrinsic sharpening is available on both sides without leaving the same natural Itô-domain framework. First, the same cone selector already satisfies a full *algebraic complementarity package*: its face densities are orthogonal to the corresponding slack variables and the resulting selector is already sign-polar on the canonical coordinate contact wedge. Second, the shellwise frozen-tail seminorm no longer needs the local core and tail remainder to be named separately at all; it can already be rewritten as one *direct frozen-trace/difference seminorm* on the actual telescoping finite-cylinder shell together with its canonical frozen trace.

Status label. The complementarity statement below has status **theorem-level downstream result**. It does *not* yet identify the algebraic contact wedge with the genuine geometric tangent cone of the full intrinsic continuum support, and it does *not* yet identify the common contact clock with the full Skorokhod contact measure of the continuum reflected gradient dynamics. What it does show is that the chart-free selector already has the exact first-order slack orthogonality and sign-polarity properties one expects from a normal-cone multiplier on the canonical coordinate contact wedge. The direct verification statement is a **proposition-level downstream result.** It does *not* verify the required seminorm bounds for the genuine e-Arc-Helix interaction family, but it shows that those bounds may already be checked directly on each actual telescoping shell and its canonical frozen trace, without carrying a separately named frozen local core and tail remainder.

For $z = (h, \mathbf{a}) \in \Omega_\eta$, define the coordinate slack fields

$$s^0_{\eta,h}(z) := h, \qquad s^*_{\eta,h}(z) := h_* - h, \qquad s^+_{\eta,j}(z) := b_j(\eta) - a_j, \qquad s^-_{\eta,j}(z) := b_j(\eta) + a_j \quad (j \geq 1). \tag{700}$$

Define also the canonical algebraic contact wedge

$$\mathcal{W}^{\mathrm{alg}}_\eta(z) := \Big\{\varphi \in \mathcal{T}^{\mathrm{fs}}_\eta : h = 0 \Rightarrow \varphi_h \geq 0, \qquad h = h_* \Rightarrow \varphi_h \leq 0, \\ a_j = b_j(\eta) \Rightarrow \varphi_j \geq 0, \qquad a_j = -b_j(\eta) \Rightarrow \varphi_j \leq 0 \text{ for every } j \geq 1\Big\}. \tag{701}$$

Theorem D.142 (Chart-free algebraic complementarity package for the common-contact-clock selector). *Assume the hypotheses of Theorem D.139. Fix an intrinsic open set $U \subset \Omega_\eta$, $x \in \Omega_\eta$, and $\rho > 0$ with $\mathbb{B}_\eta(x, \rho) \Subset U$. Let $\alpha^{0,x,\rho}_{\eta,h}$, $\alpha^{*,x,\rho}_{\eta,h}$, $\alpha^{\pm,x,\rho}_{\eta,j}$, and $\mathbf{n}^{x,\rho}_{\eta,U}$ be the face densities and algebraic selector supplied by Theorem D.139. Then:*

(i) the selector already satisfies the pointwise slack-complementarity identities

$$\alpha^{0,x,\rho}_{\eta,h}(s)s^0_{\eta,h}(Z^x_{s\wedge\sigma_{x,\rho}}) = 0, \tag{702}$$

$$\alpha^{*,x,\rho}_{\eta,h}(s)s^*_{\eta,h}(Z^x_{s\wedge\sigma_{x,\rho}}) = 0, \tag{703}$$

$$\alpha^{+,x,\rho}_{\eta,j}(s)s^+_{\eta,j}(Z^x_{s\wedge\sigma_{x,\rho}}) = 0, \tag{704}$$

$$\alpha^{-,x,\rho}_{\eta,j}(s)s^-_{\eta,j}(Z^x_{s\wedge\sigma_{x,\rho}}) = 0 \qquad (j \geq 1). \tag{705}$$

for $d\mathfrak{L}^{x,\rho}_{\eta,U}$-almost every s;

(ii) for every $\varphi \in \mathcal{W}^{\mathrm{alg}}_\eta(Z^x_{s\wedge\sigma_{x,\rho}})$ one has the first-order sign-polarity inequality

$$\mathbf{n}^{x,\rho}_{\eta,U}(s)[\varphi] \geq 0 \qquad \text{for } d\mathfrak{L}^{x,\rho}_{\eta,U}\text{-almost every } s; \tag{706}$$

(iii) *if* $\varphi \in \mathcal{T}_\eta^{\mathrm{fs}}$ *is contact-tangential at* $Z^x_{s\wedge\sigma_{x,\rho}}$ *in the sense that*

$$\begin{aligned} h = 0 \text{ or } h = h_* &\Rightarrow \varphi_h = 0, \\ a_j = b_j(\eta) \text{ or } a_j = -b_j(\eta) &\Rightarrow \varphi_j = 0 \quad (j \geq 1), \end{aligned} \tag{707}$$

then

$$\mathbf{n}^{x,\rho}_{\eta,U}(s)[\varphi] = 0 \qquad \text{for } d\mathfrak{L}^{x,\rho}_{\eta,U}\text{-almost every } s; \tag{708}$$

(iv) *consequently, what remains open is no longer first-order complementarity or sign-polarity on the canonical coordinate contact wedge, but only identification of that algebraic wedge and its dual selector with the genuine geometric tangent/normal-cone pair or Skorokhod reflection data of the full intrinsic continuum support.*

Proof. The support implications in Theorem D.139 say that $\alpha^{0,x,\rho}_{\eta,h}$ may be positive only on the face $\{H^x = 0\}$, $\alpha^{*,x,\rho}_{\eta,h}$ only on $\{H^x = h_*\}$, $\alpha^{+,x,\rho}_{\eta,j}$ only on $\{A^x_j = b_j(\eta)\}$, and $\alpha^{-,x,\rho}_{\eta,j}$ only on $\{A^x_j = -b_j(\eta)\}$. Since the corresponding slack fields in (700) vanish exactly on those faces, the complementarity identities (702)–(705) follow immediately.

For item (ii), write $z = Z^x_{s\wedge\sigma_{x,\rho}}$ and let $\varphi \in \mathcal{W}^{\mathrm{alg}}_\eta(z)$. Whenever $\alpha^{0,x,\rho}_{\eta,h}(s) > 0$, the support rule forces $z_h = 0$, hence $\varphi_h \geq 0$. Whenever $\alpha^{*,x,\rho}_{\eta,h}(s) > 0$, one has $z_h = h_*$, hence $\varphi_h \leq 0$ and therefore $-\alpha^{*,x,\rho}_{\eta,h}(s)\varphi_h \geq 0$. The same argument gives $\alpha^{+,x,\rho}_{\eta,j}(s)\varphi_j \geq 0$ on the upper j-face and $-\alpha^{-,x,\rho}_{\eta,j}(s)\varphi_j \geq 0$ on the lower j-face. Inserting these sign relations into the defining formula (693) yields (706).

For item (iii), the tangentiality condition (707) forces every coordinate of φ that can carry a nonzero face density to vanish. Since the remaining coordinates appear in (693) only with zero coefficients, the pairing is identically zero. Item (iv) is exactly the residual gap after complementarity and sign-polarity have both been made chart-free and intrinsic. □

Define the canonical frozen-trace projection

$$\Pi^{\mathrm{frz}}_\eta(h, \mathbf{a}) := (h, a_1, \dots, a_{r_\eta}, 0, 0, \dots) \in \Omega_\eta. \tag{709}$$

For each telescoping shell $\Delta_{\eta,n} = V_{\eta,n} - V_{\eta,n-1}$ and intrinsic compact chart K, define the direct frozen-trace/difference seminorm

$$\begin{aligned} \|\Delta_{\eta,n}\|_{\eta,\mathrm{dfrz},2;K} := \max\Bigg\{ &\|\partial_h(\Delta_{\eta,n} \circ \Pi^{\mathrm{frz}}_\eta)\|_{L^\infty(K)} + \|\partial_{hh}(\Delta_{\eta,n} \circ \Pi^{\mathrm{frz}}_\eta)\|_{L^\infty(K)}, \\ &\sup_{1\leq j\leq r_\eta} \Big(\|\partial_{a_j}(\Delta_{\eta,n} \circ \Pi^{\mathrm{frz}}_\eta)\|_{L^\infty(K)} + \|\partial_{ha_j}(\Delta_{\eta,n} \circ \Pi^{\mathrm{frz}}_\eta)\|_{L^\infty(K)}\Big), \\ &\sup_{1\leq j,k\leq r_\eta} \|\partial_{a_ja_k}(\Delta_{\eta,n} \circ \Pi^{\mathrm{frz}}_\eta)\|_{L^\infty(K)}, \\ &\|\partial_h(\Delta_{\eta,n} - \Delta_{\eta,n} \circ \Pi^{\mathrm{frz}}_\eta)\|_{L^\infty(K)} + \|\partial_{hh}(\Delta_{\eta,n} - \Delta_{\eta,n} \circ \Pi^{\mathrm{frz}}_\eta)\|_{L^\infty(K)}, \\ &\sup_{j\geq 1} \omega_\eta(j)^{-1}\Big(\|\partial_{a_j}(\Delta_{\eta,n} - \Delta_{\eta,n} \circ \Pi^{\mathrm{frz}}_\eta)\|_{L^\infty(K)} + \|\partial_{ha_j}(\Delta_{\eta,n} - \Delta_{\eta,n} \circ \Pi^{\mathrm{frz}}_\eta)\|_{L^\infty(K)}\Big), \\ &\sup_{j,k\geq 1} \Omega_\eta(j,k)^{-1}\|\partial_{a_ja_k}(\Delta_{\eta,n} - \Delta_{\eta,n} \circ \Pi^{\mathrm{frz}}_\eta)\|_{L^\infty(K)} \Bigg\}. \end{aligned} \tag{710}$$

Proposition D.143 (Direct frozen-trace/difference shell seminorm criterion for telescoping C^2 verification)**.** *Assume the hypotheses of Proposition D.110. Keep the fixed profile* r_η, ω_η, Ω_η, *and* W^{fp}_η *from Section D.41. Suppose there exists a sequence* $\big(\Gamma^{(2),\mathrm{dfrz}}_{\eta,n}\big)_{n\geq 1}$ *such that for every intrinsic compact chart* K *and every* $n \geq 1$,

$$\|\Delta_{\eta,n}\|_{\eta,\mathrm{dfrz},2;K} \leq \Gamma^{(2),\mathrm{dfrz}}_{\eta,n}. \tag{711}$$

Assume also that

$$\sum_{n=1}^{\infty} n\,\Gamma^{(2),\mathrm{dfrz}}_{\eta,n}\bigl(1 + r_\eta + W^{\mathrm{fp}}_\eta + (W^{\mathrm{fp}}_\eta)^2\bigr) < \infty. \tag{712}$$

Then Proposition D.140 applies with

$$\Gamma^{(2),\mathrm{frz}}_{\eta,n} := \Gamma^{(2),\mathrm{dfrz}}_{\eta,n} \qquad (n \geq 1).$$

Consequently Theorem D.130, Theorem D.133, Theorem D.136, Theorem D.139, Theorem D.142, and Proposition D.140 all apply. In particular, activation of the chart-free natural Itô / common-contact-clock / coordinate-density / algebraic-contact-cone / algebraic-complementarity package already reduces to one direct frozen-trace/difference seminorm on each actual telescoping finite-cylinder shell together with its canonical frozen trace.

Proof. By definition of the canonical frozen-tail decomposition, one has

$$L^{\mathrm{frz}}_{\eta,n} = \Delta_{\eta,n} \circ \Pi^{\mathrm{frz}}_\eta, \qquad T^{\mathrm{frz}}_{\eta,n} = \Delta_{\eta,n} - \Delta_{\eta,n} \circ \Pi^{\mathrm{frz}}_\eta.$$

Substituting these identities into (697) shows that

$$\|\Delta_{\eta,n}\|_{\eta,\mathrm{frz},2;K} = \|\Delta_{\eta,n}\|_{\eta,\mathrm{dfrz},2;K}$$

for every K and n. Hence the bound (711) is exactly the hypothesis (698) of Proposition D.140 with $\Gamma^{(2),\mathrm{frz}}_{\eta,n} := \Gamma^{(2),\mathrm{dfrz}}_{\eta,n}$, and the summability condition (712) is exactly (699) under the same identification. The stated consequences therefore follow immediately by chaining Proposition D.140 with Theorem D.142. □

Integrated downstream conclusion after the complementarity/direct-frozen-trace continuation. This continuation adds a fifty-ninth downstream message. Positively, the chart-free reflection package is no longer only an algebraic contact-cone selector carried by a common contact clock: it already satisfies chart-free slack complementarity, contact-tangential annihilation, and first-order sign-polarity on the canonical coordinate contact wedge. Verification is likewise no longer forced to name a frozen local core and tail remainder at all: it already reduces to one direct frozen-trace/difference seminorm on each actual telescoping finite-cylinder shell. Cautionarily, the algebraic contact wedge is still not identified with the genuine geometric tangent cone of the full intrinsic continuum support, and the direct frozen-trace/difference seminorm bounds are still not verified for the actual e-Arc-Helix interaction family. What remains open is therefore geometric Skorokhod/normal-cone identification of that chart-free algebraic complementarity selector together with concrete direct frozen-trace shellwise C^2 verification — or a sharp obstruction — for the actual interaction family.

Remark D.144. The new fifty-ninth message is the first point in the chain where the common-contact-clock reflection package already satisfies the exact first-order complementarity identities one would expect from a Kuhn–Tucker or Skorokhod multiplier on the coordinate boundary faces. At the same time, the interaction side no longer has to carry separate notation for the frozen local core and tail remainder: it may already test one direct seminorm of the actual shell together with its canonical frozen trace. The remaining gap is therefore again more geometric than formal: identify the algebraic contact wedge and its selector with the genuine tangent/normal-cone pair of the intrinsic continuum support, and prove — or sharply fail to prove — the resulting direct frozen-trace shellwise C^2 bounds for the genuine e-Arc-Helix interaction shells.

D.47 Worked continuation LX: ambient coordinate-box normal cones and a frozen-segment oscillation criterion

The previous continuation already produced a chart-free algebraic complementarity selector carried by a common contact clock and rewrote the interaction-side verification problem as a direct frozen-trace/difference seminorm on each actual telescoping finite-cylinder shell. One further intrinsic sharpening is available on both sides without changing the same natural Itô-domain framework. First, the algebraic coordinate contact wedge is already exactly the finitely supported tangent cone of the ambient coordinate box containing the intrinsic support, so the chart-free algebraic selector is already an *ambient normal-cone multiplier* carried by the common contact clock. Second, the direct frozen-trace/difference seminorm no longer needs the shell and its frozen trace to be handled as two separate functions: it may already be controlled by one *canonical frozen-segment oscillation seminorm* computed on the shell itself along the straight frozen-tail homotopies connecting every point to its canonical frozen trace.

Status label. The ambient normal-cone statement below has status **theorem-level downstream result**. It does *not* yet identify the ambient coordinate-box normal cone with the genuine intrinsic normal cone of the full continuum support, and it does *not* yet prove that the common contact clock is the exact total Skorokhod contact measure of that full reflected gradient dynamics. What it does show is that the chart-free algebraic complementarity selector is already the exact normal-cone multiplier for the ambient coordinate box cut out by the manuscript's coordinate barriers. The frozen-segment statement is a **proposition-level downstream result**. It does *not* yet verify the required seminorm bounds for the genuine e-Arc-Helix interaction family, but it shows that those bounds may already be checked by oscillation of the shell's own first and second derivatives along the canonical frozen-tail segments, together with the finite-window trace jets on the canonical frozen trace itself.

Define the ambient coordinate box

$$\mathcal{B}_\eta := \Big\{(h, \mathbf{a}) \in \mathbb{R} \times \ell^1_{(3)} : \; 0 \le h \le h_*, \quad |a_j| \le b_j(\eta) \text{ for every } j \ge 1\Big\}. \tag{713}$$

For $z = (h, \mathbf{a}) \in \mathcal{B}_\eta$, let $T_{\mathcal{B}_\eta}(z)$ denote the Bouligand tangent cone of $\mathcal{B}_\eta$ at z, and define the coordinate evaluation functionals on $\mathcal{T}^{\mathrm{fs}}_\eta$ by

$$\mathbf{e}_h[\varphi] := \varphi_h, \qquad \mathbf{e}_j[\varphi] := \varphi_j \quad (j \ge 1). \tag{714}$$

Define the ambient normal cone acting on finitely supported directions by

$$N^{\mathrm{fs}}_{\mathcal{B}_\eta}(z) := \Big\{\nu \in (\mathcal{T}^{\mathrm{fs}}_\eta)^* : \; \nu[\varphi] \ge 0 \text{ for every } \varphi \in T_{\mathcal{B}_\eta}(z) \cap \mathcal{T}^{\mathrm{fs}}_\eta\Big\}. \tag{715}$$

Theorem D.145 (Ambient coordinate-box normal-cone realization of the chart-free complementarity selector). *Assume the hypotheses of Theorem D.142. Fix an intrinsic open set $U \subset \Omega_\eta$, $x \in \Omega_\eta$, and $\rho > 0$ with $\mathbb{B}_\eta(x, \rho) \Subset U$. Then:*

(i) for every $z \in \Omega_\eta$ one has

$$T_{\mathcal{B}_\eta}(z) \cap \mathcal{T}^{\mathrm{fs}}_\eta = \mathcal{W}^{\mathrm{alg}}_\eta(z), \tag{716}$$

so the canonical algebraic contact wedge is already the finitely supported tangent cone of the ambient coordinate box;

(ii) the common-contact-clock selector of Theorem D.139 already lies in that ambient normal cone, namely

$$\mathbf{n}^{x,\rho}_{\eta,U}(s) \in N^{\mathrm{fs}}_{\mathcal{B}_\eta}(Z^x_{s\wedge\sigma_{x,\rho}}) \qquad \text{for } d\mathfrak{L}^{x,\rho}_{\eta,U}\text{-almost every } s; \tag{717}$$

(iii) more explicitly, for $d\mathfrak{L}^{x,\rho}_{\eta,U}$-almost every s one has the coordinate normal-cone decomposition

$$\begin{aligned} \mathbf{n}^{x,\rho}_{\eta,U}(s) = \alpha^{0,x,\rho}_{\eta,h}(s)\,\mathbf{e}_h - \alpha^{*,x,\rho}_{\eta,h}(s)\,\mathbf{e}_h \\ + \sum_{j\geq 1}\alpha^{+,x,\rho}_{\eta,j}(s)\,\mathbf{e}_j - \sum_{j\geq 1}\alpha^{-,x,\rho}_{\eta,j}(s)\,\mathbf{e}_j, \end{aligned} \tag{718}$$

where the sums are automatically finite when paired with finitely supported directions, and for every cylinder observable $F \in \mathcal{A}^{\mathrm{cyl}}_\eta(U)$,

$$\mathbf{n}^{x,\rho}_{\eta,U}[F]_t = \int_0^t \mathbf{n}^{x,\rho}_{\eta,U}(s)\big[\nabla_{\mathrm{fs}}F(Z^x_{s\wedge\sigma_{x,\rho}})\big]\,d\mathfrak{L}^{x,\rho}_{\eta,U}(s); \tag{719}$$

(iv) consequently, what remains open is no longer whether the chart-free common-contact-clock selector is a genuine normal-cone multiplier for the ambient coordinate barriers of the manuscript, but only whether that ambient coordinate-box normal cone coincides with the genuine geometric normal cone or Skorokhod reflection cone of the full intrinsic continuum support, together with identification of $\mathfrak{L}^{x,\rho}_{\eta,U}$ as the exact total contact measure of that full reflected gradient dynamics.

Proof. For a coordinate box, the Bouligand tangent directions are exactly those that point inward or tangentially at each active face. Thus $v = (v_h, \mathbf{v})$ belongs to $T_{\mathcal{B}_\eta}(z)$ if and only if

$$h = 0 \Rightarrow v_h \geq 0, \qquad h = h_* \Rightarrow v_h \leq 0, \qquad a_j = b_j(\eta) \Rightarrow v_j \geq 0, \qquad a_j = -b_j(\eta) \Rightarrow v_j \leq 0,$$

for every $j \geq 1$. Intersecting with the finitely supported directions gives exactly the definition (701) of $\mathcal{W}^{\mathrm{alg}}_\eta(z)$, proving (716).

Item (ii) is now immediate from Theorem D.142: the sign-polarity inequality (706) says precisely that $\mathbf{n}^{x,\rho}_{\eta,U}(s)$ is nonnegative on $\mathcal{W}^{\mathrm{alg}}_\eta(Z^x_{s\wedge\sigma_{x,\rho}}) = T_{\mathcal{B}_\eta}(Z^x_{s\wedge\sigma_{x,\rho}}) \cap \mathcal{T}^{\mathrm{fs}}_\eta$, hence belongs to the normal cone (715).

For item (iii), the decomposition (718) is exactly the explicit selector formula (693) rewritten using the coordinate evaluation functionals (714). The factorization (719) is then just the gradient-pairing identity (696) with this ambient normal-cone interpretation inserted. Item (iv) is the residual geometric gap after the ambient coordinate-box normal-cone structure has been made chart-free and intrinsic. □

Define the canonical frozen-tail segment homotopy by

$$H^\theta_\eta(z) := \Pi^{\mathrm{frz}}_\eta(z) + \theta\big(z - \Pi^{\mathrm{frz}}_\eta(z)\big), \qquad 0 \leq \theta \leq 1, \tag{720}$$

for every $z \in \Omega_\eta$. For every bounded scalar field G on Ω_η and every $z \in \Omega_\eta$, define the canonical frozen-segment oscillation

$$\mathrm{Osc}^{\mathrm{frz}}_{\eta,z}(G) := \sup_{0\leq\theta,\theta'\leq 1}\big|G(H^\theta_\eta(z)) - G(H^{\theta'}_\eta(z))\big|. \tag{721}$$

For each telescoping shell $\Delta_{\eta,n} = V_{\eta,n} - V_{\eta,n-1}$ and intrinsic compact chart K, define the canonical frozen-segment shell seminorm

$$\|\Delta_{\eta,n}\|_{\eta,\mathrm{oscfrz},2;K} := \max\Bigg\{ \|\partial_h(\Delta_{\eta,n}\circ\Pi_\eta^{\mathrm{frz}})\|_{L^\infty(K)} + \|\partial_{hh}(\Delta_{\eta,n}\circ\Pi_\eta^{\mathrm{frz}})\|_{L^\infty(K)},$$
$$\sup_{1\le j\le r_\eta}\Big(\|\partial_{a_j}(\Delta_{\eta,n}\circ\Pi_\eta^{\mathrm{frz}})\|_{L^\infty(K)} + \|\partial_{ha_j}(\Delta_{\eta,n}\circ\Pi_\eta^{\mathrm{frz}})\|_{L^\infty(K)}\Big),$$
$$\sup_{1\le j,k\le r_\eta}\|\partial_{a_ja_k}(\Delta_{\eta,n}\circ\Pi_\eta^{\mathrm{frz}})\|_{L^\infty(K)},$$
$$\sup_{z\in K}\mathrm{Osc}^{\mathrm{frz}}_{\eta,z}(\partial_h\Delta_{\eta,n}) + \sup_{z\in K}\mathrm{Osc}^{\mathrm{frz}}_{\eta,z}(\partial_{hh}\Delta_{\eta,n}),$$
$$\sup_{j\ge1}\omega_\eta(j)^{-1}\Big(\sup_{z\in K}\mathrm{Osc}^{\mathrm{frz}}_{\eta,z}(\partial_{a_j}\Delta_{\eta,n}) + \sup_{z\in K}\mathrm{Osc}^{\mathrm{frz}}_{\eta,z}(\partial_{ha_j}\Delta_{\eta,n})\Big),$$
$$\sup_{j,k\ge1}\Omega_\eta(j,k)^{-1}\sup_{z\in K}\mathrm{Osc}^{\mathrm{frz}}_{\eta,z}(\partial_{a_ja_k}\Delta_{\eta,n})\Bigg\}. \tag{722}$$

Proposition D.146 (Canonical frozen-segment oscillation criterion for shellwise C^2 verification). *Assume the hypotheses of Proposition D.110. Keep the fixed profile r_η, ω_η, Ω_η, and W_η^{fp} from Section D.41. Suppose there exists a sequence $(\Gamma^{(2),\mathrm{oscfrz}}_{\eta,n})_{n\ge1}$ such that for every intrinsic compact chart K and every $n\ge1$,*

$$\|\Delta_{\eta,n}\|_{\eta,\mathrm{oscfrz},2;K} \le \Gamma^{(2),\mathrm{oscfrz}}_{\eta,n}. \tag{723}$$

Assume also that

$$\sum_{n=1}^{\infty} n\,\Gamma^{(2),\mathrm{oscfrz}}_{\eta,n}\big(1+r_\eta+W_\eta^{\mathrm{fp}}+(W_\eta^{\mathrm{fp}})^2\big)<\infty. \tag{724}$$

Then Proposition D.143 applies with

$$\Gamma^{(2),\mathrm{dfrz}}_{\eta,n} := \Gamma^{(2),\mathrm{oscfrz}}_{\eta,n} \qquad (n\ge1).$$

Consequently Theorem D.130, Theorem D.133, Theorem D.136, Theorem D.139, Theorem D.142, Theorem D.145, and Proposition D.143 all apply. In particular, activation of the chart-free natural Itô / common-contact-clock / coordinate-density / algebraic-complementarity / ambient-normal-cone package already reduces to one canonical frozen-segment oscillation seminorm on each actual telescoping finite-cylinder shell.

Proof. For every $z\in\Omega_\eta$ one has $H^0_\eta(z)=\Pi_\eta^{\mathrm{frz}}(z)$ and $H^1_\eta(z)=z$. Therefore, for any scalar field G,

$$|G(z)-G(\Pi_\eta^{\mathrm{frz}}(z))| \le \mathrm{Osc}^{\mathrm{frz}}_{\eta,z}(G).$$

Apply this estimate successively to

$$G=\partial_h\Delta_{\eta,n},\ \partial_{hh}\Delta_{\eta,n},\ \partial_{a_j}\Delta_{\eta,n},\ \partial_{ha_j}\Delta_{\eta,n},\ \partial_{a_ja_k}\Delta_{\eta,n}.$$

The first three lines in (722) are exactly the finite-window trace terms appearing in (710); the last three lines dominate the corresponding difference terms because

$$\partial^\beta(\Delta_{\eta,n}-\Delta_{\eta,n}\circ\Pi_\eta^{\mathrm{frz}})(z)=\partial^\beta\Delta_{\eta,n}(z)-\partial^\beta\Delta_{\eta,n}(\Pi_\eta^{\mathrm{frz}}(z))$$

for every first- or second-order coordinate derivative ∂^β appearing in (710). Hence

$$\|\Delta_{\eta,n}\|_{\eta,\mathrm{dfrz},2;K}\le\|\Delta_{\eta,n}\|_{\eta,\mathrm{oscfrz},2;K}$$

for all K and n, so the bound (723) implies the hypothesis (711) of Proposition D.143 with $\Gamma^{(2),\mathrm{dfrz}}_{\eta,n}:=\Gamma^{(2),\mathrm{oscfrz}}_{\eta,n}$. The summability condition (724) is then exactly (712) under the same identification. The stated consequences follow immediately. □

Integrated downstream conclusion after the ambient-normal-cone/frozen-segment continuation. This continuation adds a sixtieth downstream message. Positively, the chart-free reflection package is no longer only an algebraic complementarity selector carried by a common contact clock: it is already the exact ambient normal-cone multiplier for the manuscript's coordinate barrier box. Verification is likewise no longer forced to compare the shell and its frozen trace as separate objects: it already reduces to one canonical oscillation seminorm of the shell's own first and second derivatives along the frozen-tail segments joining every point to its canonical frozen trace. Cautionarily, the ambient coordinate-box normal cone is still not identified with the genuine intrinsic normal cone of the full continuum support, and the frozen-segment oscillation bounds are still not verified for the actual e-Arc-Helix interaction family. What remains open is therefore intrinsic-support Skorokhod/normal-cone identification beyond the ambient coordinate box together with concrete canonical frozen-segment shellwise C^2 verification — or a sharp obstruction — for the actual interaction family.

Remark D.147. The new sixtieth message is the first point in the chain where the chart-free common-contact-clock reflection package is already a bona fide normal-cone multiplier for an explicit geometric state-space constraint set, namely the ambient coordinate box cut out by the barrier amplitudes. At the same time, the interaction side no longer needs to be described by a shell plus a separately named frozen-trace difference: it may already be tested by oscillation of the shell's own derivatives along the canonical straight segments to that frozen trace. The remaining gap is therefore sharper than before: replace the ambient coordinate-box normal cone by the true intrinsic normal cone/Skorokhod selector, and prove — or sharply fail to prove — the resulting canonical frozen-segment shellwise C^2 bounds for the genuine e-Arc-Helix interaction shells.

D.48 Worked continuation LXI: ambient-box Skorokhod projections and a frozen-segment envelope criterion

The previous continuation already identified the chart-free common-contact-clock selector as an ambient coordinate-box normal-cone multiplier and reduced verification of the same natural-domain package to one canonical frozen-segment oscillation seminorm on each actual telescoping finite-cylinder shell. One further closing step is available on both sides without leaving the same intrinsic Itô framework. First, the ambient normal-cone multiplier already integrates to a genuine *projected ambient-box Skorokhod compensator* on every finite cylinder projection, so the stopped finite-coordinate process already solves a finite-dimensional ambient reflected decomposition driven by the common contact clock. Second, the frozen-segment oscillation seminorm itself no longer needs pairwise oscillation to be estimated directly: it already follows from one *canonical frozen-segment envelope seminorm* controlling the same first and second shell derivatives uniformly along the straight frozen-tail homotopies.

Status label. The ambient-box Skorokhod statement below has status **theorem-level downstream result**. It does *not* yet identify the ambient coordinate-box reflection with the genuine intrinsic-support Skorokhod mechanism of the full continuum support, and it does *not* yet prove that the same finite-variation compensator closes on a non-cylinder intrinsic tangent cone beyond the ambient box. What it does show is that every finite observable projection already carries an explicit chart-free ambient-box reflected decomposition driven by the same common contact clock. The frozen-segment envelope statement is a **proposition-level downstream result**. It does *not* yet verify the required bounds for the genuine e-Arc-Helix interaction family, but it shows that the canonical frozen-segment oscillation criterion itself already follows from one direct envelope seminorm of the shell's first and second derivatives along the canonical frozen-tail segments.

Define the finite projected ambient coordinate box by

$$\mathcal{B}_\eta^{(N)} := [0, h_*] \times \prod_{j=1}^{N} [-b_j(\eta), b_j(\eta)] \subset \mathbb{R}^{N+1}, \qquad N \geq 1. \tag{725}$$

For an intrinsic open set $U \subset \Omega_\eta$, $x \in \Omega_\eta$, and $\rho > 0$ with $\mathbb{B}_\eta(x, \rho) \Subset U$, define for each $N \geq 1$ the projected ambient-box contact compensator by

$$\mathbf{K}_{\eta,N}^{x,\rho}(t) := \left(\int_0^t (\alpha_{\eta,h}^{0,x,\rho} - \alpha_{\eta,h}^{*,x,\rho})(s)\, d\mathfrak{L}_{\eta,U}^{x,\rho}(s), \int_0^t (\alpha_{\eta,1}^{+,x,\rho} - \alpha_{\eta,1}^{-,x,\rho})(s)\, d\mathfrak{L}_{\eta,U}^{x,\rho}(s), \ldots, \int_0^t (\alpha_{\eta,N}^{+,x,\rho} - \alpha_{\eta,N}^{-,x,\rho})(s)\, d\mathfrak{L}_{\eta,U}^{x,\rho}(s) \right). \tag{726}$$

Write its coordinates as

$$\mathbf{K}_{\eta,N}^{x,\rho}(t) = \big(K_{\eta,h}^{x,\rho}(t), K_{\eta,1}^{x,\rho}(t), \ldots, K_{\eta,N}^{x,\rho}(t) \big). \tag{727}$$

Theorem D.148 (Projected ambient-box Skorokhod decomposition carried by the common contact clock)**.** *Assume the hypotheses of Theorem D.145. Fix an intrinsic open set $U \subset \Omega_\eta$, $x \in \Omega_\eta$, and $\rho > 0$ with $\mathbb{B}_\eta(x, \rho) \Subset U$. Then for every $N \geq 1$ the following hold.*

(i) The process $\mathbf{K}_{\eta,N}^{x,\rho}$ has finite variation on compact time intervals, is absolutely continuous with respect to the common contact clock $\mathfrak{L}_{\eta,U}^{x,\rho}$, and satisfies

$$d\mathbf{K}_{\eta,N}^{x,\rho}(s) \in N_{\mathcal{B}_\eta^{(N)}}\big(\Pi_N Z_{s\wedge\sigma_{x,\rho}}^x\big)\, d\mathfrak{L}_{\eta,U}^{x,\rho}(s) \qquad \textit{for } d\mathfrak{L}_{\eta,U}^{x,\rho}\textit{-almost every } s. \tag{728}$$

(ii) For every cylinder observable $F = f \circ \Pi_N \in \mathcal{A}_\eta^{\mathrm{cyl}}(U)$ one has the chart-free pairing formula

$$\mathfrak{n}_{\eta,U}^{x,\rho}[F]_t = \int_0^t \nabla f\big(\Pi_N Z_{s\wedge\sigma_{x,\rho}}^x\big) \cdot d\mathbf{K}_{\eta,N}^{x,\rho}(s), \qquad t \geq 0. \tag{729}$$

In particular, every finite observable projection already solves a stopped ambient-box reflected martingale problem driven by the same common contact clock.

(iii) There exist continuous local martingales $M_{\eta,h}^{x,\rho}$ and $M_{\eta,j}^{x,\rho}$, $1 \leq j \leq N$, such that the stopped coordinate processes satisfy

$$\begin{aligned} h(Z_{t\wedge\sigma_{x,\rho}}^x) &= h(x) + M_{\eta,h}^{x,\rho}(t) + \int_0^t \Big(2 - \frac{5h(Z_{s\wedge\sigma_{x,\rho}}^x)}{h_*} - A\big(h(Z_{s\wedge\sigma_{x,\rho}}^x)\big)\partial_h V_{\eta,\infty}^{\mathrm{grad}}(Z_{s\wedge\sigma_{x,\rho}}^x) \Big) ds \\ &\quad + K_{\eta,h}^{x,\rho}(t), \end{aligned} \tag{730}$$

$$\begin{aligned} a_j(Z_{t\wedge\sigma_{x,\rho}}^x) &= a_j(x) + M_{\eta,j}^{x,\rho}(t) - \int_0^t \partial_{a_j} V_{\eta,\infty}^{\mathrm{grad}}(Z_{s\wedge\sigma_{x,\rho}}^x)\, ds \\ &\quad + K_{\eta,j}^{x,\rho}(t), \qquad 1 \leq j \leq N. \end{aligned} \tag{731}$$

Consequently $\Pi_N Z_{\cdot\wedge\sigma_{x,\rho}}^x$ already admits a chart-free ambient-box Skorokhod decomposition with drift given by the explicit continuum gradient field and reflection measure $\mathbf{K}_{\eta,N}^{x,\rho}$.

(iv) What remains open is no longer whether the common-contact-clock selector integrates to a genuine reflected finite-variation term on finite observable projections, but only whether these ambient-box projected Skorokhod decompositions are already the true intrinsic-support reflected-gradient decompositions for the full continuum process beyond the ambient coordinate box.

Proof. By construction each coordinate of $\mathbf{K}^{x,\rho}_{\eta,N}$ is an integral of a nonnegative density difference against the increasing clock $\mathfrak{L}^{x,\rho}_{\eta,U}$, so $\mathbf{K}^{x,\rho}_{\eta,N}$ has finite variation on compact intervals and is absolutely continuous with respect to that clock. For $d\mathfrak{L}^{x,\rho}_{\eta,U}$-almost every s, define

$$\dot{\mathbf{K}}^{x,\rho}_{\eta,N}(s) := \left(\alpha^{0,x,\rho}_{\eta,h}(s) - \alpha^{*,x,\rho}_{\eta,h}(s), \alpha^{+,x,\rho}_{\eta,1}(s) - \alpha^{-,x,\rho}_{\eta,1}(s), \dots, \alpha^{+,x,\rho}_{\eta,N}(s) - \alpha^{-,x,\rho}_{\eta,N}(s)\right).$$

Let $v = (v_h, v_1, \dots, v_N) \in T_{\mathcal{B}^{(N)}_\eta}(\Pi_N Z^x_{s\wedge\sigma_{x,\rho}})$. Extend v by zeros beyond the Nth mode to a finitely supported direction $\widetilde{v} \in \mathcal{T}^{\mathrm{fs}}_\eta$. By (716), $\widetilde{v} \in \mathcal{W}^{\mathrm{alg}}_\eta(Z^x_{s\wedge\sigma_{x,\rho}})$, so Theorem D.145 gives

$$\dot{\mathbf{K}}^{x,\rho}_{\eta,N}(s) \cdot v = \mathbf{n}^{x,\rho}_{\eta,U}(s)[\widetilde{v}] \geq 0.$$

Hence $\dot{\mathbf{K}}^{x,\rho}_{\eta,N}(s) \in N_{\mathcal{B}^{(N)}_\eta}(\Pi_N Z^x_{s\wedge\sigma_{x,\rho}})$ for $d\mathfrak{L}^{x,\rho}_{\eta,U}$-almost every s, which proves (728).

For item (ii), if $F = f \circ \Pi_N$, then its finitely supported cylinder gradient has coordinates $\nabla f(\Pi_N z)$ on the first $N+1$ entries and vanishes thereafter. Substituting this gradient into (719) and using the definition (726) gives exactly (729).

For item (iii), apply the chart-free local Itô/carré-du-champ package from Theorem D.118, together with the explicit gradient drift from Theorem D.88, to the coordinate observables $F_h(z) := h$ and $F_j(z) := a_j$. Their martingale parts are exactly the local martingales $M^{x,\rho}_{\eta,h}$ and $M^{x,\rho}_{\eta,j}$, while the reflection terms are given by (729) with $f(u_h, u_1, \dots, u_N) = u_h$ and $f(u_h, u_1, \dots, u_N) = u_j$. This yields (730)–(731). Item (iv) is the residual geometric gap once the ambient-box Skorokhod decomposition has been made explicit on every finite observable projection. □

For each telescoping shell $\Delta_{\eta,n} = V_{\eta,n} - V_{\eta,n-1}$ and intrinsic compact chart K, define the canonical frozen-segment envelope seminorm

$$\begin{aligned}
\|\Delta_{\eta,n}\|_{\eta,\mathrm{envfrz},2;K} := \max\Bigg\{ & \sup_{z\in K}\sup_{0\leq\theta\leq1}\Big(|\partial_h\Delta_{\eta,n}(H^\theta_\eta(z))| + |\partial_{hh}\Delta_{\eta,n}(H^\theta_\eta(z))|\Big),\\
& \sup_{1\leq j\leq r_\eta}\sup_{z\in K}\sup_{0\leq\theta\leq1}\Big(|\partial_{a_j}\Delta_{\eta,n}(H^\theta_\eta(z))| + |\partial_{ha_j}\Delta_{\eta,n}(H^\theta_\eta(z))|\Big),\\
& \sup_{1\leq j,k\leq r_\eta}\sup_{z\in K}\sup_{0\leq\theta\leq1}|\partial_{a_ja_k}\Delta_{\eta,n}(H^\theta_\eta(z))|,\\
& \sup_{j\geq1}\omega_\eta(j)^{-1}\sup_{z\in K}\sup_{0\leq\theta\leq1}\Big(|\partial_{a_j}\Delta_{\eta,n}(H^\theta_\eta(z))| + |\partial_{ha_j}\Delta_{\eta,n}(H^\theta_\eta(z))|\Big),\\
& \sup_{j,k\geq1}\Omega_\eta(j,k)^{-1}\sup_{z\in K}\sup_{0\leq\theta\leq1}|\partial_{a_ja_k}\Delta_{\eta,n}(H^\theta_\eta(z))|\Bigg\}.
\end{aligned} \tag{732}$$

Proposition D.149 (Canonical frozen-segment envelope criterion for shellwise C^2 verification). *Assume the hypotheses of Proposition D.110. Keep the fixed profile r_η, ω_η, Ω_η, and W^{fp}_η from Section D.41. Suppose there exists a sequence $(\Gamma^{(2),\mathrm{envfrz}}_{\eta,n})_{n\geq1}$ such that for every intrinsic compact chart K and every $n \geq 1$,*

$$\|\Delta_{\eta,n}\|_{\eta,\mathrm{envfrz},2;K} \leq \Gamma^{(2),\mathrm{envfrz}}_{\eta,n}. \tag{733}$$

Assume also that

$$\sum_{n=1}^{\infty} n\,\Gamma^{(2),\mathrm{envfrz}}_{\eta,n}\left(1 + r_\eta + W^{\mathrm{fp}}_\eta + (W^{\mathrm{fp}}_\eta)^2\right) < \infty. \tag{734}$$

Then Proposition D.146 applies with

$$\Gamma^{(2),\mathrm{oscfrz}}_{\eta,n} := 2\Gamma^{(2),\mathrm{envfrz}}_{\eta,n} \qquad (n \geq 1).$$

Consequently Theorem D.130, Theorem D.133, Theorem D.136, Theorem D.139, Theorem D.142, Theorem D.145, Theorem D.148, and Proposition D.146 all apply. In particular, activation of the chart-free natural Itô / common-contact-clock / ambient-box-Skorokhod package already reduces to one canonical frozen-segment envelope seminorm on each actual telescoping finite-cylinder shell.

Proof. For every scalar field G and every $z \in K$ one has

$$\mathrm{Osc}^{\mathrm{frz}}_{\eta,z}(G) \leq 2 \sup_{0\leq\theta\leq 1} |G(H^{\theta}_{\eta}(z))|.$$

Apply this estimate to

$$G = \partial_h \Delta_{\eta,n},\ \partial_{hh}\Delta_{\eta,n},\ \partial_{a_j}\Delta_{\eta,n},\ \partial_{ha_j}\Delta_{\eta,n},\ \partial_{a_j a_k}\Delta_{\eta,n}.$$

The trace terms in (722) are bounded by the same envelope because they are exactly the $\theta = 0$ slices of (732). The oscillation terms are bounded by twice the corresponding envelope suprema. Hence

$$\|\Delta_{\eta,n}\|_{\eta,\mathrm{oscfrz},2;K} \leq 2\|\Delta_{\eta,n}\|_{\eta,\mathrm{envfrz},2;K} \leq 2\Gamma^{(2),\mathrm{envfrz}}_{\eta,n}.$$

Thus Proposition D.146 applies with $\Gamma^{(2),\mathrm{oscfrz}}_{\eta,n} := 2\Gamma^{(2),\mathrm{envfrz}}_{\eta,n}$. The summability condition (734) is exactly the required oscillation summability under this identification, and the stated consequences follow. □

Integrated downstream conclusion after the ambient-box-Skorokhod/frozen-segment-envelope continuation. This continuation adds a sixty-first downstream message. Positively, the chart-free common-contact-clock reflection package no longer stops at an ambient normal-cone multiplier: every finite observable projection already carries an explicit ambient-box Skorokhod decomposition with a common-contact-clock finite-variation compensator. Verification is likewise no longer forced to compare oscillations between arbitrary pairs of segment points: it already reduces to one canonical envelope seminorm controlling the shell's own first and second derivatives uniformly along the frozen-tail segments. Cautionarily, the ambient-box Skorokhod decomposition is still not identified with the genuine intrinsic-support reflected-gradient mechanism of the full continuum process, and the frozen-segment envelope bounds are still not verified for the actual e-Arc-Helix interaction family. What remains open is therefore intrinsic-support geometric Skorokhod identification beyond the ambient box together with concrete canonical frozen-segment-envelope shellwise C^2 verification — or a sharp obstruction — for the actual interaction family.

Remark D.150. The new sixty-first message is the first point in the chain where the common contact clock already integrates to a bona fide reflected finite-variation term for every finite observable projection of the continuum lift. At the same time, the interaction side no longer needs pairwise oscillation control along the frozen-tail homotopies: a single uniform derivative envelope along those same segments already suffices. The remaining gap is therefore sharper again: identify the resulting ambient reflected decompositions with the genuine intrinsic-support Skorokhod mechanism, and verify — or sharply fail to verify — the canonical frozen-segment-envelope shellwise C^2 bounds for the genuine e-Arc-Helix interaction shells.

Integrated downstream conclusion after the intrinsic-support/terminal-criterion closure. This continuation adds a sixty-second downstream message. Positively, the reflection side no longer stops at projected ambient-box Skorokhod decompositions: because the natural state space $\Omega_\eta = \mathcal{U}_\eta$ already sits inside the same coordinate box, the common-contact-clock multiplier is now a genuine intrinsic-support normal-cone/Skorokhod selector on the manuscript's finitely supported tangent cone. Verification is likewise no longer waiting for any additional abstract reduction: once the canonical frozen-segment envelope seminorm is summably controlled on the actual telescoping finite-cylinder shells, the whole natural-domain / intrinsic-support-Skorokhod package is activated. Cautionarily, the manuscript still does not compute those terminal shell bounds for the genuine e-Arc-Helix interaction family. What remains open is therefore no longer a further structural Skorokhod theorem inside the intrinsic domain, but only direct verification — or a sharp obstruction — for that terminal shellwise C^2 criterion, together with beyond-intrinsic quantitative mixing and external dictionary questions.

Remark D.151. The new sixty-second message is the first point in the chain where the chart-free reflection package is already closed on the manuscript's own intrinsic state space rather than only on the surrounding coordinate box. At the same time, the interaction side no longer has another layer of abstract reduction left: the canonical frozen-segment envelope seminorm is now the terminal verification object. The remaining mathematical burden is therefore genuinely model-specific rather than structural: establish those terminal shellwise bounds for the actual e-Arc-Helix interaction family, or show sharply where and why they fail.

D.49 Worked continuation LXII: intrinsic-support Skorokhod closure and a terminal shell criterion

The previous continuation already produced an explicit projected ambient-box Skorokhod decomposition on every finite observable cylinder and reduced activation of the same natural-domain package to one canonical frozen-segment envelope seminorm on each actual telescoping finite-cylinder shell. One final manuscript-internal closure step is now available on both sides. First, because the natural state space itself is the admissible intrinsic subset $\Omega_\eta = \mathcal{U}_\eta \subset (0, h_*) \times \mathcal{K}_\eta^{(\infty)}$, every finitely supported intrinsic tangent direction is automatically admissible for the surrounding ambient coordinate box. Hence the common-contact-clock ambient multiplier already defines a genuine *intrinsic-support Skorokhod compensator* on the manuscript's own finitely supported tangent cone. Second, the canonical frozen-segment envelope seminorm is already terminal: once it is summably controlled on the actual telescoping finite-cylinder shells, there is no further abstract reduction left inside the manuscript's intrinsic continuation chain.

Status label. The intrinsic-support Skorokhod statement below has status **theorem-level downstream closure**. It does *not* verify the terminal shellwise bounds for the genuine e-Arc-Helix interaction family, and it does *not* address beyond-intrinsic mixing or external field-theoretic dictionaries. What it does prove is that the chart-free reflected-gradient package is now structurally closed on the manuscript's own intrinsic state space: on the finitely supported cylinder core of the natural Itô domain, the common-contact-clock multiplier already is the intrinsic-support normal-cone/Skorokhod compensator. The terminal interaction statement is a **proposition-level downstream result**. It does *not* compute the required bounds for the genuine e-Arc-Helix interaction family, but it shows that the manuscript-internal activation problem has now collapsed to one terminal canonical frozen-segment envelope seminorm on the actual shells themselves.

For $z \in \Omega_\eta$, define the finitely supported intrinsic tangent cone by

$$T^{\mathrm{fs}}_{\Omega_\eta}(z) := \left\{\varphi \in \mathcal{T}^{\mathrm{fs}}_\eta : \exists \varepsilon_0 > 0 \text{ such that } z + \varepsilon\varphi \in \Omega_\eta \text{ for all } 0 < \varepsilon < \varepsilon_0\right\}, \tag{735}$$

and the corresponding finitely supported intrinsic normal cone by

$$N^{\mathrm{fs}}_{\Omega_\eta}(z) := \left\{\Lambda \in (\mathcal{T}^{\mathrm{fs}}_\eta)^* : \Lambda[\varphi] \geq 0 \text{ for every } \varphi \in T^{\mathrm{fs}}_{\Omega_\eta}(z)\right\}. \tag{736}$$

For an intrinsic open set $U \subset \Omega_\eta$, a starting point $x \in \Omega_\eta$, and $\rho > 0$ with $\mathbb{B}_\eta(x, \rho) \Subset U$, define the functional-valued finite-variation process

$$\mathfrak{K}^{x,\rho}_{\eta,U}(t)[\varphi] := \int_0^t \mathbf{n}^{x,\rho}_{\eta,U}(s)[\varphi], d\mathfrak{L}^{x,\rho}_{\eta,U}(s), \qquad \varphi \in \mathcal{T}^{\mathrm{fs}}_\eta,\ t \geq 0, \tag{737}$$

where $\mathbf{n}^{x,\rho}_{\eta,U}$ is the algebraic contact-cone selector from Theorem D.139.

Theorem D.152 (Chart-free intrinsic-support Skorokhod closure on the natural Itô domain). *Assume the hypotheses of Theorem D.148. Fix an intrinsic open set $U \subset \Omega_\eta$, $x \in \Omega_\eta$, and $\rho > 0$ with $\mathbb{B}_\eta(x,\rho) \Subset U$. Then the following hold.*

(i) The process $\mathfrak{K}^{x,\rho}_{\eta,U}$ is well defined on $(\mathcal{T}^{\mathrm{fs}}_\eta)^*$, has finite variation on compact time intervals, and is absolutely continuous with respect to the same common contact clock $\mathfrak{L}^{x,\rho}_{\eta,U}$.

(ii) For $d\mathfrak{L}^{x,\rho}_{\eta,U}$-almost every time s,

$$\mathbf{n}^{x,\rho}_{\eta,U}(s) \in N^{\mathrm{fs}}_{\Omega_\eta}\big(Z^x_{s\wedge\sigma_{x,\rho}}\big). \tag{738}$$

Equivalently,

$$d\mathfrak{K}^{x,\rho}_{\eta,U}(s) \in N^{\mathrm{fs}}_{\Omega_\eta}\big(Z^x_{s\wedge\sigma_{x,\rho}}\big)\, d\mathfrak{L}^{x,\rho}_{\eta,U}(s) \qquad \text{for } d\mathfrak{L}^{x,\rho}_{\eta,U}\text{-almost every } s. \tag{739}$$

(iii) For every cylinder observable $F = f \circ \Pi_N \in \mathcal{A}^{\mathrm{cyl}}_\eta(U)$,

$$\mathfrak{n}^{x,\rho}_{\eta,U}[F]_t = \int_0^t \mathfrak{K}^{x,\rho}_{\eta,U}(ds)\big[\nabla_{\mathrm{fs}} F(Z^x_{s\wedge\sigma_{x,\rho}})\big], \qquad t \geq 0. \tag{740}$$

In particular, the stopped full continuum process already solves the intrinsic-support reflected-gradient martingale problem on its finitely supported cylinder core.

(iv) For every $N \geq 1$, the first $N+1$ coordinate projection of $\mathfrak{K}^{x,\rho}_{\eta,U}$ coincides with the projected ambient-box compensator $\mathbf{K}^{x,\rho}_{\eta,N}$ of (726). Thus the projected ambient-box Skorokhod decompositions from Theorem D.148 are precisely the finite-coordinate shadows of one chart-free intrinsic-support Skorokhod compensator.

(v) What remains open is no longer any further structural Skorokhod or intrinsic-normal-cone identification on the manuscript's own natural Itô domain, but only concrete verification — or a sharp obstruction — for the terminal shellwise interaction criterion stated below, together with beyond-intrinsic quantitative mixing and external dictionary questions.

Proof. Item (i) is immediate from (737) because $\mathbf{n}^{x,\rho}_{\eta,U}(s)$ is linear on $\mathcal{T}^{\mathrm{fs}}_\eta$ and is integrated against the increasing clock $\mathfrak{L}^{x,\rho}_{\eta,U}$. For item (ii), let $\varphi \in T^{\mathrm{fs}}_{\Omega_\eta}(z)$. Since $\Omega_\eta = \mathcal{U}_\eta \subset (0,h_*) \times \mathcal{K}^{(\infty)}_\eta$, every sufficiently small positive displacement $z + \varepsilon\varphi$ remains inside the ambient coordinate box, so the sign constraints defining the canonical algebraic wedge (701) are automatically satisfied. Hence

$$T^{\mathrm{fs}}_{\Omega_\eta}(z) \subset \mathcal{W}^{\mathrm{alg}}_\eta(z) \qquad (z \in \Omega_\eta).$$

Applying Theorem D.145 to the path value $z = Z^x_{s\wedge\sigma_{x,\rho}}$ yields

$$\mathbf{n}^{x,\rho}_{\eta,U}(s)[\varphi] \geq 0 \qquad \text{for every } \varphi \in T^{\mathrm{fs}}_{\Omega_\eta}(Z^x_{s\wedge\sigma_{x,\rho}})$$

for $d\mathfrak{L}^{x,\rho}_{\eta,U}$-almost every s, which is exactly (738). Formula (739) is simply the measure-valued restatement of the same inclusion.

For item (iii), let $F = f \circ \Pi_N$. By Theorem D.148,

$$\mathfrak{n}^{x,\rho}_{\eta,U}[F]_t = \int_0^t \nabla f\big(\Pi_N Z^x_{s\wedge\sigma_{x,\rho}}\big) \cdot d\mathbf{K}^{x,\rho}_{\eta,N}(s).$$

But $\nabla_{\mathrm{fs}} F$ has exactly these first $N+1$ coordinates and vanishes thereafter, while (737) defines $\mathfrak{K}^{x,\rho}_{\eta,U}$ by the same common-contact-clock density. This yields (740). Item (iv) is then immediate by testing $\mathfrak{K}^{x,\rho}_{\eta,U}$ on the coordinate basis vectors of $\mathcal{T}^{\mathrm{fs}}_\eta$: the resulting first $N+1$ coordinates are exactly the components of $\mathbf{K}^{x,\rho}_{\eta,N}$. Item (v) records the only remaining non-structural burden once the intrinsic-support Skorokhod compensator and the terminal criterion below are both in place. ☐

Proposition D.153 (Terminal canonical frozen-segment envelope criterion for full intrinsic activation). *Assume the hypotheses of Proposition D.149. Then Theorem D.152 applies as soon as the shellwise bounds* (733) *and the summability condition* (734) *hold. Equivalently, the entire manuscript-internal natural-domain / intrinsic-support-Skorokhod package now activates from the single terminal requirement*

$$\sum_{n=1}^{\infty} n\,\Gamma_{\eta,n}^{(2),\mathrm{envfrz}}\big(1 + r_\eta + W_\eta^{\mathrm{fp}} + (W_\eta^{\mathrm{fp}})^2\big) < \infty \tag{741}$$

for a sequence controlling the canonical frozen-segment envelope seminorm of every actual telescoping finite-cylinder shell through (733)*. There is no further abstract reduction inside the manuscript beyond this terminal shellwise criterion.*

Proof. Proposition D.149 shows that the hypotheses (733) and (734) already activate the full natural Itô / common-contact-clock / ambient-box-Skorokhod package. Theorem D.152 then upgrades that ambient-box package to the intrinsic-support Skorokhod compensator on the manuscript's own natural state space. Thus the only remaining activation hypothesis is exactly the canonical frozen-segment envelope control already encoded in (741). No additional structural reduction remains. □

References

Correspondence email: arcman@arcii.org
ORCID: 0009-0002-7399-6243

[1] Meng, Frank F. [Arcman]. *The Arc Theory: The non-spacetime general relativity about the essence of nature.* Beijing: Modern Education Press (China Publishing Group), 2008. ISBN 978-7-80196-633-9.

[2] Arcman. *Arc Geometry I: Differential and Topological Structure of Arc Helices.* Preprint, Zenodo, 2026. DOI: https://doi.org/10.5281/zenodo.18869377.

[3] Arcman. *Arc Geometry II: Variational Structures and Energies of Arc Helices.* Preprint, Zenodo, 2026. DOI: https://doi.org/10.5281/zenodo.18905574.

[4] Arcman. *Arc Geometry III: Spectral Stability and Resonant Bifurcation from the q-Fold Circle.* Preprint, Zenodo, 2026. DOI: https://doi.org/10.5281/zenodo.18918603.

[5] Arcman. *On the Pure Geometric Origin of the Fine-Structure Constant via Spontaneous Topological Doubling.* Preprint, Zenodo, 2026. DOI: https://doi.org/10.5281/zenodo.18973743.

[6] National Institute of Standards and Technology (NIST), *CODATA Value: inverse fine-structure constant*, 2022 CODATA recommended values, accessed March 2026. Available at https://www.physics.nist.gov/cgi-bin/cuu/Value?alphinv=.

[7] M. P. do Carmo, *Differential Geometry of Curves and Surfaces*, Prentice-Hall, 1976.

[8] D. Rolfsen, *Knots and Links*, AMS Chelsea Publishing, 1976. Reprinted by the American Mathematical Society, 2003.

[9] J. Langer and D. A. Singer, "The total squared curvature of closed curves," *Journal of Differential Geometry* **20**(1) (1984), 1–22. DOI: 10.4310/jdg/1214438990.

[10] J. Langer and D. A. Singer, "Knotted elastic curves in $\mathbb{R}^3$," *Journal of the London Mathematical Society* **s2-30**(3) (1984), 512–520. DOI: 10.1112/jlms/s2-30.3.512.

Publisher's Cataloging-in-Publication Data
provided by Five Rainbows Cataloging Services

Names: Arcman, author.
Title: e-Arc-Helix : volume 1, geometry, rational-closure sectors,
and upper continuations / Arcman.
Description: Ontario, CA : International Institute of Arc Theory, 2026. |
Series: Arc theory, vol 1.
Identifiers: LCCN 2026914656 (print) | ISBN 979-8-9956273-1-9 (paperback) |
ISBN 979-8-9956273-6-4 (ebook)
Subjects: LCSH: Academic writing. | Professional education. | Geometry. |
Differential geometry. | Mathematical logic. | Decision making. |
BISAC: MATHEMATICS / Geometry / General. |
MATHEMATICS / Mathematical Analysis. | MATHEMATICS / Number Theory.
Classification: LCC QA641 .E37 (print) | LCC QA641 (ebook) |
DDC 516--dc23.

Errata and Version-Control Notice

This volume is released as a print-on-demand archival edition of a large technical manuscript. Any typographical, formatting, or formula-display corrections identified after publication will be recorded in the official errata and version-control record maintained by the International Institute of Arc Theory. Such corrections do not alter the claim status of the volume unless explicitly stated in a later edition or formal revision notice.

Cover Design: Chloe Chen
Cover Illustration: Alex Ren

www.ingramcontent.com/pod-product-compliance
Lightning Source LLC
LaVergne TN
LVHW071208210926
848482LV00015B/751